# Breeding, Genetics and Safety Production of Dairy Cattle

# Breeding, Genetics and Safety Production of Dairy Cattle

Editors

**Zhi Chen**
**Cong Li**

MDPI • Basel • Beijing • Wuhan • Barcelona • Belgrade • Manchester • Tokyo • Cluj • Tianjin

*Editors*
Zhi Chen
Yangzhou University
Yangzhou, China

Cong Li
Northwest A&F University
Yangling, China

*Editorial Office*
MDPI
St. Alban-Anlage 66
4052 Basel, Switzerland

This is a reprint of articles from the Special Issue published online in the open access journal *Agriculture* (ISSN 2077-0472) (available at: https://www.mdpi.com/journal/agriculture/special_issues/Genetics_Dairy_Cattle).

For citation purposes, cite each article independently as indicated on the article page online and as indicated below:

LastName, A.A.; LastName, B.B.; LastName, C.C. Article Title. *Journal Name* **Year**, *Volume Number*, Page Range.

**ISBN 978-3-0365-8246-7 (Hbk)**
**ISBN 978-3-0365-8247-4 (PDF)**

# Contents

# About the Editors

**Zhi Chen**

Zhi Chen, Ph.D., is an Associate Professor at the College of Animal Science and Technology of Yangzhou University, Yangzhou City, Jiangsu Province, China. From October 2021 to the present, Dr. Chen has served as Associate Professor in the College of Animal Science and Technology, Yangzhou University, and from September 2017 to October 2021, he served as a Lecturer in the College of Animal Science and Technology, Yangzhou University. He obtained a doctorate degree from Northwest A&F University, Shaanxi Provincial Key Laboratory of Molecular Biology for Agriculture, from September 2013 to June 2017, obtained a master's degree from the College of Animal Science, Guizhou University, from September 2010 to June 2013, and obtained a bachelor's degree from the College of Animal Science and Technology, Yangzhou University, from September 2006 to June 2010. He is active in many fields of milk fat traits, gene regulation network construction, gene expression, epigenetics, circRNA, lincRNA, miRNA, and mRNA.

**Cong Li**

Cong Li, Ph.D., is an Associate Professor at the College of Animal Science and Technology (Northwest A&F University, Yangling, 712100, China). He obtained a doctorate in Animal Genetics, Breeding and Reproduction from China Agricultural University from September 2010 to June 2016, and joined a joint training of functional genomics at the University of Illinois at Urbana–Champaign from February 2014 to February 2015; Dr. Li obtained a Bachelor's degree of Animal Science from Inner Mongolia University for Nationalities from September 2006 to June 2010. He is active in many fields of gene regulation, functional genomics, genetic polymorphisms, genetic heritability, gwas, bioinformatics, milk component, fatty acids, and proteins.

*agriculture*

*Editorial*

# Genetics and Production of Safe, High-Quality Milk by Dairy Cattle

**Sitian Yang [1], Xiang Cao [1], Yu Wang [1], Cong Li [2,*] and Zhi Chen [1,*]**

1 College of Animal Science and Technology, Yangzhou University, Yangzhou 225009, China; 221902425@stu.yzu.edu.cn (S.Y.); mx120210843@yzu.edu.cn (X.C.); shanheyicheng@163.com (Y.W.)
2 Key Laboratory of Animal Genetics, Breeding and Reproduction of Shaanxi Province, College of Animal Science and Technology, Northwest A&F University, Xianyang 712100, China
* Correspondence: congl@nwafu.edu.cn (C.L.); zhichen@yzu.edu.cn (Z.C.)

A crucial part of the livestock industry is dairy cattle which contribute significantly to the livestock economy. Milk-fat traits of dairy cattle are related to the efficiency and quality of dairy cattle production, and their genetic breeding is also vital for the production of dairy cattle products. Presently, the dairy industry, particularly in developing countries, is shifting from quantity-based scale production to a quality- and efficiency-based scale. As a result, the level of dairy production and milk quality will inevitably be improved, leading to higher standards for the quality of dairy products. In view of the current situation and future trends in dairy development, it is of great significance to attach importance to the research on high-quality milk-production technology, meat-quality regulation and product quality and safety control.

The most common disease on dairy farms is bovine mastitis, which reduces milk production and affects milk quality. The infection occurs when pathogenic bacteria infiltrate the mammary gland through infection or environmental means. In actual samples, *E. coli* was found to have high rates of infection and prevalence, although it could be suppressed effectively by some antibiotics, such as gentamicin and tetracycline. However, *E. coli* isolates have shown severe resistance to many β-lactams, including penicillin, cephalothin and amoxicillin [1]. It is therefore important to explore other ingredients that can effectively inhibit Gram-induced inflammation and oxidative stress response. Accordingly, Liu et al. turned their attention to traditional Chinese herbal ingredients, the most prominent being Lycium barbarum polysaccharides (LBP), which exhibit antioxidant and anti-inflammatory properties both in vivo and in vitro. Through its activation of cytokine signaling-3 (SOCS3) and inhibition of mitogen-activated protein kinase (MAPK) signaling, LBP effectively alleviated *E. coli*-induced inflammation and oxidation in primary bovine mammary epithelial cells (pbMECs) [2]. These findings will enable us to conduct better research on the replacement of antibiotic-based treatments, and provide new ideas, technologies, and materials for the prevention and treatment of cow mastitis.

At the cellular level, microRNAs (miRNAs) are a class of endogenous non-coding RNAs. They have been shown to play an important role in cell differentiation, apoptosis, gene-expression regulation, and tumorigenesis. The inflammatory response is a complex process of inflammatory reactions that occurs in the body to various stimuli, including the release of inflammatory cytokines and chemokines, as well as the activation of corresponding signaling pathways. Additionally, miRNAs are vital to the immune system in that they can participate in and regulate the immune response, and suppress it by binding with target genes. With increasing research, miR-29 has emerged as a key regulator of liver lipid metabolism. The transient transfection of bta-miRNA29d-3p mimics and inhibitors in bovine mammary epithelial cells (BMECs) revealed that bta-miRNA29d-3p could regulate fatty-acid metabolism and TAG synthesis by regulating lipid-metabolism-related genes in BMECs and targeting glycerol-3-phosphate acyltransferase (GPAM) [3]. Meanwhile,

**Citation:** Yang, S.; Cao, X.; Wang, Y.; Li, C.; Chen, Z. Genetics and Production of Safe, High-Quality Milk by Dairy Cattle. *Agriculture* **2023**, *13*, 1348. https://doi.org/10.3390/agriculture13071348

Received: 19 June 2023
Revised: 26 June 2023
Accepted: 30 June 2023
Published: 4 July 2023

based on mammary transcriptome analysis, a kind of miRNA, bta-miR-125a, has been identified as controlling bovine-milk fat production by targeting the 3′ untranslated region (UTR) of serum amyloid A-3 (SAA1) mRNA. This miRNA can act as a positive regulator of lipid synthesis in mammary epithelial cells by targeting *SAA1* gene expression [4]. N6-methyladenosine ($m_6A$) methylation, a methylation modification found on mRNA molecules, has also been suggested to play a role in the development of lipid metabolism since it can change the mRNA levels of genes related to milk-fat metabolism. More importantly, genes and signaling pathways related to milk-fat metabolism have also been shown to be closely related to $m_6A$ methylation. The above studies are expected to provide references for further research into mammary-gland development and milk-fat metabolism in ruminants [5].

A study by Zhang et al. investigated the effects of two metabolic-syndrome pathway genes, interleukin 6 receptor (*IL-6*) and leptin receptor (*LEPR*), on reproductive performance in dairy cows. Among the six SNPs they examined, five were significantly associated with at least one reproductive trait, including age at the first service, age at the first calving, the interval from calving to first insemination, and the calving ease in heifers and cows. Thus, IL6R and LEPR are considered to be important molecular markers for the genetic selection of reproductive traits in high-yielding dairy cattle [6]. In addition, *DDIT3* is also considered to be a novel gene that can regulate the mechanism of milk production traits, thus being used to regulate milk protein and fat traits of dairy cows. On the other hand, as a result of *DDIT3* silencing, inflammatory markers such as *IL-6* and *IL6R* were up-regulated, which means that lower *DDIT3* expression can prompt increased lipid accumulation and apoptosis through the up-regulation of *IL-6* expression [7].

The remaining concepts are discussed from the perspectives of molecular genetics, development, and selective breeding. In future production practices, it is important to pay special attention to mutations associated with abnormalities in dairy-cattle populations, such as defects in bovine leukocyte adhesion, complex vertebral malformations, and short spinal or short spinal lethal syndrome, as representative examples of inherited diseases [8]. Meanwhile, two variants of cofilin, cofilin-1 (CFL1, non-muscular type) and cofilin-2 (CFL2, muscular type), have been found in mammals, with a dual function on skeletal muscle fibers [9]. The acute phase response (APR) is a non-specific systemic response in which acute phase protein (APP) production changes. Hence, it is possible to use APP as a biomarker of inflammation. The rapid dynamics of serum amyloid A (SAA), one of the main APPs in dairy cows, also calls for special attention. Therefore, the breeding and reproduction stages should be controlled specifically with regard to genetics and gene selection [10]. In addition, beta-oxidation, an important metabolic process involving multiple steps, produces energy via the breakdown of fatty-acid molecules. Future studies can benefit from summarizing the importance of peroxisome β oxidation [11].

Taking common diseases of dairy cows as a starting point, this research enables the further analysis of the susceptibility of dairy cows to disease and the methods of mitigation. It also comprises an examination of some important regulatory factors that contribute to mRNA regulation, which provides suggestions for the improvement and enhancement of milk-fat traits. A greater emphasis is being placed on livestock production practices from the perspectives of genetics, reproduction, and breeding selection. In addition, fatty acid oxidation is summarized and the regulatory mechanism of fatty acids is analyzed in depth, providing new insights into future research on the genetics of milk-fat traits.

**Author Contributions:** Writing—review and editing, S.Y. and X.C.; validation, Y.W.; supervision, Z.C. and C.L.; project administration, Z.C. All authors have read and agreed to the published version of the manuscript.

**Funding:** This work was supported by the Postgraduate Research and Practice Innovation Program of Jiangsu Province (KYCX22_3533); independent Innovation in Jiangsu Province of China (CX(21)3119); "Qing Lan Project" and the "High-end talent support program" of Yangzhou University, China.

**Conflicts of Interest:** The authors declare no conflict of interest.

## References


1. Xu, T.; Cao, W.; Huang, Y.; Zhao, J.; Wu, X.; Yang, Z. The Prevalence of *Escherichia coli* Derived from Bovine Clinical Mastitis and Distribution of Resistance to Antimicrobials in Part of Jiangsu Province, China. *Agriculture* **2023**, *13*, 90. [CrossRef]
2. Liu, R.; Zhu, H.; Zhao, J.; Wu, X.; Lu, X.; Xu, T.; Yang, Z. Lycium barbarum Polysaccharide Inhibits *E. coli*-Induced Inflammation and Oxidative Stress in Mammary Epithelial Cells of Dairy Cows via SOCS3 Activation and MAPK Suppression. *Agriculture* **2022**, *12*, 598. [CrossRef]
3. Zhao, X.; Li, J.; Zhao, S.; Chen, L.; Zhang, M.; Ma, Y.; Yao, D. Regulation of bta-miRNA29d-3p on Lipid Accumulation via GPAM in Bovine Mammary Epithelial Cells. *Agriculture* **2023**, *13*, 501. [CrossRef]
4. Cui, X.; Yuan, T.; Fang, Z.; Feng, J.; Wu, C. Bta-miR-125a Regulates Milk-Fat Synthesis by Targeting SAA1 mRNA in Bovine Mammary Epithelial Cells. *Agriculture* **2022**, *12*, 344. [CrossRef]
5. Xin, Z.; Zhang, T.; Lu, Q.; Yang, Z.; Chen, Z. Progress of m6A Methylation in Lipid Metabolism in Humans and Animals. *Agriculture* **2022**, *12*, 1683. [CrossRef]
6. Zhang, H.; Sammad, A.; Shi, R.; Dong, Y.; Zhao, S.; Liu, L.; Guo, G.; Xu, Q.; Liu, A.; Wang, Y. Genetic Polymorphism and mRNA Expression Studies Reveal IL6R and LEPR Gene Associations with Reproductive Traits in Chinese Holsteins. *Agriculture* **2023**, *13*, 321. [CrossRef]
7. Cui, X.; Li, C.; Wei, Z.; Meng, H.; Zhang, F.; Liu, Y.; Wu, C.; Yang, S. DDIT3 Governs Milk Production Traits by Targeting IL-6 to Induce Apoptosis in Dairy Cattle. *Agriculture* **2023**, *13*, 117. [CrossRef]
8. Marzanova, S.N.; Devrishov, D.A.; Turbina, I.S.; Marzanov, N.S.; Griffin, D.K.; Romanov, M.N. Genetic Load of Mutations Causing Inherited Diseases and Its Classification in Dairy Cattle Bred in the Russian Federation. *Agriculture* **2023**, *13*, 299. [CrossRef]
9. Sun, Y.; Ma, Y.; Wu, X.; Zhao, T.; Lu, L.; Yang, Z. Functional and Comparative Analysis of Two Subtypes of Cofilin Family on Cattle Myoblasts Differentiation. *Agriculture* **2022**, *12*, 1420. [CrossRef]
10. Trela, M.; Domańska, D.; Witkowska-Piłaszewicz, O. Diagnostic Use of Serum Amyloid A in Dairy Cattle. *Agriculture* **2022**, *12*, 459. [CrossRef]
11. Lu, Q.; Zong, W.; Zhang, M.; Chen, Z.; Yang, Z. The Overlooked Transformation Mechanisms of VLCFAs: Peroxisomal β-Oxidation. *Agriculture* **2022**, *12*, 947.

*agriculture*

*Article*

# Bta-miR-125a Regulates Milk-Fat Synthesis by Targeting *SAA1* mRNA in Bovine Mammary Epithelial Cells

**Xiaogang Cui, Tianqi Yuan, Zhengyu Fang, Jiao Feng and Changxin Wu ***

Key Laboratory of Medical Molecular Cell Biology of Shanxi Province, Institute of Biomedical Sciences, Shanxi University, Taiyuan 030006, China; helloxiaogangcui@sxu.edu.cn (X.C.); 202023105015@email.sxu.edu.cn (T.Y.); 202023118003@email.sxu.edu.cn (Z.F.); fengjiao2018@sxu.edu.cn (J.F.)
* Correspondence: cxw20@sxu.edu.cn

**Abstract:** The nutritional value of cow milk mainly depends on its fatty acid content and protein composition. The identification of genes controlling milk production traits and their regulatory mechanisms is particularly important for accelerating genetic progress in the breeding of dairy cows. On the basis of mammary gland transcriptome analyses, in this study we identified an miRNA, bta-miR-125a, that could control bovine milk-fat production by targeting the 3′ untranslated region (UTR) of the serum amyloid A-1 (SAA1) mRNA. The presence of synthetic bta-miR-125a (i.e., an miR-125a mimic) significantly down-regulated the expression of luciferase from mRNAs containing the binding sequence for bta-miR-125a in the 3′-UTR in a dual-luciferase reporter assay. Furthermore, the presence of the miR-125a mimic decreased the steady-state level of the *SAA1* protein, but increased the accumulation of triglycerides and cholesterol content in bovine mammary epithelial cells (MAC-Ts). Blocking the function of bta-miR-125a using a specific inhibitor decreased the level of triglycerides and cholesterol content in the cells. These results indicate that bta-miR-125a can serve as a positive regulator of lipid synthesis in mammary epithelial cells, which acts by targeting *SAA1* gene expression.

**Keywords:** mammary epithelial cells; milk-fat synthesis; miR-125a; *SAA1*; MAC-T

**Citation:** Cui, X.; Yuan, T.; Fang, Z.; Feng, J.; Wu, C. Bta-miR-125a Regulates Milk-Fat Synthesis by Targeting *SAA1* mRNA in Bovine Mammary Epithelial Cells. *Agriculture* **2022**, *12*, 344. https://doi.org/10.3390/agriculture12030344

Academic Editors: Zhi Chen and Cong Li

Received: 28 January 2022
Accepted: 25 February 2022
Published: 28 February 2022



## 1. Introduction

Cow milk is regarded as a basic food in many diets worldwide [1], and contains nearly all of the essential elements required for healthy human nutrition. Milk fat is a milk quality indicator, and a determining element of the nutritional value of milk. It is composed of lipid droplets, which mainly consist of triacylglycerides (TAGs), which are synthesized and released from the bovine mammary epithelial cells. The milk-fat levels and composition are affected by various factors, including heredity, nutrition, physiological conditions and the environment. More studies have revealed some miR-NAs affecting lipid metabolism by targeting genes, e.g., ATP binding cassette transporter A1 (*ABCA1*), with Chen et al. reporting that miR-106b functions through ABCA1 by mRNA and protein levels [2]; however, the underlying mechanisms remain poorly understood [3–5].

Serum amyloid A (SAA) is the most prominent and highly up-regulated protein during acute inflammation [6]. The SAA protein is synthesized in the liver, and has been shown to be involved in the metabolism of lipids [7]. Four isoforms of SAA have been identified through amino-acid sequence analysis: *SAA1*, *2*, *3* and *4* [8]. The *SAA1* and *SAA2* isoforms are mainly synthesized in hepatocytes, and are primarily associated with high-density lipoprotein (HDL) [9]. In addition, the expression of *SAA1* and *SAA2* in non-hepatic tissues has also been reported [10]. *SAA1* is involved in the development of the mammary gland through NF-κB-dependent signaling, while the over-expression of *SAA1* may accelerate apoptosis and suppress mammary epithelial cell growth [11]. Furthermore, seven SNPs (single nucleotide polymorphisms) in the *SAA1* gene have been significantly associated with milk yield and composition traits [12]. Unlike *SAA1* and *SAA2*, *SAA3* isoform is present in a lipid-poor form, not associated with HDL [7]. The *SAA4* isoform is an HDL-associated apolipoprotein, and constitutively expressed at relatively low levels in both human and mouse liver [13]. To date, the SAA4 isoform function is largely unknown [14].

MicroRNAs (miRNAs) are small, non-coding RNAs that regulate gene expression at the post-transcriptional level in various biological processes [15]. It has been estimated that the expression of 30% of protein-coding genes is regulated by miRNAs [16,17].

The mammary gland is a uniquely specialized organ in humans and mammals. The specialized tissues, including the mammary epithelial cells in the mammary gland, undergo proliferation, differentiation and apoptosis. Various roles of miRNAs in the development of the mammary gland in humans [18,19], the maintenance of mammary epithelial progenitor cells in mice [20] and the proliferation and differentiation of mammary epithelial cells in humans [21,22] have been uncovered, as have those in the outgrowth of epithelial ducts in mice [23]. In mammary epithelial cells, several miRNAs have been found to be involved in the production of milk fat by targeting different genes. miR-224 can inhibit the secretion of milk fat by down-regulating the expression of the acyl-coenzyme A dehydrogenase (*ACADM*) and aldehyde dehydrogenase2 (*ALDH2*) genes [24]. The over-expression of miR-224 has been associated with a decrease in triglycerides in mammary epithelial cells [24]. Contrary to miR-224, miR-21-3p promotes triglyceride production in cow mammary epithelial cells—through the inhibition of the elongation of the very long chain fatty acids protein 5 (*Elovl5*) gene, which is a gene important in lipid metabolism—by catalyzing the elongation of fatty acids [25]. Furthermore, miR-15a inhibits the vitality and lactation of mammary epithelial cells by targeting the *GHR* gene [26], which is associated with milk composition. Moreover, miR-27a-3p can inhibit milk-fat synthesis by dairy cow mammary epithelial cells (MAC-Ts) by targeting peroxisome proliferator-activated receptor gamma (PPARG), which enhances the synthesis of monounsaturated fatty acids by controlling stearoyl-CoA desaturase [4].

In our previous studies, we analyzed the mammary gland epithelial tissues of four lactating Holstein cows with extremely high and low milk-protein (PP) and fat percentages (FP), using RNA sequencing (RNA-seq) and small RNA-seq [27,28]. We identified 21 differentially expressed genes as potential targets for some of the 71 differentially expressed miRNAs, including the *SAA1* gene and bta-miR-125a, respectively [29]. Based on those preliminary results, we hypothesized that bta-miR-125a might regulate the expression of the *SAA1* gene. In this study, this hypothesis was tested using different techniques, including dual-luciferase reporter assays, quantitative reverse-transcription PCR (qRT-PCR) and oil red O staining assays, in order to investigate the functional relevance of bta-miR-125a for the production of milk fat in dairy cow mammary epithelial cells (MAC-Ts). Our results shed new light on the network of miRNAs involved in the production of milk.

## 2. Materials and Methods

### 2.1. Cell Cultures

Human HEK293T and bovine MAC-T cells were grown in Dulbecco's modified Eagle's medium (DMEM; Boster, Wuhan, China) containing 10% fetal bovine serum (Gibco, Waltham, MA, USA), 100 IU/mL penicillin and 100 μg/mL streptomycin (Solarbio, Beijing, China) at 37 °C under 5% $CO_2$.

### 2.2. Plasmid Construction and Site-Directed Mutagenesis

The binding sequence of bta-miR-125a was predicted to be in the 3′ untranslated region (UTR) of the *SAA1* mRNA. Thus, the DNA sequence for this region was amplified by PCR, using bovine mammary gland DNAs as templates and a primer pair (p*SAA1* 3′UTR Forward: 5′-GCTGCCTCTCTCTCTGCTCAG-3′; p*SAA1* 3′UTR Reverse: 5′-TTTTGTTTGACCCAAATATAGTGAGGATAAAGGT-3′; Sangon, Shanghai, China). The PCR conditions were as follows: 94 °C for 5 min, followed by 35 cycles of 94 °C for 30 s; 60 °C for 30 s; and 72 °C for 45 s; and a final extension at 72 °C for 10 min. The 683 bp PCR products were digested with PmeI and XbaI, and then inserted into the plasmid pmirGLO Dual-Luciferase miRNA Target Expression Vector (PmiRGLO, Promega, Madison, WI, USA), in order to obtain the recombinant plasmid PmiR-*SAA1*–3′UTR-wild type (WT) (Figure 1A). The seed sequences recognized by miR-125a in the plasmid PmiR-*SAA1*–3′UTR-wild type (WT) were deleted using a QuikChange site-directed mutagenesis kit (Stratagene, La Jolla, CA, USA) to generate the plasmid PmiR-*SAA1*–3′UTR-mutant (MUT), which was used as a negative control (Figure 1B). The plasmids were all confirmed by DNA sequencing and used in the luciferase reporter assay.

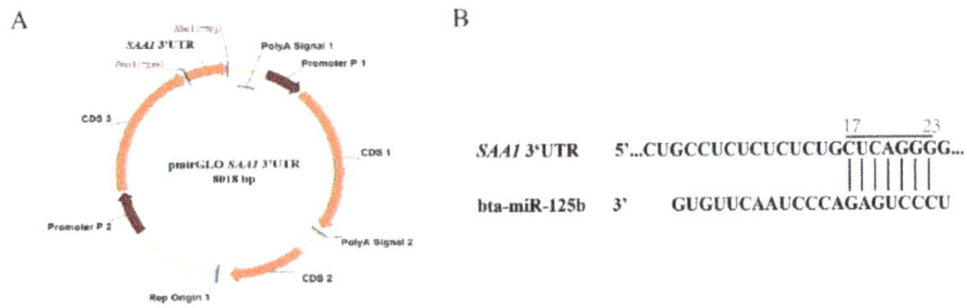


**Figure 1.** (**A**) The pmirGLO vectors with the predicted 3′UTR target sequences of the expressed gene, pmirGLO-SAA1-3′UTR; (**B**) Locations and sequences of the miRNA target sites in the 3′UTR of the expressed gene. The sequences of the miRNAs are indicated, along with mutations introduced in the target sites (underlined nucleotides) for generating the mutated reporter constructs.

### 2.3. Luciferase Reporter Assays

Synthetic bta-miR-125a (miR mimic), bta-miR-125a-specific inhibitor miRNA (miR inhibitor) and non-specific negative control miRNA (NC) were synthesized and purchased from GenePharma (Shanghai, China). Luciferase reporter assays were performed in HEK293T cells. The cells were plated into six-well plates at $5.0 \times 10^5$ cells/well for 24 h. Then, 30 ng of the empty PmiRGLO vector, PmiRGLO-*SAA1*–3′UTR-wild type (WT) or PmiR-*SAA1*–3′UTR-mutant (MUT) was co-transfected with a 30 nM concentration (final concentration) of the miRNA mimic, miRNA inhibitor or NC (GenePharma) in various combinations in each well, using 1 μL of Lipofectamine 3000 (Invitrogen, Waltham, MA, USA). At 24 h post-transfection, the relative activities of firefly luciferase were measured using a TECAN Infinite 200 multifunctional microplate reader (Tecan, Männedorf, Switzerland), and normalized to the relative activity of *Renilla* luciferase. The experiments were performed in triplicate, and the presented data are averages from three independent experiments.

### 2.4. Transient Transfection in MAC-T Cells

MAC-T cells were seeded in 24-well cell culture plates at a density of approximately $4 \times 10^5$ cells per well, with quadruplicate wells per group. For each well, a 50 nM concentration (final concentration) of the miR-125a miRNA mimic, miRNA-inhibitor or NC (GenePharma) was mixed with 1 μL of Lipofectamine 3000 (Invitrogen) in 50 μL of opti-MEM for 15 min at room temperature. Then, the mixture was added into each well. At 24 h, 48 h and 72 h post-transfection, the cells were collected and processed for further analysis.

### 2.5. RNA Isolation and Quantitative RT-PCR (qRT-PCR)

Total RNA was isolated from transfected MAC-T cells using TRIzol reagent (Invitrogen, Carlsbad, CA, USA), following the manufacturer's instructions. The concentration and purity of the total RNA were determined using a Nanodrop 2000 spectrophotometer (Thermo, Waltham, MA, USA). RNA samples with optical density values between 1.8 and 2.0 at 260/280 nm (OD 260/280) were used in the qRT-PCR analysis. The first-strand complementary DNA (cDNA) was synthesized using an miRcutePlus miRNA First-Strand cDNA Kit (Tiangen, Beijing, China), following the manufacturer's protocol. qRT-PCR was performed using TB Green Premix Ex Taq II (SYBR Green, TAKARA) and a miRcute Plus miRNA qPCR Kit (SYBR Green) (Tiangen, Beijing, China) on a LightCycler 480 II Real-time RT-PCR System (LightCycler, Indianapolis, IN, USA). The levels of the *GAPDH* mRNA or U6 RNA were tested as endogenous controls. The $2^{-\Delta\Delta CT}$ method was applied to calculate the relative expression of the indicated genes. The results are representative of at least three independent experiments. The primers used in the qRT-PCR are listed in Tables 1 and 2.

**Table 1.** Primer sequences for qPCR.

| Name | RT Primer (5′ to 3′) | Forward Primer (5′ to 3′) | Reversed Primer (5′ to 3′) | Tm (°C) |
|---|---|---|---|---|
| bta-miR-125a | CTCAACTGGTGTCGTGGAG TCGGCAATTCAG TTGAG CACAGGTT | GGGCTTCCCTGA GACCCTTT | CTCAACTGGTGTCGTG GAGTC | 60 |
| U6 | | TIANGEN: CD201–0145 | | 60 |

**Table 2.** Primer sequences used for qRT-PCR.

| Gene Name | Forward Primer Sequence | Reverse Primer Sequence | Amplicon (bp) | Tm (°C) |
|---|---|---|---|---|
| *SAA1* | AGTCCACAGCCAGTGGATGT | ATCTCTGAA TATTTTCTCTGGCATC | 2433 | 60 |
| *GAPDH* | AGATGGTGAAGGTCGGAGTG | CGTTCTCTGCCTT GACTGTG | 189 | 60 |
| *MARVELD1* | GGCCAGCTGTAAGATCATCACA | TCTGATCACAGA CAGAGCACCAT | 100 | 60 |
| *FABP3* | GAACTCGACTCCCAGCTTGAA | AAGCCTAC CACAATCATCGAAG | 103 | 59 |
| *FABP4* | TGGATAGTGCAGCCAGTGTGA | TCCAG TGTGATGCGGTGTGTA | 109 | 60 |
| *SCD* | TCCTGTTGTTGTGCTTCATCC | GGCATAACG GAATAAGGTGGC | 101 | 60 |
| *APOA1* | CGGCGGCTTCTCTTGTATAGC | TTCAA GCGTGAGCTGAAACG | 83 | 60 |
| *LPL* | ACACAGCTGAGGACACTTGCC | GCCATGGATCAC CACAAAGG | 101 | 58 |
| *PPARG* | CCAAATATCGGTGGGAGTCG | ACAGCG AAGGGCTCACTCTC | 101 | 58 |
| *SLC27A1* | GTACCAGCACGAAAGGCTCAA | ATCACAC GGCGCTCTTCAA | 120 | 58 |
| *SLC27A4* | CACGGAGGAACTTCAGATGTGA | GGCCCCGC TATACTGACTATGA | 127 | 59 |

### 2.6. Western Blot Analysis

Total proteins were extracted from MAC-T cells using radio-immunoprecipitation assay (RIPA; Solarbio, Beijing, China) lysis buffer containing 1% phenylmethanesulfonyl fluoride (PMSF; Solarbio, Beijing, China) at 24 h, 48 h and 72 h post-transfection. The protein extracts were quantified using a bicinchoninic acid protein kit (BCA; Solarbio, Beijing, China). Approximately 20 µg of total protein was separated by polyacrylamide gel electrophoresis in a 15% SDS-PAGE gel, and transferred to nitrocellulose membranes at 300 mA, which were then probed with an *SAA1*-specific antibody (ABclonal, Wuhan, China) or a β-actin antibody (ABclonal, Wuhan, China). The membrane was then washed three times with Tris-buffered saline and Tween 20, and then probed with a horseradish peroxidase (HRP)-conjugated secondary antibody (Bioss, Beijing, China) at a 1:5000 dilution. Chemiluminescence detection was performed using a SuperKing™ Hypersensitive luminescent ELC solution (Abbkine, Beijing, China).

*2.7. Flow Cytometry*

An Annexin V/PI Kit (Solarbio, Beijing, China) was applied to detect the apoptosis of MAC-T cells following the manufacturer's instructions. 24 h post-transfection, MAC-T cells were washed with PBS, digested with trypsin and collected by centrifuging at 176× $g$ at room temperature for 5 min. The cells were re-suspended in binding buffer and stained sequentially with Annexin V and propidium iodide.

*2.8. Oil Red O Staining*

The lipid droplets in MAC-T cells were stained at 72 h post-transfection using an oil red O staining kit (Solarbio, Beijing, China). The MAC-T cells were washed with PBS, and then fixed and stained with oil red O dye for 30 min. After the oil red O dye was washed away with distilled water, the nucleus was stained with a hematoxylin staining solution for 1 min. Finally, the hematoxylin stain was washed off and the MAC-T cells were covered with distilled water. Lipid droplets were observed and photographed under an inverted microscope.

*2.9. Triglyceride Assay*

The levels of cellular TAG in MAC-T cells were measured using a Triglyceride Assay Kit (Applygen, Beijing, China). All the experiments were performed according to the manufacturer's instructions.

*2.10. Data Analysis*

Statistical analyses were performed using the SPSS Statistics 19.0 statistical software (SPSS Inc., Chicago, IL, USA). All the data are expressed as means ± standard errors (SEs). Student's $t$-test was used to determine the statistical significance of the difference between two groups. ImageJ software was used to analyze the relative content of lipid droplets. $p < 0.05$ was considered to indicate statistical significance, and $p < 0.01$ indicated high statistical significance.

**3. Results**

*3.1. Bta-miR-125a Mimic Suppressed the Gene Expression of SAA1 mRNA by Targeting a Specific Sequence in Its 3′-UTR*

We first ran TargetScan (http://www.targetscan.org/vert_71/, accessed on 27 June 2016) and MiRanda (http://www.microrna.org/microrna/home.do, accessed on 27 June 2016), predicting that the 3′-UTR of the *SAA1* mRNA was a target of bta-miR-125a. To validate the regulatory effect of bta-miR-125a on the expression of *SAA1*, we performed a dual-luciferase reporter assay using a plasmid containing the 3′-UTR sequence of *SAA1* mRNA fused to the open reading frame (ORF) of luciferase. The synthetic bta-miR-125a (miR-mimic) was co-transfected with the plasmid into HEK293 cells. The synthetic inhibitor of bta-miR-125a (miR-inhibitor) and a synthetic control miRNA (NC) were also co-transfected with the plasmid. As shown in Figure 2A, at 24 h post-transfection, the luciferase level in the HEK293 cells transfected with miR-125a mimic decreased by 48%, relative to that in cells with the miRNA control ($p < 0.05$). By contrast, the miR-125a inhibitor yielded the same luciferase level as the negative control (Figure 2A). However, when the predicted binding sites of bta-miR-125a were mutated, the luciferase activity was effectively restored to the control level (Figure 2B). These results suggest that bta-miR-125a may inhibit the expression of *SAA1* by targeting its 3′-UTR in MAC-T cells.

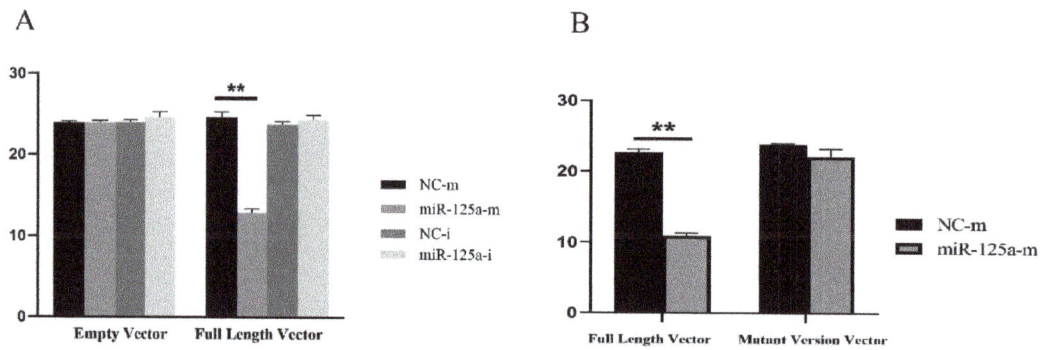


**Figure 2.** MicroRNAs suppressed the expression of *SAA1* by binding to the 3′-UTR target sequence. Luciferase activity in HEK293 cells co-transfected with miRNA mimic, miRNA inhibitor, miRNA control and empty vector for the *SAA1* 3′-UTR. Luciferase activity was assayed 24 h after transfection. All luciferase values were normalized to *Renilla* luciferase. Black columns represent the luciferase activity upon co-transfection with miRNA-mimic control. Middle gray columns represent the luciferase activity upon co-transfection with miRNA-inhibitor control. Dark gray columns represent the luciferase activity upon co-transfection with miRNA mimic. Light gray columns represent the luciferase activity upon co-transfection with miRNA inhibitor. (**A**) represents the luciferase activity of *SAA1* after over- or under-expression of miR-125a compared with controls; (**B**) represents the luciferase activity of *SAA1* after transfecting the mutant vector and miR-125a, compared with control. ** Very significant difference between the control and the treatment ($p < 0.01$).

### 3.2. Bta-miR-125a Inhibits SAA1 Expression in MAC-T Cells

After the transfection of cells with the bta-miR-125a mimic and inhibitor, our flow cytometry findings indicated that there were no obvious differences in the apoptosis rates of MAC-T cells among the groups (Figure 3A).

Our qRT-PCR results showed that the expression of bta-miR-125a was significantly up-regulated in the mimic group, compared with negative-control group (NC; $p < 0.01$), while the opposite trend was found for the expression of bta-miR-125a upon transfection with the inhibitor (Figure 3B). These results indicate that bta-miR-125a was successfully over-expressed and inhibited in the corresponding groups.

In addition, transfection with the miR-125a mimic significantly down-regulated the expression of the *SAA1* gene, compared with that in the NC group ($p < 0.01$); by contrast, transfection with miR-125a inhibitors significantly elevated the expression of the *SAA1* gene, compared with that in the NC-i group ($p < 0.01$); Figure 4A,B).

The results of Western blotting showed that, compared with that in the negative-control group (miR-NC-m), the protein expression of *SAA1* decreased gradually in MAC-T cells transfected with the miR-125a mimic at 48 and 72 h post-transfection, while MAC-T cells transfected with miR-125a inhibitors showed higher levels of the *SAA1* protein than cells in the NC-i group at 48 and 72 h post-transfection (Figure 4C).

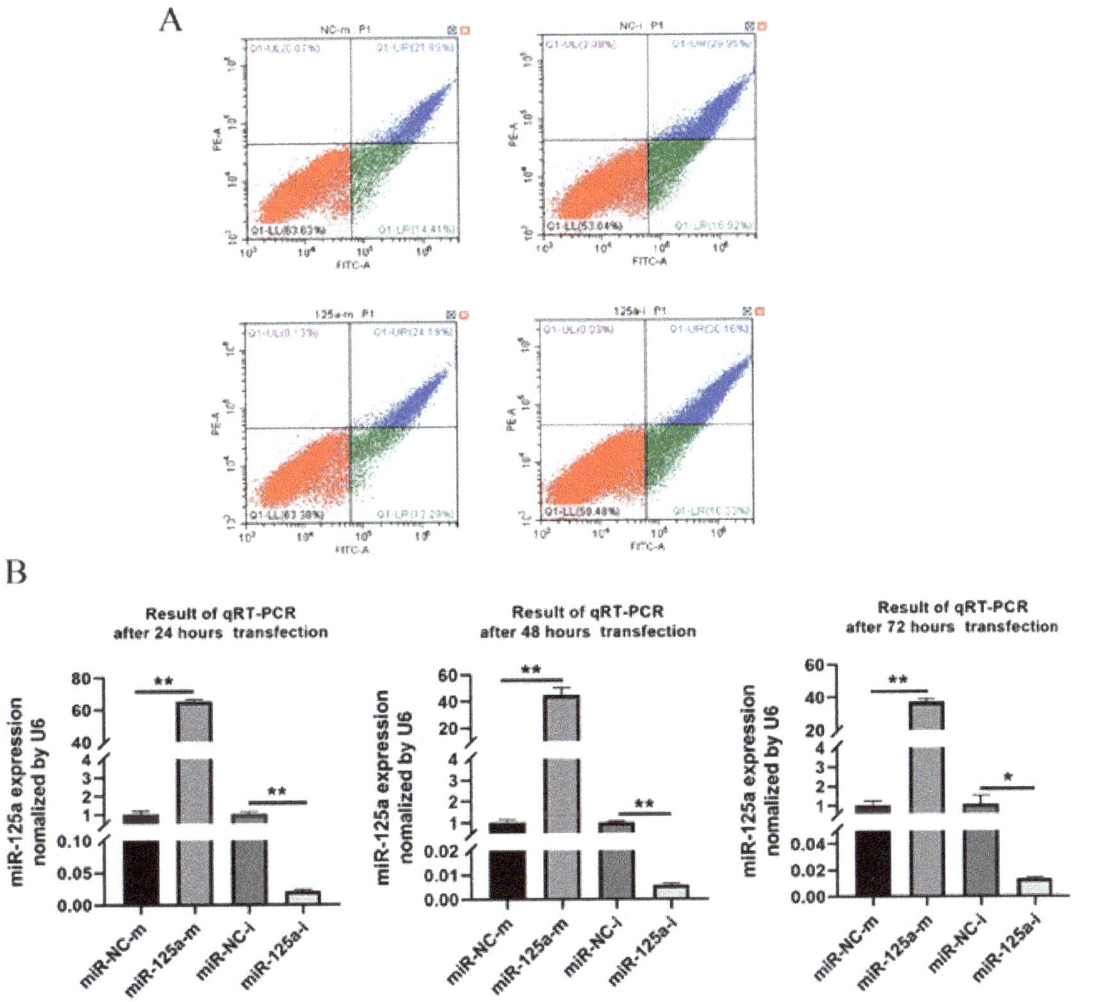


**Figure 3.** (**A**) After transfection with miR-125a mimic and inhibitor for 72 h, flow cytometry was used to detect the apoptosis rate of MAC-T cells; (**B**) Results of qRT-PCR. The expression levels of miRNAs at 24, 48 and 72 h after transfection. * Significant difference between the control and the treatment ($p < 0.05$); ** Very significant difference between the control and the treatment ($p < 0.01$).

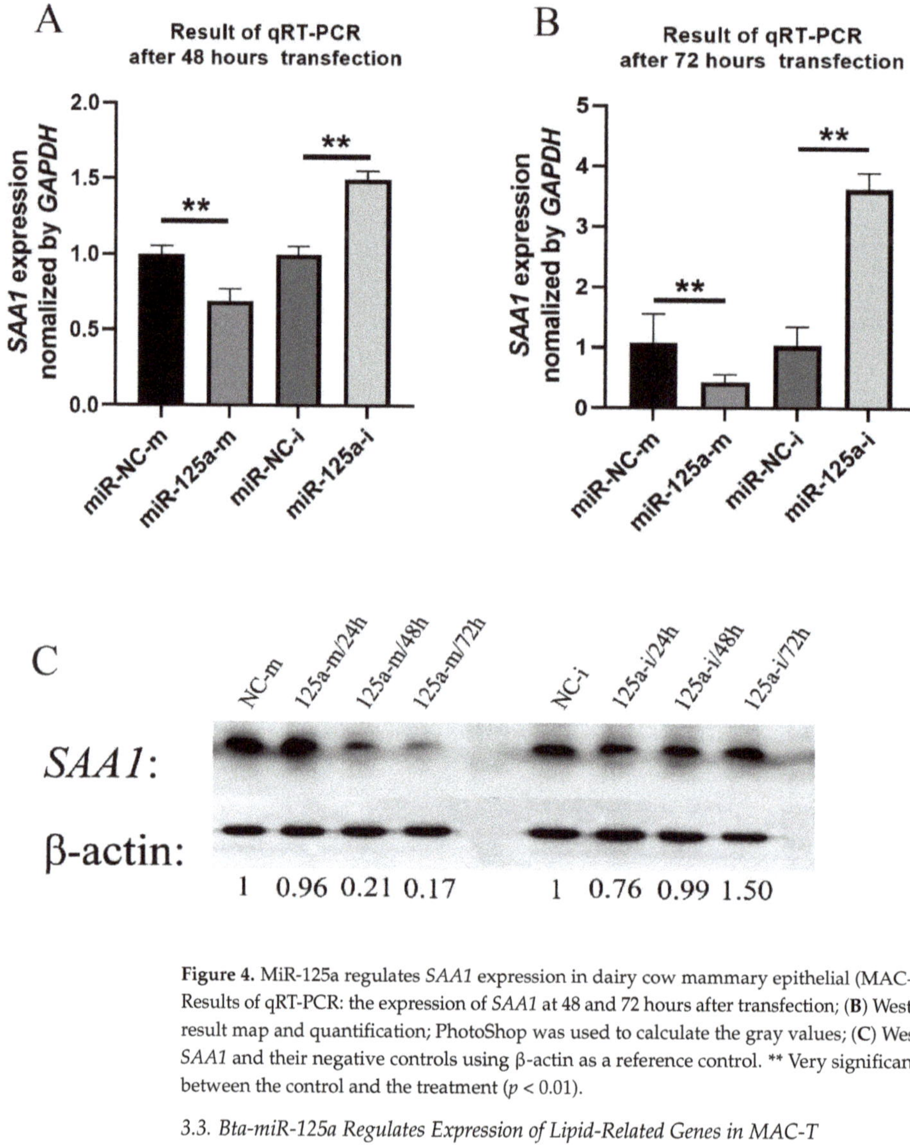


**Figure 4.** MiR-125a regulates *SAA1* expression in dairy cow mammary epithelial (MAC-T) cells. (**A**) Results of qRT-PCR: the expression of *SAA1* at 48 and 72 hours after transfection; (**B**) Western blotting result map and quantification; PhotoShop was used to calculate the gray values; (**C**) Western blot of *SAA1* and their negative controls using β-actin as a reference control. ** Very significant difference between the control and the treatment ($p < 0.01$).

### 3.3. *Bta-miR-125a Regulates Expression of Lipid-Related Genes in MAC-T*

The real-time qPCR results revealed that, relative to the control, the ectopic over-expression of bta-miR-125a strongly up-regulated the expression of *SLC27A1* ($p < 0.05$), *FABP3* ($p < 0.01$), *FABP4* ($p < 0.01$), *LPL* ($p < 0.01$), *PPARG* ($p < 0.01$) and *SLC27A4* ($p < 0.01$); see Figure 5A. By contrast, cells transfected with the bta-miR-125a inhibitor displayed marked down-regulation of *LPL* ($p < 0.05$), *SLC27A1* ($p < 0.05$) and *PPARG* ($p < 0.05$); see Figure 5B.

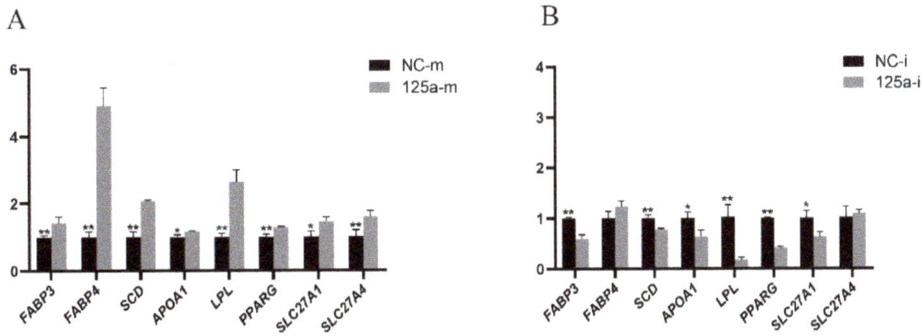


**Figure 5.** (**A**) Expression of lipid-related genes after miR-125a-mimic transfection; (**B**) Expression of lipid-related genes after miR-125a-inhibitor transfection. * Significant difference between the control and the treatment ($p < 0.05$); ** Very significant difference between the control and the treatment ($p < 0.01$).

### *3.4. Bta-miR-125a Enhances Lipogenesis in Bovine Mammary Epithelial Cells*

The results of oil red O staining showed that transfection with the miR mimic increased the aggregation of lipid droplets, whereas bta-miR-125a knockdown suppressed the aggregation of lipid droplets (Figure 6A). Using the ImageJ software to analyze the relative content of lipid droplets, we found that the content of lipid droplets in cells transfected with the miR mimic was higher than that in the NC group and in the group transfected with miR inhibitors (Figure 6B). In addition, the triglyceride assay revealed that transfection with miR mimics increased the triglyceride content, while transfection with miR inhibitors had the opposite effect ($p < 0.01$; Figure 6C).

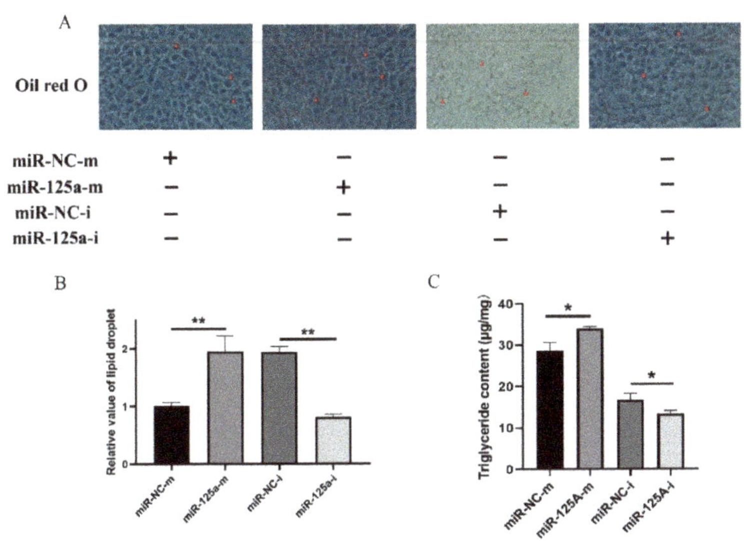


**Figure 6.** (**A**) Oil red O staining: staining at 72 h after transfection, with microscopic examination; (**B**) Oil red O staining: staining at 72 h after transfection, with quantitative analysis of images; (**C**) Triglyceride production by MAC-T cells after miR-125a-mimic and inhibitor transfection for 72 h. All experiments were performed in triplicate. * Significant difference between the control and the treatment ($p < 0.05$); ** Very significant difference between the control and the treatment ($p < 0.01$).

## 4. Discussion

Mammary epithelial cells are the principal milk-producing epithelial cells [30], and are a critical research object for understanding lipid metabolism in dairy cows [31]. Mammary alveolar cell-T (MAC-T) samples were obtained after the transfection of mammary epithelial cells with simian virus-40 (SV-40) large T-antigen, which makes them immortal and stable cells, even after 350 serial passages [32]. Contrary to other established MEC cell lines, MAC-T cells possess the unique characteristic of secreting milk products, and have been widely used in research related to lipid metabolism [33,34]. In the udder, the milk-producing organ in dairy cows, the number of MAC-T cells is considered an essential determinant of milk yield. Therefore, we selected MAC-T cells as an in-vitro model in which to explore milk-fat synthesis. However, compared with cow mammary epithelial primary cells, MAC-T cells have defects, in terms of a relatively insufficient lactation capacity, which we will consider in future research.

Previously, using RNA-seq and small RNA-seq analysis, our group demonstrated that the *SAA1* gene and miR-125a were differentially expressed in the mammary gland between groups of lactating Holstein cows with high and low levels of milk protein and fat percentage [28]. Numerous studies have reported the various effects of the *SAA1* gene in humans [35]; it promotes the release of pro-inflammatory cytokines [36], may induce lipolysis [36,37] and serves as an apolipoprotein of the high-density lipoprotein (HDL) group, which is present in acute-phase serum [38,39]. Previously, we identified *SAA1* as a target for miR-125a [28], which is known to be involved in the differentiation of adipocytes, as well as triglyceride synthesis [40]. In our previous study, *SAA1* and miR-125a were found to be differentially expressed in mammary glands of groups with high and low milk-fat percentages (FP) [28]. Therefore, in this study, we aimed to assess the function of miR-125a and *SAA1* in milk-fat synthesis.

Triglycerides are crucial components of milk fat, representing approximately 98% of the total milk lipids. The content of TAG is an important quality evaluation index for milk [41]. Milk-fat synthesis is affected by multiple physiological, environmental and genetic factors. In cattle, *SAA1* encodes the acute-phase protein serum amyloid A (SAA), which is primarily produced in the liver during the inflammatory response [39]. However, another study has reported a good correlation between amyloid A and mammary inflammation, with a reduced correlation in cows with high SAA [42]. Previous studies have shown that *SAA1* might be involved in mammary gland development through the NF-κB signaling pathway [43], and that *SAA1* over-expression may suppress the growth of mammary epithelial cells [44]. In addition, *SAA1* is a gene essential for milk production in dairy cattle, which negatively regulates milk-fat traits and has been suggested as a genetic marker [12]. In this study, our qRT-PCR and Western blotting results indicate that miR-125a inhibits the expression of *SAA1*. Furthermore, the TAG results reveal that *SAA1* inhibits milk-fat synthesis in MAC-T cells [45]. Therefore, our results demonstrate that *SAA1* may inhibit milk-fat synthesis in MAC-T cells through miR-125a.

MicroRNAs are post-transcriptional regulatory factors that participate in many biological processes, mainly by binding to the 3′ untranslated regions (3′-UTRs) of their target mRNAs and regulating gene expression [46]. Recent studies have reported that miRNAs play an essential regulatory role in milk-fat synthesis. In MAC-T cells, the inhibition of miR-27a-3p, which targets peroxisome proliferator-activated receptor (*PPARG*), may restore the LPS inhibition of milk-fat synthesis. Furthermore, miR-27a-3p inhibition can reverse the LPS-induced down-regulation of *PPARG* expression in LPS-stimulated MAC-T cells [47]. miR-34b has been found to be involved in lipid metabolism, and reduces the accumulation of TAG in primary bovine mammary epithelial cells (BMECs) by targeting lipid-metabolism genes, including *PPARγ*, *C/EBPα*, *FABP4* and *FASN* [48]. Moreover, miR-34b–DCP1A (decapping enzyme 1A) might be an essential axis for milk-fat synthesis in BMECs and the production of beneficial milk components [48]. Furthermore, miR-221 has the ability to inhibit the proliferation of mammary gland epithelial cells by targeting *STAT5a* and *IRS1*, which are considered critical genes in the PI3K–Akt/mTOR and JAK–

STAT signaling pathways, respectively [49]. It has recently been shown that changes in the expression of miR-24 in goat mammary epithelial cells (GMECs) may increase the unsaturated fatty acids and change the TAG content. In ruminant mammary cells, the fatty acid synthase (*FASN*) gene appears to be a direct target of miR-24, where miR-24/*FASN* have exhibited a potential role in controlling lipid metabolism [50]. Moreover, miR-125a-5p was expressed at lower levels in the adipose tissues of mice fed a high-fat diet than in mice fed standard chow. miR-125a-5p expression has also been found to be strongly up-regulated (nearly five-fold), when 3T3-L1 pre-adipocytes were induced to differentiate into mature adipocytes. Functional analysis has indicated that the over-expression of miR-125a-5p promoted the proliferation of 3T3-L1 pre-adipocytes and inhibited their differentiation [51]. Most of the miRNAs in the above-mentioned studies were shown to be involved in milk-fat synthesis. The contrasting results might be attributable to the complex regulatory network of miR-125a, which impairs the expression of genes involved in the lipid synthesis pathway (e.g., *FABP3*, *LPL* and *PPARG*). Our study demonstrated that miR-125a altered lipid accumulation and TAG content, consistent with the results of previous studies on MAC-T and GMECs [52]. The observed increase in cellular TAG was consistent with the increase in milk-fat-related genes (e.g., *PPARG*, *LPL*, *FABP4* and *FABP3*). *PPARG* is a member of the nuclear peroxisome proliferator-activated receptor (PPAR) family, which regulates the transcription of multiple genes and consists of three sub-types: PPAR-$\alpha$, -$\beta$ and -$\gamma$. *PPARG* promotes cellular processes that involve lipid accumulation [53]. In the bovine mammary gland, the mRNA expression level of *PPARG* during lactation is remarkably changeable, suggesting its prominent role in bovine milk-fat synthesis [54,55]. The increased utilization of fatty acids (FAs) in *PPARG*-knockout mice increased the synthesis of inflammatory lipids; hence, the production of toxic pro-inflammatory milk is related to a lack of TAG synthesis [3]. Data from an IPA analysis, indicating that the over-expression of PPARG down- or up-regulated these upstream transcription factors, further supports our previous hypothesis that PPARG regulates the gene network related to fatty acid metabolism, in either a direct or indirect manner [56]. *SAA1*, *PPARG*, *FABP3*, *FABP4*, *SCD*, *APOA1*, *LPL*, *SLC27A1* and *SLC27A4* are enriched in the PPAR signaling pathway, according to the KOBAS/DAVID software. Furthermore, the up-regulation of *PPARG* expression by *FABP3* significantly promoted the accumulation of lipid droplets [47]. Consequently, our results reveal the ability of miR-125a to control lipid accumulation and TAG content through the expression of *SAA1* and other fat-metabolism-related genes in MAC-T. We speculate that miR-125a promotes the formation of lipid droplets in MAC-T cells by targeting *SAA1*.

In this study, we investigated the miR-125a-based regulatory mechanisms of *SAA1* at the cellular level. Our results indicate that miR-125a promotes the formation of lipid droplets in MAC-T cells by targeting *SAA1*. Overall, the results in this paper show that miR-125a can control the synthesis of milk fat in MAC-T cells by targeting *SAA1*.

## 5. Conclusions

Our previous research revealed the genes related to milk-fat metabolism and their corresponding miRNAs. This study reveals that miR-125a significantly down-regulates the expression of *SAA1* through binding to a specific target sequence in its 3′-UTR. The results indicate that these molecules may play critical roles in the regulation of milk-fat metabolism in dairy cattle. More in-depth investigations are required, in order to validate the biological mechanisms of *SAA1* and miR-125a in the formation of milk production traits in dairy cattle.

**Author Contributions:** X.C., Z.F. and T.Y. performed the RNA-related experiments and data analysis; T.Y. and J.F. helped to interpret the results; and X.C. and C.W. wrote the manuscript and participated in the experimental design. All authors have read and agreed to the published version of the manuscript.

**Funding:** This research was funded by the Scientific and Technological Innovation Programs of Higher Education Institutions in Shanxi, China (Grant No. 2019L0037), and Shanxi Province Science Foundation for Youths, China (Grant No. 201801D221285).

**Institutional Review Board Statement:** Not applicable.

**Informed Consent Statement:** Not applicable.

**Data Availability Statement:** All the relevant data are available within the manuscript.

**Conflicts of Interest:** The authors declare no conflict of interest.

## References

1. Melfsen, A.; Holstermann, M.; Haeussermann, A.; Molkentin, J.; Susenbeth, A.; Hartung, E. Accuracy and application of milk fatty acid estimation with diffuse reflectance near-infrared spectroscopy. *J. Dairy Res.* **2018**, *85*, 212–221. [CrossRef] [PubMed]
2. Chen, Z.; Chu, S.; Wang, X.; Fan, Y.; Zhan, T.; Arbab, A.A.I.; Li, M.; Zhang, H.; Mao, Y.; Loor, J.J.; et al. MicroRNA-106b Regulates Milk Fat Metabolism via ATP Binding Cassette Subfamily A Member 1 (ABCA1) in Bovine Mammary Epithelial Cells. *J. Agric. Food Chem.* **2019**, *67*, 3981–3990. [CrossRef] [PubMed]
3. Wan, Y.; Saghatelian, A.; Chong, L.W.; Zhang, C.L.; Cravatt, B.F.; Evans, R.M. Maternal PPAR gamma protects nursing neonates by suppressing the production of inflammatory milk. *Genes Dev.* **2007**, *21*, 1895–1908. [CrossRef] [PubMed]
4. Wang, M.Q.; Zhou, C.H.; Cong, S.; Han, D.X.; Wang, C.J.; Tian, Y.; Zhang, J.B.; Jiang, H.; Yuan, B. Lipopolysaccharide inhibits triglyceride synthesis in dairy cow mammary epithelial cells by upregulating miR-27a-3p, which targets the PPARG gene. *J. Dairy Sci.* **2021**, *104*, 989–1001. [CrossRef]
5. Yu, L.; Wu, W.K.; Li, Z.J.; Liu, Q.C.; Li, H.T.; Wu, Y.C.; Cho, C.H. Enhancement of doxorubicin cytotoxicity on human esophageal squamous cell carcinoma cells by indomethacin and 4-[5-(4-chlorophenyl)-3-(trifluoromethyl)-1H-pyrazol-1-yl]benzenesulfonamide (SC236) via inhibiting P-glycoprotein activity. *Mol. Pharmacol.* **2009**, *75*, 1364–1373. [CrossRef] [PubMed]
6. Digre, A.; Nan, J.; Frank, M.; Li, J.P. Heparin interactions with apoA1 and SAA in inflammation-associated HDL. *Biochem. Biophys. Res. Commun.* **2016**, *474*, 309–314. [CrossRef]
7. Yamazaki, K.; Kuromitsu, J.; Tanaka, I. Microarray analysis of gene expression changes in mouse liver induced by peroxisome proliferator- activated receptor alpha agonists. *Biochem. Biophys. Res. Commun.* **2002**, *290*, 1114–1122. [CrossRef] [PubMed]
8. Ray, B.K.; Ray, A. Rabbit serum amyloid a gene: Cloning, characterization and sequence analysis. *Biochem. Biophys. Res. Commun.* **1991**, *180*, 1258–1264. [CrossRef]
9. Li, D.; Xie, P.; Zhao, S.; Zhao, J.; Yao, Y.; Zhao, Y.; Ren, G.; Liu, X. Hepatocytes derived increased SAA1 promotes intrahepatic platelet aggregation and aggravates liver inflammation in NAFLD. *Biochem. Biophys. Res. Commun.* **2021**, *555*, 54–60. [CrossRef]
10. Gan, X.W.; Wang, W.S.; Lu, J.W.; Ling, L.J.; Zhou, Q.; Zhang, H.J.; Ying, H.; Sun, K. De novo Synthesis of SAA1 in the Placenta Participates in Parturition. *Front. Immunol.* **2020**, *11*, 1038. [CrossRef]
11. Siegmund, S.V.; Schlosser, M.; Schildberg, F.A.; Seki, E.; De Minicis, S.; Uchinami, H.; Kuntzen, C.; Knolle, P.A.; Strassburg, C.P.; Schwabe, R.F. Serum Amyloid a Induces Inflammation, Proliferation and Cell Death in Activated Hepatic Stellate Cells. *PLoS ONE* **2016**, *11*, e0150893. [CrossRef]
12. Yang, S.; Gao, Y.; Zhang, S.; Zhang, Q.; Sun, D. Identification of Genetic Associations and Functional Polymorphisms of SAA1 Gene Affecting Milk Production Traits in Dairy Cattle. *PLoS ONE* **2016**, *11*, e0162195. [CrossRef] [PubMed]
13. Whitehead, A.S.; de Beer, M.C.; Steel, D.M.; Rits, M.; Lelias, J.M.; Lane, W.S.; de Beer, F.C. Identification of novel members of the serum amyloid A protein superfamily as constitutive apolipoproteins of high density lipoprotein. *J. Biol. Chem.* **1992**, *267*, 3862–3867. [CrossRef]
14. De Buck, M.; Gouwy, M.; Wang, J.M.; Van Snick, J.; Opdenakker, G.; Struyf, S.; Van Damme, J. Structure and Expression of Different Serum Amyloid A (SAA) Variants and their Concentration-Dependent Functions During Host Insults. *Curr. Med. Chem.* **2016**, *23*, 1725–1755. [CrossRef]
15. Del Pozo-Acebo, L.; Hazas, M.; Tomé-Carneiro, J.; Gil-Cabrerizo, P.; San-Cristobal, R.; Busto, R.; García-Ruiz, A.; Dávalos, A. Bovine Milk-Derived Exosomes as a Drug Delivery Vehicle for miRNA-Based Therapy. *Int. J. Mol. Sci.* **2021**, *22*, 1105. [CrossRef] [PubMed]
16. Lewis, B.P.; Shih, I.H.; Jones-Rhoades, M.W.; Bartel, D.P.; Burge, C.B. Prediction of mammalian microRNA targets. *Cell* **2003**, *115*, 787–798. [CrossRef]
17. Lim, L.P.; Glasner, M.E.; Yekta, S.; Burge, C.B.; Bartel, D.P. Vertebrate microRNA genes. *Science* **2003**, *299*, 1540. [CrossRef]
18. Silveri, L.; Tilly, G.; Vilotte, J.L.; Le Provost, F. MicroRNA involvement in mammary gland development and breast cancer. *Reprod. Nutr. Dev.* **2006**, *46*, 549–556. [CrossRef] [PubMed]
19. Wang, X.; Zhang, L.; Jin, J.; Xia, A.; Wang, C.; Cui, Y.; Qu, B.; Li, Q.; Sheng, C. Comparative transcriptome analysis to investigate the potential role of miRNAs in milk protein/fat quality. *Sci. Rep.* **2018**, *8*, 6250. [CrossRef]
20. Ibarra, I.; Erlich, Y.; Muthuswamy, S.K.; Sachidanandam, R.; Hannon, G.J. A role for microRNAs in maintenance of mouse mammary epithelial progenitor cells. *Genes Dev.* **2007**, *21*, 3238–3243. [CrossRef]

21. Li, M.; Li, Q.; Gao, X. Expression and function of leptin and its receptor in dairy goat mammary gland. *J. Dairy Res.* **2010**, *77*, 213–219. [CrossRef] [PubMed]
22. Liu, S.; Patel, S.H.; Ginestier, C.; Ibarra, I.; Martin-Trevino, R.; Bai, S.; McDermott, S.P.; Shang, L.; Ke, J.; Ou, S.J.; et al. MicroRNA93 regulates proliferation and differentiation of normal and malignant breast stem cells. *PLoS Genet.* **2012**, *8*, e1002751. [CrossRef]
23. Ucar, A.; Vafaizadeh, V.; Jarry, H.; Fiedler, J.; Klemmt, P.A.; Thum, T.; Groner, B.; Chowdhury, K. miR-212 and miR-132 are required for epithelial stromal interactions necessary for mouse mammary gland development. *Nat. Genet.* **2010**, *42*, 1101–1108. [CrossRef] [PubMed]
24. Shen, B.; Pan, Q.; Yang, Y.; Gao, Y.; Liu, X.; Li, W.; Han, Y.; Yuan, X.; Qu, Y.; Zhao, Z. miR-224 Affects Mammary Epithelial Cell Apoptosis and Triglyceride Production by Downregulating ACADM and ALDH2 Genes. *DNA Cell Biol.* **2017**, *36*, 26–33. [CrossRef] [PubMed]
25. Li, X.; Jiang, P.; Yu, H.; Yang, Y.; Xia, L.; Yang, R.; Fang, X.; Zhao, Z. miR-21-3p Targets Elovl5 and Regulates Triglyceride Production in Mammary Epithelial Cells of Cow. *DNA Cell Biol.* **2019**, *38*, 352–357. [CrossRef] [PubMed]
26. Li, H.M.; Wang, C.M.; Li, Q.Z.; Gao, X.J. MiR-15a decreases bovine mammary epithelial cell viability and lactation and regulates growth hormone receptor expression. *Molecules* **2012**, *17*, 12037–12048. [CrossRef] [PubMed]
27. Cui, X.; Hou, Y.; Yang, S.; Xie, Y.; Zhang, S.; Zhang, Y.; Zhang, Q.; Lu, X.; Liu, G.E.; Sun, D. Transcriptional profiling of mammary gland in Holstein cows with extremely different milk protein and fat percentage using RNA sequencing. *BMC Genom.* **2014**, *15*, 226. [CrossRef]
28. Cui, X.; Zhang, S.; Zhang, Q.; Guo, X.; Wu, C.; Yao, M.; Sun, D. Comprehensive MicroRNA Expression Profile of the Mammary Gland in Lactating Dairy Cows with Extremely Different Milk Protein and Fat Percentages. *Front. Genet.* **2020**, *11*, 548268. [CrossRef] [PubMed]
29. Cai, X.; Liu, Q.; Zhang, X.; Ren, Y.; Lei, X.; Li, S.; Chen, Q.; Deng, K.; Wang, P.; Zhang, H.; et al. Identification and analysis of the expression of microRNA from lactating and nonlactating mammary glands of the Chinese swamp buffalo. *J. Dairy Sci.* **2017**, *100*, 1971–1986. [CrossRef]
30. Perruchot, M.H.; Arévalo-Turrubiarte, M.; Dufreneix, F.; Finot, L.; Lollivier, V.; Chanat, E.; Mayeur, F.; Dessauge, F. Mammary Epithelial Cell Hierarchy in the Dairy Cow Throughout Lactation. *Stem Cells Dev.* **2016**, *25*, 1407–1418. [CrossRef]
31. Xia, L.; Zhao, Z.; Yu, X.; Lu, C.; Jiang, P.; Yu, H.; Li, X.; Yu, X.; Liu, J.; Fang, X.; et al. Integrative analysis of miRNAs and mRNAs revealed regulation of lipid metabolism in dairy cattle. *Funct. Integr. Genom.* **2021**, *21*, 393–404. [CrossRef]
32. Rejman, J.J.; Oliver, S.P.; Muenchen, R.A.; Turner, J.D. Proliferation of the MAC-T bovine mammary epithelial cell line in the presence of mammary secretion whey proteins. *Cell Biol. Int. Rep.* **1992**, *16*, 993–1001. [CrossRef]
33. Johnson, T.L.; Fujimoto, B.A.; Jiménez-Flores, R.; Peterson, D.G. Growth hormone alters lipid composition and increases the abundance of casein and lactalbumin mRNA in the MAC-T cell line. *J. Dairy Res.* **2010**, *77*, 199–204. [CrossRef]
34. Wang, T.; Lee, H.; Zhen, Y. Responses of MAC-T Cells to Inhibited Stearoyl-CoA Desaturase 1 during cis-9, trans-11 Conjugated Linoleic Acid Synthesis. *Lipids* **2018**, *53*, 647–652. [CrossRef] [PubMed]
35. Jéru, I.; Hayrapetyan, H.; Duquesnoy, P.; Cochet, E.; Serre, J.L.; Feingold, J.; Grateau, G.; Sarkisian, T.; Jeanpierre, M.; Amselem, S. Involvement of the modifier gene of a human Mendelian disorder in a negative selection process. *PLoS ONE* **2009**, *4*, e7676. [CrossRef] [PubMed]
36. Yang, R.Z.; Lee, M.J.; Hu, H.; Pollin, T.I.; Ryan, A.S.; Nicklas, B.J.; Snitker, S.; Horenstein, R.B.; Hull, K.; Goldberg, N.H.; et al. Acute-phase serum amyloid A: An inflammatory adipokine and potential link between obesity and its metabolic complications. *PLoS Med.* **2006**, *3*, e287. [CrossRef] [PubMed]
37. Wang, Y.C.; Kuo, W.H.; Chen, C.Y.; Lin, H.Y.; Wu, H.T.; Liu, B.H.; Chen, C.H.; Mersmann, H.J.; Chang, K.J.; Ding, S.T. Docosahexaenoic acid regulates serum amyloid A protein to promote lipolysis through down regulation of perilipin. *J. Nutr. Biochem.* **2010**, *21*, 317–324. [CrossRef] [PubMed]
38. Benditt, E.P.; Eriksen, N. Amyloid protein SAA is associated with high density lipoprotein from human serum. *Proc. Natl. Acad. Sci. USA* **1977**, *74*, 4025–4028. [CrossRef] [PubMed]
39. Sun, L.; Ye, R.D. Serum amyloid A1: Structure, function and gene polymorphism. *Gene* **2016**, *583*, 48–57. [CrossRef]
40. Du, J.; Xu, Y.; Zhang, P.; Zhao, X.; Gan, M.; Li, Q.; Ma, J.; Tang, G.; Jiang, Y.; Wang, J.; et al. MicroRNA-125a-5p Affects Adipocytes Proliferation, Differentiation and Fatty Acid Composition of Porcine Intramuscular Fat. *Int. J. Mol. Sci.* **2018**, *19*, 501. [CrossRef] [PubMed]
41. Perretta, L.; Ouldibbat, L.; Hagadorn, J.I.; Brumberg, H.L. High versus low medium chain triglyceride content of formula for promoting short-term growth of preterm infants. *Cochrane Database Syst. Rev.* **2021**, *2*, Cd002777. [CrossRef] [PubMed]
42. O'Mahony, M.C.; Healy, A.M.; Harte, D.; Walshe, K.G.; Torgerson, P.R.; Doherty, M.L. Milk amyloid A: Correlation with cellular indices of mammary inflammation in cows with normal and raised serum amyloid A. *Res. Vet. Sci.* **2006**, *80*, 155–161. [CrossRef]
43. Kho, Y.; Kim, S.; Yoon, B.S.; Moon, J.H.; Kim, B.; Kwak, S.; Woo, J.; Oh, S.; Hong, K.; Kim, S.; et al. Induction of serum amyloid A genes is associated with growth and apoptosis of HC11 mammary epithelial cells. *Biosci. Biotechnol. Biochem.* **2008**, *72*, 70–81. [CrossRef] [PubMed]
44. Huang, W.; Khatib, H. Comparison of transcriptomic landscapes of bovine embryos using RNA-Seq. *BMC Genom.* **2010**, *11*, 711. [CrossRef] [PubMed]
45. Wang, Y.; Cao, F.; Wang, Y.; Yu, G.; Jia, B.L. Silencing of SAA1 inhibits palmitate- or high-fat diet induced insulin resistance through suppression of the NF-κB pathway. *Mol. Med.* **2019**, *25*, 17. [CrossRef]

46. HafezQorani, S.; Lafzi, A.; de Bruin, R.G.; van Zonneveld, A.J.; van der Veer, E.P.; Son, Y.A.; Kazan, H. Modeling the combined effect of RNA-binding proteins and microRNAs in post-transcriptional regulation. *Nucleic Acids Res.* **2016**, *44*, e83. [CrossRef] [PubMed]
47. Liang, M.Y.; Hou, X.M.; Qu, B.; Zhang, N.; Li, N.; Cui, Y.J.; Li, Q.Z.; Gao, X.J. Functional analysis of FABP3 in the milk fat synthesis signaling pathway of dairy cow mammary epithelial cells. *Vitr. Cell. Dev. Biol. Anim.* **2014**, *50*, 865–873. [CrossRef] [PubMed]
48. Wang, Y.; Guo, W.; Tang, K.; Wang, Y.; Zan, L.; Yang, W. Bta-miR-34b regulates milk fat biosynthesis by targeting mRNA decapping enzyme 1A (DCP1A) in cultured bovine mammary epithelial cells1. *J. Anim. Sci.* **2019**, *97*, 3823–3831. [CrossRef]
49. Jiao, B.L.; Zhang, X.L.; Wang, S.H.; Wang, L.X.; Luo, Z.X.; Zhao, H.B.; Khatib, H.; Wang, X. MicroRNA-221 regulates proliferation of bovine mammary gland epithelial cells by targeting the STAT5a and IRS1 genes. *J. Dairy Sci.* **2019**, *102*, 426–435. [CrossRef]
50. Wang, M.; Li, L.; Liu, R.; Song, Y.; Zhang, X.; Niu, W.; Kumar, A.K.; Guo, Z.; Hu, Z. Obesity-induced overexpression of miRNA-24 regulates cholesterol uptake and lipid metabolism by targeting SR-B1. *Gene* **2018**, *668*, 196–203. [CrossRef]
51. Xu, Y.; Du, J.; Zhang, P.; Zhao, X.; Li, Q.; Jiang, A.; Jiang, D.; Tang, G.; Jiang, Y.; Wang, J.; et al. MicroRNA-125a-5p Mediates 3T3-L1 Preadipocyte Proliferation and Differentiation. *Molecules* **2018**, *23*, 317. [CrossRef] [PubMed]
52. Kang, Y.; Hengbo, S.; Jun, L.; Jun, L.; Wangsheng, Z.; Huibin, T.; Huaiping, S. PPARG modulated lipid accumulation in dairy GMEC via regulation of ADRP gene. *J. Cell. Biochem.* **2015**, *116*, 192–201. [CrossRef] [PubMed]
53. Deng, K.; Ren, C.; Fan, Y.; Liu, Z.; Zhang, G.; Zhang, Y.; You, P.; Wang, F. miR-27a is an important adipogenesis regulator associated with differential lipid accumulation between intramuscular and subcutaneous adipose tissues of sheep. *Domest. Anim. Endocrinol.* **2020**, *71*, 106393. [CrossRef] [PubMed]
54. Bionaz, M.; Loor, J.J. Gene networks driving bovine milk fat synthesis during the lactation cycle. *BMC Genom.* **2008**, *9*, 366. [CrossRef]
55. Kadegowda, A.K.; Bionaz, M.; Piperova, L.S.; Erdman, R.A.; Loor, J.J. Peroxisome proliferator-activated receptor-gamma activation and long-chain fatty acids alter lipogenic gene networks in bovine mammary epithelial cells to various extents. *J. Dairy Sci.* **2009**, *92*, 4276–4289. [CrossRef] [PubMed]
56. Minuti, A.; Bionaz, M.; Lopreiato, V.; Janovick, N.A.; Rodriguez-Zas, S.L.; Drackley, J.K.; Loor, J.J. Prepartum dietary energy intake alters adipose tissue transcriptome profiles during the periparturient period in Holstein dairy cows. *J. Anim. Sci. Biotechnol.* **2020**, *11*, 1–14. [CrossRef]

agriculture

Article

# *Lycium barbarum* Polysaccharide Inhibits *E. coli*-Induced Inflammation and Oxidative Stress in Mammary Epithelial Cells of Dairy Cows via SOCS3 Activation and MAPK Suppression

**Run Liu [1], Hao Zhu [1], Jingwen Zhao [1], Xinyue Wu [1], Xubin Lu [1], Tianle Xu [2] and Zhangping Yang [1,2,*]**

[1] College of Animal Science and Technology, Yangzhou University, Yangzhou 225009, China; mx120200803@stu.yzu.edu.cn (R.L.); mx120210848@stu.yzu.edu.cn (H.Z.); mz120191038@stu.yzu.edu.cn (J.Z.); mz120201356@stu.yzu.edu.cn (X.W.); dx120180094@yzu.edu.cn (X.L.)

[2] Joint International Research Laboratory of Agriculture and Agri-Product Safety of Ministry of Education of China, Yangzhou University, Yangzhou 225009, China; tl-xu@yzu.edu.cn

* Correspondence: yzp@yzu.edu.cn

**Abstract:** *Escherichia coli* (*E. coli*) is one of the main causative agents of mastitis in dairy cows. *Lycium barbarum* polysaccharide (LBP) has a variety of physiological effects as it has antioxidants, it is hypoglycemic, it has anti-aging properties, it is neuroprotective, immune boosting, and it has anti-inflammatory effects in vivo and in vitro. In this study, we examined whether LBP affects the expression of pro-inflammatory factors, and the mitogen-activated protein kinase (MAPK) signaling pathway via activation of the suppressor of cytokine signaling-3 (SOCS3) in *E. coli*-induced primary bovine mammary epithelial cell (pbMEC) inflammatory responses. The experiment was designed with the control group (**NC**), cells were treated with *E. coli* for 6 h as the *E. coli* group (**E. coli**), and cells were pretreated with 100 μg/mL or 300 μg/mL of LBP for 24 h, followed by the addition of *E. coli* for 6 h as the *E. coli* + low level (E + LL) or *E. coli* + high level (E + HL) groups. The addition of LBP did not alter the cell viability of pbMEC in a dose-dependent assay. Pretreatment with LBP significantly decreased the expression of pro-inflammatory genes (IL1B, MAPK14, COX-2, iNOS) and proteins (COX-2, IL-1β, TNF-α) in the cells challenged by *E. coli* as compared with the control group ($p < 0.05$). *E. coli* stimulation significantly increased the production of reactive oxygen species (ROS) and malondialdehyde (MDA) in pbMEC, and decreased the antioxidants' capacity with regard to decreased superoxide dismutase (SOD) and total antioxidant capacity (T-AOC); however, pretreatment with LBP reversed the oxidative stress and inhibition of antioxidants in cells challenged by *E. coli*. Moreover, LBP reversed the upregulated expression of the components of the MAPK pathway (increased phosphorylation level of p38, JNK, and ERK), followed by *E. coli* stimulation. Consistently, cells exposed to *E. coli* strengthened the staining of p38, whereas pretreatment of LBP weakened the staining of p38 in cells challenged by *E. coli*. Notably, the expression of SOCS3 was increased by LBP added to the cells in a dose-dependent manner. Additionally, the level of decreased expression of proinflammatory factors (IL-1β, TNF-α, and COX-2) was higher in the E + LL group than in the E + HL group. These results indicate that LBP pretreatment is effective in the alleviation of *E. coli*-induced inflammatory and oxidative responses in pbMEC through activation of SOCS3 and depression of MAPK signaling. As such, this might help us to develop molecular strategies for mitigating the detrimental effects of clinical bovine mastitis.

**Keywords:** *Lycium barbarum* polysaccharide; bovine mammary epithelial cells; *E. coli*; MAPK signaling; SOCS3

**Citation:** Liu, R.; Zhu, H.; Zhao, J.; Wu, X.; Lu, X.; Xu, T.; Yang, Z. *Lycium barbarum* Polysaccharide Inhibits *E. coli*-Induced Inflammation and Oxidative Stress in Mammary Epithelial Cells of Dairy Cows via SOCS3 Activation and MAPK Suppression. *Agriculture* **2022**, *12*, 598. https://doi.org/10.3390/agriculture12050598

Academic Editor: M. D. (Mo) Salman

Received: 11 March 2022
Accepted: 20 April 2022
Published: 24 April 2022



## 1. Introduction

Bovine mastitis is one of the serious diseases that causes huge economic losses for dairy farming. This is mainly because the invasion of pathogenic microorganisms such as *E. coli* into mammary tissue leads to a storm of inflammation and tissue damage [1–4]. Studies have shown that *E. coli*-induced inflammation of the mammary gland has a large effect on cows' milk production and milk quality [5]. At present, antibiotics are still one of the most effective in the treatment for bovine mastitis; however, the drug resistance of pathogens, as well as drug residues in dairy products, have been an issue for the dairy industry and public health. Hence, looking for alternatives such as natural substances to treat bovine mastitis is of great importance [6,7].

The primary bovine mammary epithelial cells (pbMEC) constitute an important line of defense against the invasion of pathogens [8,9]. Moreover, pbMEC tries to express a variety of pathogen patterns, identify receptors, and activate downstream signaling pathways to trigger the production of inflammatory cytokines [10,11]. Cytokines such as COX-2, IL6, and TNF-α have been reported to be induced by activating the nuclear factor kappa-B (NF-κB) and MAPK signaling pathways [12]. Thus, the suppression of inflammation is usually mediated by inhibiting the NF-κB and MAPK signaling pathways [13]. Nevertheless, SOCS family cytokine signal transduction inhibition is a negative regulatory protein that may reduce the overactivation of MAPK and NF-κB [14,15]. Furthermore, it has been revealed that SOCS1 and SOCS3 reduce the inflammatory responses in BV-2 microglia by inhibiting the activation of the TLR/NF-κB [16–18].

LBP is a major active ingredient and water-soluble polysaccharide extracted from Goji berries [19,20]. A study has shown the anti-inflammatory, antioxidant, and antiapoptotic properties of LBP [21]. Studies have also reported that it appears to protect human lens epithelial cells and mouse tissues from oxidative damage [22,23]. Recent studies with bovine hepatocytes have shown that LBP inhibits the activation of NF-κB to decrease chemokines [24]. Preliminary research in our laboratory has shown the anti-inflammatory potential of LBP. On the basis of available data, we hypothesized that LBP might be effective at protecting cells against inflammatory responses and oxidative stress induced by *E. coli* in pbMEC. Hence, the aim of this study was to investigate the protective effect of LBP against *E. coli*-induced inflammation responses and the oxidative stress of pbMEC through SOCS3 activation and MAPK suppression. The study may provide new insights into therapeutic strategies for combating clinical mastitis using LBP as diet supplement or medical agent and highlighted the link between SOCS3 and regulation of mammary inflammation.

## 2. Materials and Methods

### *2.1. Experimental Design*

We optimized the concentration and time aspects of the treatment of *E. coli* or LBP, cells were cultured with *E. coli* ($1 \times 10^7$/mL) for 6 h as positive control (**E. coli**); cells were pre-treated with (100 μg/mL) LBP for 24 h, then washed, followed by the addition of *E. coli* exposure for 6 h (E + LL group, **E + LL**). Similarly, for cells pretreated with (300 μg/mL) LBP for 24 h, followed by the addition of *E. coli* and exposure for 6 h (E + HL group, **E + HL**), an equal amount of PBS was added to the culture as the negative control (**NC**). The concentration screening of LBP and *E. coli* has been previously studied in this laboratory. Before adding LBP and *E. coli* ($1 \times 10^7$/mL), cells were incubated using serum-free medium for 24 h.

### *2.2. Chemicals*

LBP was purchased from Solarbio (P7850, Solarbio Life Sciences, Beijing, China) with a purity of > 50%, and *E. coli* was isolated and identified as a B1 type from the milk of clinical bovine mastitis.

### *2.3. Cell Culture Conditions*

pbMEC were isolated and obtained by our laboratory from bovine mammary tissue of three lactating cows without signs of clinical disease. The pbMEC in good growth condition were inoculated in 6 well plates ($2.5 \times 10^5$/well) using complete medium (90% RPMI 1640, (Gibco, Temecula, CA, USA), 10% fetal bovine serum (Thermo Fisher Scientific, Waltham,

MA, USA), and antibiotics (penicillin 100 IU/mL; streptomycin 100 μg/mL). The medium was replaced with a fresh medium, free of antibiotics, during *E. coli* stimulation. Cells were incubated at conditions of 5% $CO_2$ and 37 °C.

### 2.4. Cell Viability

Cell viability was measured using a cell counting kit-8 (CCK-8) (Vazyme, Nanjing, China). When the cell density was appropriate ($1 \times 10^6$), different concentrations of LBP were added at 0, 25, 100, 300, 500, and 1000 μg/mL for 24 h, then 10 μL of CCK-8 solution was added for 2 h, and the absorbance was measured at 450 nm wavelength.

### 2.5. RNA Extraction and Quantitative RT-PCR

Total RNA from cells was extracted using RNA-easy Isolation Reagent (Vazyme, Nanjing, China). The extracted RNA was performed using a HiScript III 1st Strand cDNA Synthesis Kit (Vazyme, Nanjing, China) to reverse transcribe cDNA. qRT-PCR was performed using HiScript II One Step qRT-PCR SYBR Green Kit (Vazyme, Nanjing, China) on an Applied Biosystems 7300 Real-Time PCR System with Primer 5.0 to design primers for IL1B, MAPK14, COX2, iNOS, IL6, TNFA UXT, RPS9, and GAPDH, whose sequences are listed in Supplementary Table S1. The CT values were recorded by RT-PCR using AceQ® Probe Master Mix, and the results were analyzed using the $2^{-\Delta\Delta Ct}$ method.

### 2.6. ROS Determination

The intracellular ROS levels were measured using a Reactive Oxygen Species Assay Kit (Beyotime Biotechnology, Nantong, China). The 2,7-Dichlorodihydrofluorescein diacetate (DCFH-DA) probe was diluted to 10 μM/L according to a ratio of 1:1000, cells were collected and suspended in the DCFH-DA solution, and incubated in the incubator at 37 °C for 20 min. The mixture was then centrifuged at 1000 rpm for 5 min. Subsequently, pellets were resuspended in PBS and measured at 488 nm excitation and 525 nm emission.

### 2.7. Antioxidant Enzyme Activity

The cells were collected to fully disrupt and release the antioxidants therein, centrifuged at 4 °C and 12,000× *g* for 5 min, and the supernatant was taken for subsequent determination. According to the manufacturer's protocol (Shanghai Beyotime Biotechnology, Shanghai, China), MDA levels, T-AOC, and SOD were detected with biochemical kits (Shanghai Beyotime Biotechnology, Shanghai, China). For all kits, after adding the working solution, it was mixed well and centrifuged accordingly, as indicated in the instructions of the kits. Absorbance values were measured and used to calculate output by the formulas according to the protocols.

### 2.8. Immunofluorescence

Cell crawls were placed in 12 well plates and inoculated with ($2 \times 10^4$) cells per well. The medium was discarded, washed in PBS, fixed in 4% paraformaldehyde for 15 min, washed 3 times in PBS, incubated in a 0.5% Triton X-100 permeation solution for 15 min at RT, closed in a 5% bovine serum albumin (BSA) blocking buffer for 1 h at room temperature, washed three times in PBS, and incubated with a primary antibody at 4 °C overnight. It was then washed three times with PBS, and incubated for 1 h in the dark at 37 °C using a goat anti-rabbit Cy3-labeled secondary antibody. Nuclei were stained with diamidinophenyl-indole (DAPI) (1 μg/mL) (D8417, Sigma-Aldrich, St. Louis, MO, USA) for 5 min, washed three times in PBS, transferred to slides, and sealed with a fluorescent quencher. Images were then captured using Leica DMi8 fluorescent microscopy.

### 2.9. Western Blot

Cells in 6-well plates were placed on ice with radio immunoprecipitation assay (RIPA) lysate (Beyotime, Shanghai, China) for total protein extraction. Then, 10% SDS-PAGE gel was used to separate each target proteins. The blots were incubated for 2 h at RT with horseradish peroxidase coupled (HRP) secondary antibodies, followed by the incubation of primary antibodies overnight at 4 °C, which were then washed 3 times with 1 × TBS + Triton mixed solution (TBST). Differences in protein transfer efficiency between blots were normalized with a Histone-H3 quantification. The gray values of the bands of each target protein were quantified using ImageJ system analysis software (National Institutes of Health, Bethesda, MD, USA). Primary antibodies (p65, p-p65, MAPK, p-MAPK, ERK, p-ERK, JNK, p-JNK, Socs3, IL-1β, TNF-α, Histone-H3) raised from rabbit were purchased from Cell Signaling Technology (Danvers, MA, USA; #8242, #3033, #8690, #4631, #4695, #4370, #9252, #4671, #52,113, #12,703, #6945, #4499), and were diluted 1:1000 for incubation.

### *2.10. Statistics*

Data are expressed as the mean ± standard error of the means (mean ± SEM). All data were analyzed using one-way ANOVA with Dunnett's post-test by SPSS Statistics for Win (IBM Inc., New York, NY, USA). Differences with $p$ values < 0.05 were considered statistically significant. Experiments were performed in triplicate.

## 3. Results

### *3.1. The Effect of Cell Viability of LBP on pbMEC*

pbMEC were treated with different concentrations of LBP at 0, 25, 100, 300, 500, and 1000 μg/mL for 24 h. The effect of different doses of LBP on pbMEC viability was measured by CCK-8. Compared with the control group (0 μg/mL), cell viability was not affected by pretreatment with LBP from 25 to 1000 μg/mL. As the concentration of LBP increased, it did not cause cell apoptosis ($p > 0.05$, Figure 1).

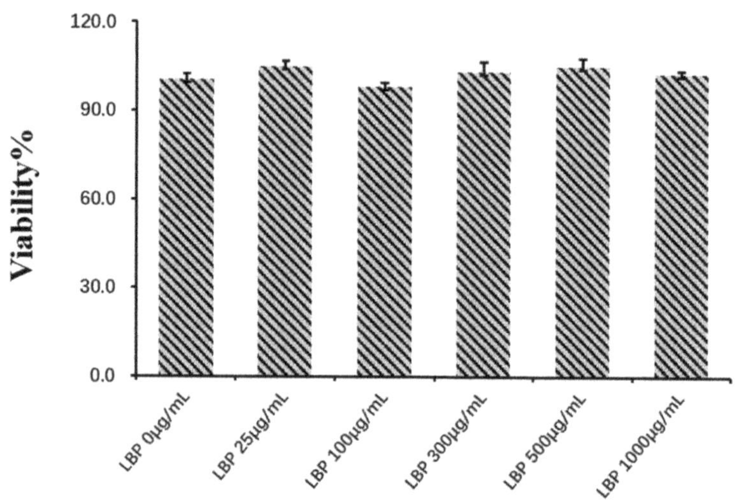


**Figure 1.** Determination of cell viability in different doses of LBP via CCK-8 assay. Effects of LBP on pbMEC viability. pbMEC were treated with different concentrations of LBP at 0, 25, 100, 300, 500, and 1000 μg/mL for 24 h. Determination of cell viability in different doses of LBP via Cell Counting Kit-8 assay. Results are presented as means ± SEM. Superscripts represent that the difference between groups was significant according to statistical analysis ($p < 0.05$).

### *3.2. The Effect of LBP on the Antioxidant Capacity of pbMEC*

The results showed that *E. coli* treatment upregulated the content of ROS and MDA, whereas LBP reversed the upregulation of ROS and MDA in a dose-dependent manner. Moreover, LBP pretreatment at 100 μg/mL did not reduce the level of MDA as compared with the *E. coli* group. We found *E. coli* treatment decreased the activity of SOD and T-AOC. As a result, the decreased T-AOC level was reversed in a similar manner to that in cells pretreated with LBP, and the effect of LBP treatment is dose dependent ($p = 0.0002$, $p = 0.005$, $p = 0.128$; Figure 2C). Compared with the *E. coli* group, 100 μg/mL of LBP altered the SOD level, whereas 300 μg/mL of LBP significantly increased the SOD activity ($p = 0.007$, $p = 0.0001$; Figure 2D).

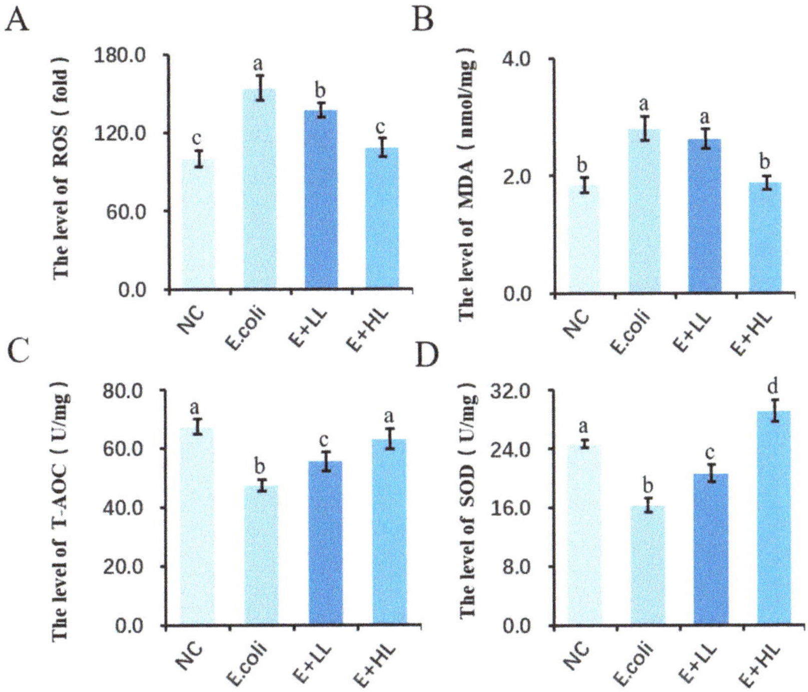


**Figure 2.** The effect of LBP on the antioxidant capacity of pbMEC. Different concentrations of LBP at 100 μg/mL or 300 μg/mL were added to pbMEC, followed by a challenge with $1 \times 10^7$/mL *E. coli* for an additional 6 h. (**A**) Determination of ROS content in pbMEC. (**B**) Detection of MDA content in pbMEC. (**C**) Detection of T-AOC content in pbMEC. (**D**) Detection of SOD content in pbMEC. Results are presented as means ± SEM. Letters in superscripts represent that the difference between groups was significant according to statistical analysis ($p < 0.05$).

### *3.3. Effect of LBP on the Expression of Genes Related to Inflammatory Responses in E. coli Stimulated pbMEC*

In this study, LBP pretreatment attenuated *E. coli*-induced proinflammatory cytokine gene expression in pbMEC. The LBP-induced alteration of COX2, IL1B, IL6, TNFA, and iNOS gene expression in pbMEC stimulated by *E. coli* was examined by RT-PCR. Compared with the control group, *E. coli* treatment upregulated the mRNA abundance of COX2, IL1B, IL6, TNFA, and iNOS. pbMEC were treated with different doses of LBP at 100 μg/mL or 300 μg/mL, and markedly downregulated the proinflammatory genes that were induced by *E. coli* challenge. Moreover, the inhibitory effect of LBP on inflammatory gene expression was more effective for cells pretreated with the increased doses of LBP, as a result of a lower expression of proinflammatory genes in E + HH than those in the E + LL group. Similarly, *E. coli* treatment upregulated the mRNA abundance of MAPK14, whereas pretreatment with different doses of LBP at 100 μg/mL or 300 μg/mL, markedly attenuated the elevation of MAPK14 induced by *E. coli* ($p = 0.007$, $p = 0.002$, $p = 0.018$; Figure 3D).

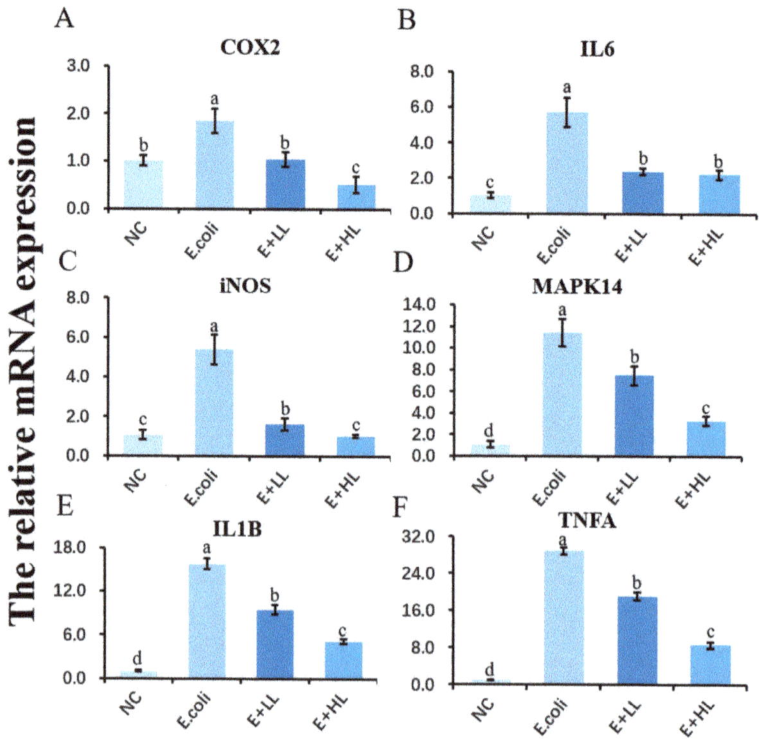


**Figure 3.** The effect of LBP on mRNA expression related to inflammatory responses. The inflammation injury of pbMEC challenged by *E. coli*. pbMEC were treated with different doses of LBP at 100 μg/mL or 300 μg/mL), followed by a challenge with $1 \times 10^7$/mL *E. coli* for an additional 6 h. (**A**) mRNA quantitative expression of COX2. (**B**) mRNA quantitative expression of IL6. (**C**) mRNA quantitative expression of iNOS. (**D**) mRNA quantitative expression of MAPK14. (**E**) mRNA quantitative expression of IL1B. (**F**) mRNA quantitative expression of TNFA. The biological replicates for PCR running were triplicate, along with three unique experiments. Results are presented as means ± SEM. Superscripts represent that the difference between groups was significant according to statistical analysis ($p < 0.05$).

### 3.4. Pretreatment with LBP Activated SOCS3 in pbMEC Challenged with E. coli

*E. coli* significantly decreased the protein expression of SOCS3, whereas pretreatment with LBP in the E + LL and E + HH groups restored the inhibited expression of SOCS3 ($p < 0.05$; Figure 4A,B). Moreover, LBP pretreatment at 100 and 300 μg/mL raised the abundance of SOCS3 to approximately twofold that of the control group. Consistently, gene expression of SOCS3 was also affected by pretreatment of LBP, as a result of the reversed expression of SOCS3 that was downregulated by *E. coli* stimulation ($p = 0.011$; Figure 4C). Additionally, results in the immunofluorescence assay agreed with the protein expression generated by Western blot. The weakened staining of SOCS3 in the *E. coli* group was restored by the pretreatment of LBP in the E + LL and E + HH groups (Figure 4D).

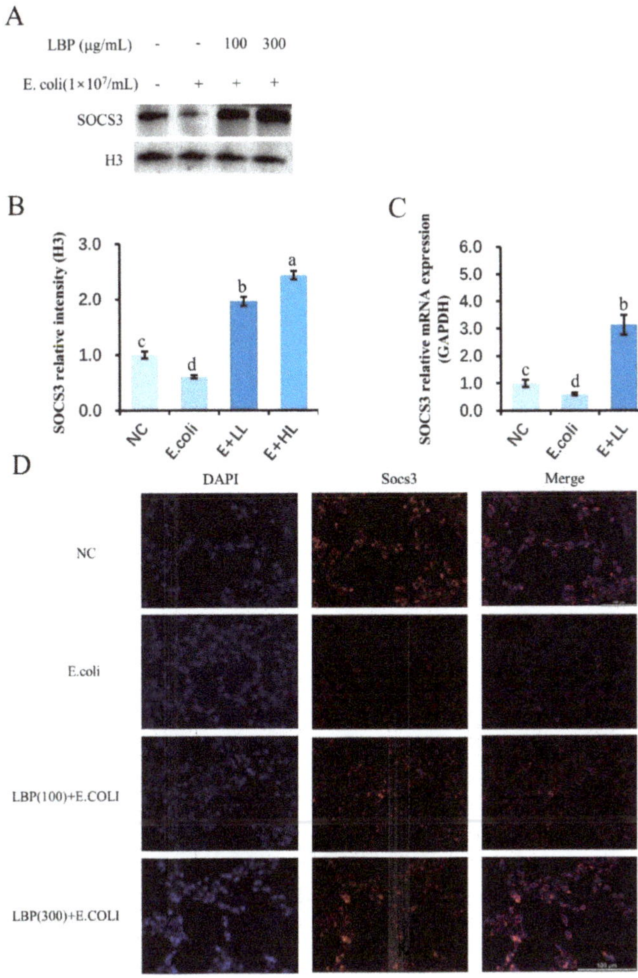


**Figure 4.** Effects of LBP on SOCS3 protein expression. pbMEC were treated with different doses of LBP at 100 μg/mL or 300 μg/mL followed by a challenge with $1 \times 10^7$/mL *E. coli* for an additional 6 h. (**A**) Western blot analyses of SOCS3. (**B**) Relative protein abundance of SOCS3 to Histone H3. (**C**) Relative mRNA expression level of SOCS3. (**D**) Fluorescence for detecting SOCS3 abundance. Antibodies against SOCS3 (Cy3 labeled secondary antibody, red) and counterstained with DAPI (blue) were conducted for target protein and nuclear staining respectively. The biological replicates for PCR running were triplicate, along with triple unique experiments. Results are presented as means ± SEM. Superscripts represent that the difference between groups was significant according to statistical analysis ($p < 0.05$).

### *3.5. Effect of LBP on the MAPK Pathway-Related Protein Expression of pbMEC under Different Treatments*

MAPK is a classic signaling pathway in charge of inflammatory responses. We investigated the components of the MAPK signaling by Western blot and immunofluorescence. When compared with the control group, *E. coli* treatment elevated the proportion of p-p38, p-JNK, and p-ERK to total p38, JNK, and ERK protein ($p = 0.005$, $p = 0.002$, $p = 0.012$, respectively; Figure 5B). Compared with the *E. coli*-treated group, 100 μg/mL LBP treatment did not reduce the ratio of p-p38 to p38 ($p = 0.062$). Instead, 100 μg/mL LBP treatment significantly reduced the ratio of p-JNK to JNK and p-ERK to ERK ($p = 0.032$, $p = 0.046$, respectively; Figure 5B). In contrast, as the pretreatment of LBP with a concentration of 300 μg/mL, the ratio of p-p38 to p38, p-JNK to JNK, and p-ERK to ERK decreased significantly in cells treated with *E. coli* ($p < 0.05$; Figure 5A,B). Consistently, immunofluorescence staining results revealed that cells exposed to *E. coli* strengthened the staining of p38, whereas LBP pretreatment attenuated the staining of p38 in nuclei. Notably, the expression of phosphorylated p38 was decreased in a dose-dependent manner of LBP added to the cells, indicating that the effect of LBP at a concentration of 300 μg/mL is better than that of 100 μg/mL (Figure 5C).

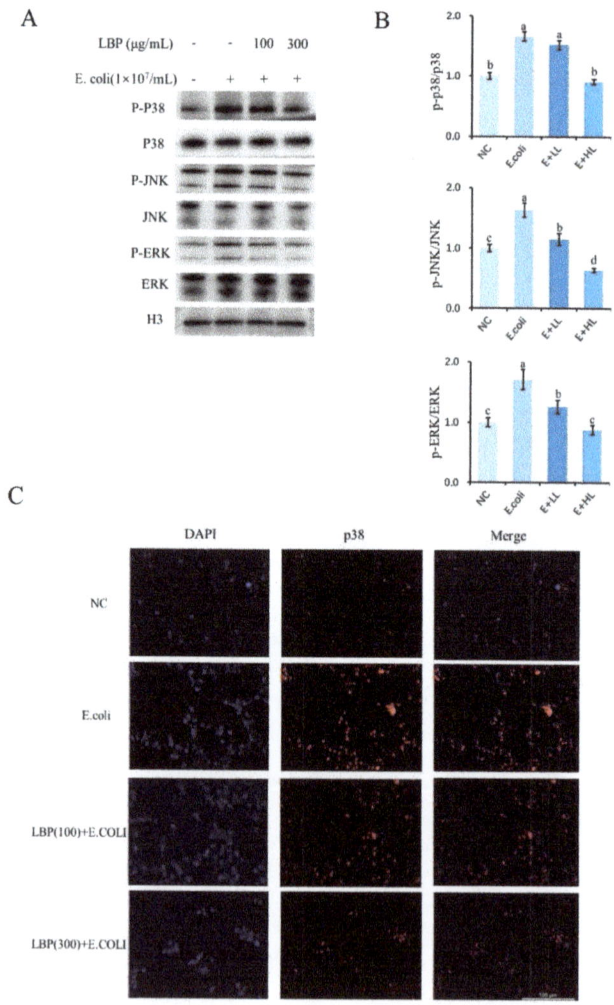


**Figure 5.** Expression of proteins related to MAPK signaling. Different doses of LBP at 100 μg/mL or 300 μg/mLwere added to pbMEC, followed by a challenge with $1 \times 10^7$/mL *E. coli* for an additional 6 h. (**A**) Western blots of the components of MAPK signaling. (**B**) The ratio of p-p38, p-JNK and p-ERK to total p38, JNK and ERK abundance. (**C**) Fluorescence detection for SOCS3 (Cy3 labeled, red) was performed, and the nucleus was stained with dye DAPI (blue). Results are expressed as means ± SEM. Results are presented as means ± SEM. Superscripts represent that the difference between groups was significant according to statistical analysis ($p < 0.05$).

### 3.6. LBP Pretreatment Inhibits E. coli-Induced Activation of Inflammatory Genes

We investigated the effect of pretreated LBP on pro-inflammatory factors by using Western blot. Compared with the control group, the *E. coli* challenge increased the protein abundance of COX-2, IL-1β, and TNF-α, and 100 μg/mL, and LBP pretreatment did not reduce the protein expression of COX-2 or IL-1β ($p = 0.287$, $p = 0.094$, respectively; Figure 6A,B); however, 300 μg/mL of LBP supplemented to cells significantly inhibited the expression of COX-2 and IL-1β. Furthermore, the abundance of TNF-α was partially inhibited by LBP pretreatment in a dose-dependent manner followed by *E. coli* stimulation ($p < 0.05$; Figure 6A,B).

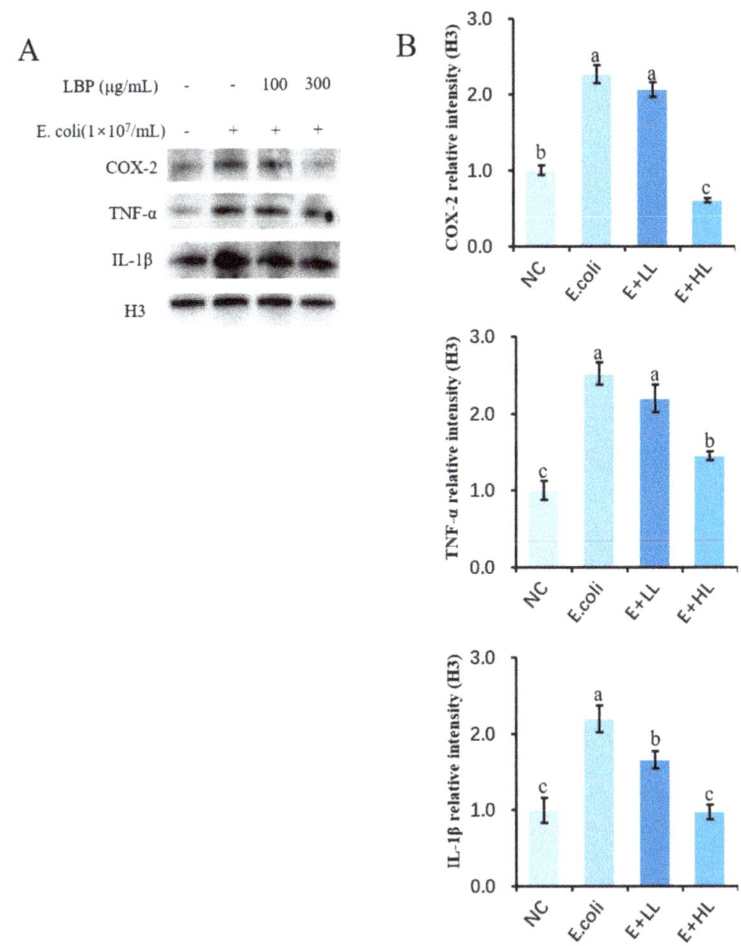


**Figure 6.** Effects of LBP on inflammatory factor protein expression. pbMEC were treated with different doses of LBP at 100 μg/mL or 300 μg/mL, followed by a challenge with $1 \times 10^7$/mL *E. coli* for an additional 6 h. (**A**) Western blots of COX-2, IL-1β, and TNF-α. (**B**) Relative protein abundance of COX-2, IL-1β, and TNF-α to Histone H3. Results are expressed as means ± SEM. Results are presented as means ± SEM. Superscripts represent that the difference between groups was significant according to statistical analysis ($p < 0.05$).

## 4. Discussion

Bovine mastitis is a common infectious disease caused mainly by microbial infections. *E. coli* is one of the most common pathogens, mainly causing subclinical mastitis [25]. Several studies agree with the upregulation of TNFA, IL6, and IL1B mRNA abundance in the mammary gland during coliform mastitis, when cows are in the transition period [26,27]. It is well known that antibiotics are currently the main treatment for mastitis, as traditional antibiotics develop resistance during treatment [6]. Hence, research on developing novel and effective antibacterial drugs might be an alternative strategy for the treatment of mastitis. LBP has been widely used for defending oxidative stress and alleviating inflammation [23,28]. In this study, we investigated the anti-inflammatory effects and mechanism of LBP on *E. coli*-stimulated pbMEC inflammatory responses via activation of SOCS3 and inactivation of MAPK pathway.

### *4.1. LBP Attenuates the E. coli-Induced Inflammatory and Oxidative Responses in pbMEC*

Since LBP did not reveal a cytotoxic effect on pbMEC at doses ranging from 0 to 1000 μg/mL, we selected the concentration of LBP at either 100 or 300 μg/mL in the formal experiment according to reported studies [29]. LBP has been shown to reduce the expression of TNFA, IL6, and IL1B in *E. coli*-induced mouse macrophages [30]. In a mouse model of steatohepatitis, LBP treatment affects the secretion of NF-κB-mediated inflammatory cytokines and inflammasome, such as NLRP3/6 [31]. LPS comprises endotoxins of Gram-negative bacteria that envelop *E. coli*; thus, it is widely used to establish a cellular inflammation model for microbial infections in vitro [32]. In this study, treatment with LPS increases the gene expression of TNFA, IL6, and IL1B in pbMEC, whereas LBP pretreatment reversed the elevated pro-inflammatory gene expression. Additionally, LPS is a typical potent stimulus of the immune response, which leads to an increase in the expression of iNOS and COX2 in macrophages [33,34]. Notably, in the current study, LBP pretreatment lowered the upregulation of iNOS and COX2 gene expression in pbMEC induced by *E. coli*. This result indicates that LBP suppresses the *E. coli*-induced expression of proinflammatory gene expression and supports the protection of inflammation cascades.

ROS is a byproduct of the physiological metabolism of mitochondria which can participate in the transduction of signaling pathways, tissue damage, and related physiological processes [35]. When ROS are produced in large quantities, it may cause oxidative stress in cells and lead to autophagy or apoptosis [36]. LBP is a well-known powerful antioxidant; the study has found that increasing doses of LBP can improve the SOD activity of mouse blood serum, liver, and brain tissues, while reducing the content of MDA [37]. It has also recently been demonstrated that LBP can reduce oxidative stress in HepG2 cells [38]; therefore, we determined whether the antioxidant capacity of LBP is related to the elimination of ROS in the cellular environment by detecting the activity of antioxidant enzymes. SOD is a specific enzyme that can scavenge free radicals in the biological antioxidant system [39]. The current study showed that cells treated with *E. coli* produced a large amount of ROS, whereas LBP pretreatment significantly attenuated MDA content and restored oxidative capacity by increasing SOD activity and T-AOC in pbMEC. The results are indicative of the role of LBP in reducing the oxidative responses induced by *E. coli* and stabilizing the homeostasis of cells with regard to the defense of oxidative damage. Although LBP affects the immune and oxidative response in vitro, the limitations of the current study regarding the reflection of the real situation in the infected udder using LBP, could be a potential therapy until more robust tests are made.

### *4.2. LBP Inhibits Inflammatory and Oxidative Responses in pbMEC through SOCS3 and MAPK Molecules*

SOCS3 is a protein that negatively regulates the cytokine signal transduction pathway [40]. Recent studies have shown that SOCS3 can be modulated by a variety of inflammatory stimuli, and that it blocks the signal interactions of immune molecules. For example, the MAPK signaling pathway plays a key role in the bacteria-induced gene expression of SOCS3 in primary human macrophages [41]. Interestingly, the study found that SOCS3 interference blocked the upregulation of TNFA and IL1B in the LPS-induced 3T3-L1 adipocyte inflammation model [42]; however, the role of SOCS3 in the regulation of inflammatory responses in bovine mastitis is still unclear. In this study, we found that pretreatment with LBP reversed the decrease in SOCS3 mRNA and protein expression induced by *E. coli* challenge; hence, this result indicates the involvement of SOCS3 in *E. coli*-stimulated pbMEC and its potential role in the modulation of inflammation by LBP.

It is well known that the NF-κB and MAPK signaling pathways play a crucial role in the innate immune response [43]. We have reported in a previous study that LBP inhibits LPS-induced NF-κB activation in pbMEC; however, the MAPK signaling pathway may act as an indispensable regulator to further study the potential mechanism of LBP that inhibits *E. coli*-induced inflammation in pbMEC. In mammalian cells, three MAPK families have been clearly characterized: extracellular signal-regulated kinase (ERK), C-Jun N-terminal kinase (JNK), and p38 kinase [44,45]. The current study shows that LBP reverses the activation of inflammatory factors through its suppression of MAPK signaling pathways. The activation of p38, ERK, and JNK was also restrained in a dose-dependent fashion by LBP. The immunofluorescence results emphasized that the activation of the p38 protein by *E. coli* stimulation is inhibited by pretreatment with LBP. Several studies have confirmed that the activation of the MAPK signaling pathway results in the phosphorylation of transcription regulators, which indirectly or directly modulates the secretion of cytokines such as TNF-α and IL-1β [46,47]. The fact that pretreatment of LBP suppressed the abundance of COX-2, IL-1β, and TNF-α protein in pbMEC stimulated by *E. coli*, is in line with the inhibitory effects of LBP on MAPK signaling.

Overall, our results indicated that LBP suppressed the expression of proinflammatory cytokines by preventing MAPK activation in *E. coli*-stimulated pbMEC, showing that the anti-inflammatory mechanism of LBP might be associated with SOCS3-mediated suppressing in MAPK signaling pathways. The results highlight the potential role of LBP as a therapeutic extract in combating clinical bovine mastitis.

**Supplementary Materials:** The following supporting information can be downloaded at: https://www.mdpi.com/article/10.3390/agriculture12050598/s1, Table S1: Primers used for real-time qPCR [11,48].

**Author Contributions:** R.L., T.X. and H.Z. analyzed and processed the data. R.L., X.L., J.Z. and X.W. performed the experiment, and T.X. and Z.Y. were major contributors to the drafting of the manuscript. Z.Y. and T.X. provided financing throughout the experiment and generated the outlines of the research. All authors have read and agreed to the published version of the manuscript.

**Funding:** This study was funded by the National Natural Science Foundation of China (32,102,731; 31,872,324).

**Institutional Review Board Statement:** The animal study protocol was approved by the Ethics Committee of Yangzhou University (YZU-2021003-168) on 10 July 2021.

**Informed Consent Statement:** Not applicable.

**Data Availability Statement:** The data presented in this study are available on request from the corresponding author.

**Acknowledgments:** The authors would like to thank all those who helped us during this project for their valuable advice and support.

**Conflicts of Interest:** The authors declare no conflict of interest.

## References

1. Merriman, K.E.; Powell, J.L.; Santos, J.E.P.; Nelson, C.D. Intramammary 25-hydroxyvitamin D 3 treatment modulates innate immune responses to endotoxin-induced mastitis. *J. Dairy Sci.* **2018**, *101*, 7593–7607. [CrossRef] [PubMed]
2. Tomazi, T.; Coura, F.M.; Goncalves, J.L.; Heinemann, M.B.; Santos, M.V. Antimicrobial susceptibility patterns of Escherichia coli phylogenetic groups isolated from bovine clinical mastitis. *J. Dairy Sci.* **2018**, *101*, 9406–9418. [CrossRef] [PubMed]

3. Guo, M.; Gao, Y.; Xue, Y.; Liu, Y.; Zeng, X.; Cheng, Y.; Ma, J.; Wang, H.; Sun, J.; Wang, Z.; et al. Bacteriophage Cocktails Protect Dairy Cows Against Mastitis Caused by Drug Resistant Escherichia coli Infection. *Front. Cell. Infect. Microbiol.* **2021**, *11*, 690377. [CrossRef] [PubMed]
4. Guerra, S.T.; Paula, C.; Bolaos, C.; Hernandes, R.T.; Ribeiro, M.G. Virulence factors of Escherichia coli: An overview of animal and human infections with emphasis in bovine mastitis. *Semin. Cienc. Agrar.* **2019**, *40*, 2087. [CrossRef]
5. Blum, S.E.; Dan, E.H.; Jacoby, S.; Krifuks, O.; Leitner, G. Physiological response of mammary glands to Escherichia coli infection: A conflict between glucose need for milk production and immune response. *Sci. Rep.* **2020**, *10*, 9602. [CrossRef]
6. Seal, B.S.; Lillehoj, H.S.; Donovan, D.M.; Gay, C.G. Alternatives to antibiotics: A symposium on the challenges and solutions for animal production. *Anim. Health Res. Rev.* **2013**, *14*, 78–87. [CrossRef]
7. Mingmongkolchai, S.; Panbangred, W. Bacillus probiotics: An alternative to antibiotics for livestock production. *J. Appl. Microbiol.* **2018**, *124*, 1334–1346. [CrossRef]
8. Jedrzejczak, M.; Szatkowska, I. Bovine mammary epithelial cell cultures for the study of mammary gland functions. *In Vitro Cell. Dev. Biol.-Anim.* **2014**, *50*, 389–398. [CrossRef]
9. Kehrli, M.E.; Shuster, D.E. Factors Affecting Milk Somatic Cells and Their Role in Health of the Bovine Mammary Gland. *J. Dairy Sci.* **1994**, *77*, 619–627. [CrossRef]
10. Gomes, F.; Henriques, M. Control of Bovine Mastitis: Old and Recent Therapeutic Approaches. *Curr. Microbiol.* **2016**, *72*, 377–382. [CrossRef]
11. Wang, J.; Guo, C.; Wei, Z.; He, X.; Kou, J.; Zhou, E.; Yang, Z.; Fu, Y. Morin suppresses inflammatory cytokine expression by downregulation of nuclear factor-κB and mitogen-activated protein kinase (MAPK) signaling pathways in lipopolysaccharide-stimulated primary bovine mammary epithelial cells. *J. Dairy Sci.* **2016**, *99*, 3016–3022. [CrossRef] [PubMed]
12. Ren, X.; Han, L.; Li, Y.; Zhao, H.; Zhang, Z.; Zhuang, Y.; Zhong, M.; Wang, Q.; Ma, W.; Wang, Y. Isorhamnetin attenuates TNF-α-induced inflammation, proliferation, and migration in human bronchial epithelial cells via MAPK and NF-κB pathways. *Anat Rec (Hoboken).* **2021**, *304*, 901–913. [CrossRef]
13. Yang, Z.; Yin, R.; Cong, Y.; Yang, Z.; Zhou, E.; Wei, Z.; Liu, Z.; Cao, Y.; Zhang, N. Oxymatrine Lightened the Inflammatory Response of LPS-Induced Mastitis in Mice Through Affecting NF-κB and MAPKs Signaling Pathways. *Inflammation* **2014**, *37*, 2047–2055. [CrossRef] [PubMed]
14. Shu, F.; Junfu, W.; Xiaoqun, X.; Xuemei, C.; Junwen, L.; Qinghong, S.; Meng, L.; Huali, W.; Changsheng, Z. The Expression of SOCS and NF-κB p65 in Hypopharyngeal Carcinoma. *Iran. J. Public Health* **2018**, *47*, 1874–1882.
15. Elliott, J.; Johnston, J.A. SOCS: Role in inflammation, allergy and homeostasis. *Trends Immunol.* **2004**, *25*, 434–440. [CrossRef]
16. Souza, J.A.C.d.; Nogueira, A.V.B.; Souza, P.P.C.d.; Kim, Y.J.; Lobo, C.S.; Oliveira, G.J.P.L.d.; Cirelli, J.A.; Garlet, G.P.; Rossa, C.; López-Collazo, E. SOCS3 Expression Correlates with Severity of Inflammation, Expression of Proinflammatory Cytokines, and Activation of STAT3 and p38 MAPK in LPS-Induced Inflammation In Vivo. *Mediat. Inflamm.* **2013**, *2013*, 650812.
17. Delgado-Ortega, M.; Melo, S.; Meurens, F. Expression of SOCS1-7 and CIS mRNA in porcine tissues. *Vet. Immunol. Immunopathol.* **2011**, *144*, 493–498. [CrossRef]
18. Hu, J.; Wang, W.; Hao, Q.; Zhang, T.; Yin, H.; Wang, M.; Zhang, C.; Zhang, C.; Zhang, L.; Zhang, X.; et al. Suppressors of cytokine signalling (SOCS)-1 inhibits neuroinflammation by regulating ROS and $TLR_4$ in $BV_2$ cells. *Inflamm. Res.* **2020**, *69*, 27–39. [CrossRef]
19. Mingzhao, D.; Xinyu, H.; Ling, K.; Baohai, Z.; Chaopu, Z. Lycium barbarum Polysaccharide Mediated the Antidiabetic and Antinephritic Effects in Diet-Streptozotocin-Induced Diabetic Sprague Dawley Rats via Regulation of NF-κB. *BioMed Res. Int.* **2016**, *2016*, 3140290.
20. Xiao-rui, Z.; Chun-hui, Q.; Jun-ping, C.; Gang, L.; Lin-juan, H.; Zhong-fu, W.; Wen-xia, Z.; Yong-xiang, Z. Lycium barbarum polysaccharide LBPF4-OL may be a new Toll-like receptor 4/MD2-MAPK signaling pathway activator and inducer. *Int. Immunopharmacol.* **2014**, *19*, 132–141.
21. Péter, F.; Ildikó, S.; Hajnalka, L.; Mariann, H.; Sándor, S.; György, P. Association of chemerin with oxidative stress, inflammation and classical adipokines in non-diabetic obese patients. *J. Cell. Mol. Med.* **2014**, *18*, 1313–1320.
22. Qi, B.; Ji, Q.; Wen, Y.; Lian, L.; Guo, X.; Hou, G.; Wang, G.; Zhong, J.; Ferenc, G. Lycium barbarum Polysaccharides Protect Human Lens Epithelial Cells against Oxidative Stress–Induced Apoptosis and Senescence. *PLoS ONE* **2014**, *9*, e110275. [CrossRef] [PubMed]
23. Li, X.M.; Ma, Y.L.; Liu, X.J. Effect of the Lycium barbarum polysaccharides on age-related oxidative stress in aged mice. *J. Ethnopharmacol.* **2007**, *111*, 504–511. [CrossRef] [PubMed]
24. Sun, Y. Biological activities and potential health benefits of polysaccharides from Poria cocos and their derivatives. *Int. J. Biol. Macromol.* **2014**, *68*, 131–134. [CrossRef]
25. Sargeant, J.M.; Gillespie, J.R.; Oberst, R.D.; Phebus, R.K.; Hyatt, D.R.; Bohra, L.K.; Galland, J.C. Results of a longitudinal study of the prevalence of Escherichia coli O157:H7 on cow-calf farms. *Am. J. Vet. Res.* **2000**, *61*, 1375–1379. [CrossRef]
26. Shuster, D.E.; Kehrli, M.E.; Stevens, M.G. Cytokine production during endotoxin-induced mastitis in lactating dairy cows. *Am. J. Vet. Res.* **1993**, *54*, 80–85.
27. Larson, M.A.; Weber, A.; Mcdonald, T.L. Bovine serum amyloid A3 gene structure and promoter analysis: Induced transcriptional expression by bacterial components and the hormone prolactin. *Gene* **2006**, *380*, 104–110. [CrossRef]

28. Xiao, J.; Xing, F.; Huo, J.; Fung, M.L.; Liong, E.C.; Ching, Y.P.; Xu, A.; Chang, R.; So, K.F.; Tipoe, G.L. Lycium barbarum polysaccharides therapeutically improve hepatic functions in non-alcoholic steatohepatitis rats and cellular steatosis model. *Sci. Rep.* **2014**, *4*, 5587. [CrossRef]
29. Cuadrado, A.; Nebreda, A. Mechanisms and functions of p38 MAPK signalling. *Biochem. J.* **2010**, *429*, 403–417. [CrossRef]
30. Peng, Q.; Liu, H.; Shi, S.; Li, M. Lycium ruthenicum polysaccharide attenuates inflammation through inhibiting TLR4/NF-κB signaling pathway. *Int. J. Biol. Macromol.* **2014**, *67*, 330–335. [CrossRef]
31. Xiao, J.; Wang, F.; Liong, E.C.; So, K.F.; Tipoe, G.L. Lycium barbarum polysaccharides improve hepatic injury through NFkappa-B and NLRP3/6 pathways in a methionine choline deficient diet steatohepatitis mouse model. *Int. J. Biol. Macromol.* **2018**, *120*, 1480–1489. [CrossRef] [PubMed]
32. Granucci, F. The Family of LPS Signal Transducers Increases: The Arrival of Chanzymes. *Immunity* **2018**, *48*, 4–6. [CrossRef] [PubMed]
33. Dang, Y.; Xu, Y.; Wu, W.; Li, W.; Sun, Y.; Yang, J.; Zhu, Y.; Zhang, C. Tetrandrine Suppresses Lipopolysaccharide-Induced Microglial Activation by Inhibiting NF-κB and ERK Signaling Pathways in $BV_2$ Cells. *PLoS ONE* **2014**, *9*, e102522. [CrossRef] [PubMed]
34. Qian, G.; Li, Y.; He, M.; Guo, W.; Kan, X.; Xu, D.; Liu, J.; Fu, S. Peiminine Protects against Lipopolysaccharide-Induced Mastitis by Inhibiting the AKT/NF-κB, ERK1/2 and p38 Signaling Pathways. *Int. J. Mol. Sci.* **2018**, *19*, 2637.
35. Wang, Y.; Branicky, R.; Noë, A.; Hekimi, S. Superoxide dismutases: Dual roles in controlling ROS damage and regulating ROS signaling. *J. Cell Biol.* **2018**, *217*, 1915–1928. [CrossRef]
36. Willson, J.A.; Arienti, S.; Sadiku, P.; Reyes, L.; Coelho, P.; Morrison, T.; Rinaldi, G.; Dockrell, D.H.; Whyte, M.K.B.; Walmsley, S.R. Neutrophil HIF-1α stabilization is augmented by mitochondrial ROS produced via the glycerol 3-phosphate shuttle. *Blood* **2022**, *139*, 281–286. [CrossRef]
37. Tian, X.; Liang, T.; Liu, Y.; Ding, G.; Zhang, F.; Ma, Z. Extraction, Structural Characterization, and Biological Functions of Lycium Barbarum Polysaccharides: A Review. *Biomolecules* **2019**, *9*, 389. [CrossRef]
38. Tuo, X.U.; Ling, H.Y.; Long, J.; Jian-Qin, H.E.; Yang, S.S.; Zhu, Z.M.; Yan, C.Q.; Feng, S.D.; Physiology, D.O. Ameliorative effects and the mechanism of lycium barbarum polysaccharide on insulin resistance of $HepG_2$ cell. *Chin. J. Appl. Physiol.* **2017**, *33*, 568–571.
39. Kwon, K.; Jung, J.; Sahu, A.; Tae, G. Nanoreactor for cascade reaction between SOD and CAT and its tissue regeneration effect. *J. Control Release* **2022**, *344*, 160–172. [CrossRef]
40. Tamiya, T.; Kashiwagi, I.; Takahashi, R.; Yasukawa, H.; Yoshimura, A. Suppressors of cytokine signaling (SOCS) proteins and JAK/STAT pathways: Regulation of T-cell inflammation by SOCS1 and SOCS3. *Arter. Thromb Vasc Biol* **2011**, *31*, 980–985. [CrossRef]
41. Latvala, S.; Miettinen, M.; Kekkonen, R.A.; Korpela, R.; Julkunen, I. Lactobacillus rhamnosus GG and Streptococcus thermophilus induce suppressor of cytokine signalling 3 (SOCS3) gene expression directly and indirectly via interleukin-10 in human primary macrophages. *Clin. Exp. Immunol.* **2011**, *165*, 94–103. [CrossRef]
42. Liu, Z.; Gan, L.; Zhou, Z.; Jin, W.; Sun, C. SOCS3 promotes inflammation and apoptosis via inhibiting JAK2/STAT3 signaling pathway in 3T3-L1 adipocyte. *Immunobiology* **2015**, *220*, 947–953. [CrossRef] [PubMed]
43. Liu, L.; Ye, L.; Liu, L.; Bian, Y.; Li, Z.; Gao, X.; Li, Q. 14-3-3γ Regulates Lipopolysaccharide-Induced Inflammatory Responses and Lactation in Dairy Cow Mammary Epithelial Cells by Inhibiting NF-κB and MAPKs and Up-Regulating mTOR Signaling. *Int. J. Mol. Sci.* **2015**, *16*, 16622–16641. [CrossRef] [PubMed]
44. Cargnello, M.; Roux, P.P. Activation and Function of the MAPKs and Their Substrates, the MAPK-Activated Protein Kinases. *Microbiol. Mol. Biol. Rev.* **2011**, *75*, 50–83. [CrossRef] [PubMed]
45. Junttila, M.R.; Li, S.P.; Westermarck, J. Phosphatase-mediated crosstalk between MAPK signaling pathways in the regulation of cell survival. *FASEB J.* **2008**, *22*, 954–965. [CrossRef] [PubMed]
46. Zahra, A.; Vaiva, K.; Volodymyr, S. Dual control of MAPK activities by AP2C1 and MKP1 MAPK phosphatases regulates defence responses in Arabidopsis. *J. Exp. Bot.* **2022**, *73*, 2369–2384.
47. Liu, K.; Liu, E.; Lin, L.; Hu, Y.; Yuan, Y.; Xiao, W. L-Theanine mediates the p38MAPK signaling pathway to alleviate heat-induced oxidative stress and inflammation in mice. *Food Funct.* **2022**, *13*, 2120–2130. [CrossRef]
48. Liu, M.; Song, S.; Li, H.; Jiang, X.; Yin, P.; Wan, C.; Liu, X.; Liu, F.; Xu, J. The protective effect of caffeic acid against inflammation injury of primary bovine mammary epithelial cells induced by lipopolysaccharide. *J. Dairy Sci.* **2014**, *97*, 2856–2865. [CrossRef]

*agriculture*

*Article*

# Functional and Comparative Analysis of Two Subtypes of Cofilin Family on Cattle Myoblasts Differentiation

**Yujia Sun [1,2], Yaoyao Ma [1,2], Xinyi Wu [1,2], Tianqi Zhao [1,2], Lu Lu [2] and Zhangping Yang [1,2,*]**

[1] Joint International Research Laboratory of Agriculture and Agri-Product Safety, the Ministry of Education of China, Institutes of Agricultural Science and Technology Development, Yangzhou University, Yangzhou 225009, China
[2] Key Laboratory of Animal Genetics & Breeding and Molecular Design of Jiangsu Province, Yangzhou University, Yangzhou 225009, China
* Correspondence: yzp@yzu.edu.cn; Tel.: +86-0514-87979230

**Abstract:** Agricultural meat composition and quality are not independent of the effects of skeletal muscle growth and development in animals. Cofilin is distributed extensively in muscle and non-muscle cells, and its function is tightly regulated in the cell. Cofilin has two variants in mammals, cofilin-1 (CFL1, non-muscle type) and cofilin-2 (CFL2, muscle type), and has a dual function on skeletal muscle fibers. Our study examined the expression pattern of CFL1 and CFL2 in different fetal bovine, calf, and adult cattle tissues. The content of the CFL2 gene increased significantly with the increase in cattle age in muscle tissues; CFL1 showed the opposite trend. In muscle tissues, DNA methylation levels of CFL1 and CFL2 were high in fetal bovine, and the mRNA level of CFL2 was significantly lower compared to CFL1. However, DNA methylation levels of CFL2 were lower than CFL1, and the mRNA level of CFL2 was remarkably higher compared to CFL1 in adult cattle. Overexpression of CFL1 or knockdown CFL2 reduced the expression levels of muscle differentiation markers, i.e., MYOD, MYOG, and MYH3. Overexpression of CFL2 or knockdown CFL1 stimulated the expression of these marker genes. Therefore, CFL2 may be superior to CFL1 as a candidate gene for subsequent research on cattle genetics and breeding.

**Keywords:** cofilin-1; cofilin-2; expression pattern; DNA methylation; myoblast differentiation

**Citation:** Sun, Y.; Ma, Y.; Wu, X.; Zhao, T.; Lu, L.; Yang, Z. Functional and Comparative Analysis of Two Subtypes of Cofilin Family on Cattle Myoblasts Differentiation. *Agriculture* **2022**, *12*, 1420. https://doi.org/10.3390/agriculture12091420

Academic Editor: Michael Schutz

Received: 28 July 2022
Accepted: 7 September 2022
Published: 8 September 2022



## 1. Introduction

Animal muscle is one of the main sources of human protein, and its basic unit is muscle fiber. In recent years, beef quality has become a hot and difficult issue for cattle genetics and breeding researchers. It is very important to study the molecular genetic mechanisms of meat quality traits [1]. The composition and structure of muscle, as well as the histological properties of muscle fibers, are tightly correlated with meat quality traits. Actin is the main component of muscle cells and cytoskeleton and participates in a series of important physiological activities in organisms, such as muscle contraction, cell movement, cytokinesis, and so on. These processes require not only actin but also regulatory proteins bound to actin. Cofilin is one of the 21 kD actin-binding proteins widely distributed in organisms from yeast to mammals [2]. Cofilin was originally identified in pig brain extracts [3]; its basic function is to bind and depolymerize F-actin in cells, inhibit the polymerization of G-actin, and synergistically participate in a series of important physiological activities in the organism.

The cofilin family is essential in eukaryotes, known to play critical roles in normal muscle function and regeneration, and has highly complex and interesting regulatory patterns. In mammals, the ADF/cofilin protein family has three members: ADF (Actin Depolymerizing Factor), CFL1 (non-muscle-type cofilin, NM-type cofilin, cofilin-1), and CFL2 (muscle-type cofilin, M-type cofilin, cofilin-2), Although they overlap in function, their distribution and expression in tissues vary. ADF is mainly expressed in neurons and epithelial cells, CFL1 is expressed in non-muscle cells of various tissues, and CFL2 is mainly found in muscle cells. In addition, some cells can express the above three members at the same time [4,5]. In terms of function, ADF is more effective in the segmentation of actin monomers, while CFL1/CFL2 mainly acts on the nucleation and cleavage of actin filaments [6]. Activated CFL1 can cleave actin fibers (Fibrous action, F-actin) to form short actin fragments with more nucleation and aggregation activities, thereby regulating the dynamic balance between actin depolymerization and promoting polymerization and maintaining cytoskeleton function [7]. In addition to directly cleaving actin, CFL1 also affects gene expression, cell proliferation, and apoptosis and plays an important role in maintaining cell homeostasis. The CFL2 gene is highly expressed in the muscle tissue of domestic animals, which may be closely related to its major part in muscle growth and development [8,9]. Studies have found that the precursor proteins encoded by the mammalian CFL2 gene all contain ADF domains, and ADF deletion can lead to mouse corneal developmental diseases, such as actin cytoskeleton disorder, epithelial hyperplasia, and even blindness [10,11]. Deletion of the CFL2 gene kills 8-day-old mice due to severe muscle defects and abnormal actin accumulation, indicating that the CFL2 gene plays an important role in muscle development and maintenance of normal muscle morphology [12].

CFL1 can achieve actin movement by promoting actin depolymerization and severing myofilaments. It is expressed in various non-muscle tissues and is more expressed in the brain and liver, while CFL2 is mainly expressed in the mammalian heart and muscle. It is expressed in tissues and is considered to be an actin assembler in muscle tissue and has an irreplaceable role in myofiber formation and muscle regeneration [13]. During myogenesis in healthy mice, a cofilin transition occurs from non-muscle type CFL1 to muscle type CFL2 [9]. Both CFL1 and CFL2 are expressed in embryonic skeletal muscle. With the progress of muscle development, CFL1 gene expression decreases, and CFL1 expression disappears in terminally differentiated myoblasts, while CFL2 gene expression increases. Therefore, CFL1 is almost undetectable in adult skeletal muscle, which mainly expresses CFL2, of which 30–50% is in the phosphorylated form [11]. The existing form of the cofilin family is usually closely related to function and shows different molecular subtypes in cells and tissues according to functional needs.

This study attempted to understand further the role of CFL1 and CFL2 in the growth and development of skeletal muscle cells and to explore the interaction between CFL1 and CFL2 by studying the expression rules of CFL1 and CFL1 genes in Qinchuan (QC) cattle muscle tissue and their function on muscle cell differentiation in order to provide a theoretical basis for the subsequent molecular genetic breeding of cattle; it may also provide some new ideas for mammalian genetic breeding.

## 2. Materials and Methods

All experimental animal procedures were approved by the Institutional Animal Care and Use Committee of Yangzhou University (License number: SYXK [Su] 2017-0044).

### *2.1. Collection of Tissue Samples*

The tissue samples of the cattle were sampled immediately after slaughter, frozen in liquid nitrogen, brought back to the laboratory, and stored in a −80 °C refrigerator for future use. Tissue samples were collected from female cattle at three key stages of muscle formation and maturation: fetal bovine (FB, embryonic day 90), calf (1 month postnatal), and adult cattle (AC, 24 months). Heart tissues, liver tissues, spleen tissues, lung tissues, kidney tissues, fat tissues, and muscle tissues were collected from three QC cattle for RNA extraction. Fresh longissimus dorsi muscle tissue of fetal bovine was sampled, stored in PBS containing 1% streptomycin and penicillin, and brought back to the laboratory for isolation and culture of primary bovine myoblasts (PBMs).

All cattle were obtained from the reserved farm of Qinchuan cattle (Fufeng country, Shaanxi Province, P.R. China) and the fineness breeding center of Qinchuan cattle (Yangling, Shaanxi province, P.R. China). They were raised at the same level.

### 2.2. RNA Extraction and cDNA Synthesis

Total RNA from tissues and cellular were obtained using TRIzol reagent (TaKaRa, Dalian, China) following the manufacturer's protocol, cDNA was synthesized from an equal amount RNA (500 ng) using PrimeScript RT reagent Kit (TaKaRa, Dalian, China) according to manufacturer's instruction.

### 2.3. Cell Culture

PBMs were isolated and cultured using collagenase I digestion. First, the muscle tissue samples were washed three times with sterile PBS with 1% double antibody, and then the epidermal tissue was cut along the fetal bovine dorsal spine. In a 6 cm petri dish of PBS, scissors were used to cut the muscle mass as much as possible, and then it was collected into a 50 mL centrifuge tube; collagenase I was added to a water bath at 37 °C to digest for 1.5 h, it was filtered and centrifuged at 1000 r/min for 10 min, and the precipitated cells were collected. The cells were resuspended in DMEM/F12 complete medium supplemented with 15% FBS and 1% double-antibody. After counting the cells, the cell suspension was inoculated in a 6 cm culture dish in proportion, then cultured in a 37 °C, 5% $CO_2$ incubator for 2 h, and the upper layer was aspirated. The culture medium continued to be cultured in a new 6 cm culture dish until the cell density reached about 80–90%, and the cells were passaged or frozen for subsequent experiments. HEK293 and C2C12 cell lines were provided by our laboratory. All cells were cultured at 37 °C containing 5% $CO_2$.

### 2.4. Induce Cell Differentiation and Transfection

To investigate myocyte differentiation, the growth medium (GM) was changed to differentiation medium (DM, 2% horse serum) after cell confluence reached approximately 80%. Meanwhile, CFL1-CMV, CFL2-CMV, pAdTrack-CMV vector (control group), shCFL1-2, shCFL2-1, and siRNA negative control (NC) was transfected into PBMs using Lipofectamine$^{TM}$ 2000 (Invitrogen, Carlsbad, CA, USA) to confirm the regulatory mechanism of the CFL1 and CFL2 genes on myoblast differentiation. C2C12 cells were cultured for 1, 2, 4, 6, and 8 days, and PBMs were harvested for RNA extraction at 1, 3, 5, and 7 days. Recombinant vectors (overexpression vector CFL1-CMV and CFL2-CMV, interference vector shCFL1-1, shCFL1-2, shCFL2-1, and shCFL2-2) are detailed in the Supplementary Material. Four interference vectors were transfected into HEK293 cells, and shCFL1-2 and shCFL2-1 were found to have a better interference effect (Figure S1). The entire process for adenovirus generation and proliferation was performed as previously described [14].

### 2.5. Quantitative Real-Time PCR (qRT-PCR) and Western Blot

Total RNA was extracted from tissues and cells using TRIzol reagent (TaKaRa, Dalian, China), and then 500 ng RNA was transcribed to cDNA using the PrimeScript RT reagent kit (TaKaRa, Dalian, China) for use in qRT-PCR with SYBR Green Master Mix reagent kit (GenStar, Beijing, China). Gene-specific primers for the CFL1 gene (CFL1-F/R) and CFL2 (CFL2-F/R) gene were designed based on published mRNA sequences using Primer Premier 5.0 software (PREMIER Biosoft International, USA). Transcription levels were normalized to the levels of the housekeeping genes GAPDH (GAPDH-F/R). The primers are listed in Table 1. The total protein extracted from HEK293 cells is listed in the Supplementary Material. Western blot results are shown in Figure S1. In previous work, we selected GAPDH as the housekeeping gene [15,16], the stability data expression of GAPDH is shown in Table S2.

### 2.6. Bisulfite Sequencing Polymerase Chain Reaction (BSP)

Longissimus dorsi DNA from fetal bovine (FB, 90-day embryo, 4 male individuals) and adult bovine (AB, 24 months old, 4 male individuals) was converted using the methylSEQr Bisulfite Conversion Kit (Applied Biosystems, Foster City, CA, USA) following the manufacturer's protocol. The BSP primers of CFL2 DMR (differentially methylated region) were synthesized by the online software MethPrimer [17]. BSP sequencing was analyzed by the online software QUMA. The nucleotide sequences of CFL1 DMR and CFL2 DMR (DMR, differentially methylated region) are shown in Table 1. We performed three independent replicates and sequenced five clones from each independent amplification and cloning set.

**Table 1.** Primer information for qRT-PCR.

| Name | Primer (5′-3′) |
|---|---|
| CFL1-F | GTGTGGCTGTCTCTGATG |
| CFL1-R | CGCTTCTTCACTTCCTCTG |
| CFL2-F | GGTGACATTGGTGATACTG |
| CFL2-R | CATATCGGCAATCATTCAGA |
| GAPDH-F | AGATAGCCGTAACTTCTGTGC |
| GAPDH-R | ACGATGTCCACTTTGCCAG |
| CFL1-DMR-F | TGTTTTAATAAGGATATTTAGGGTATTT |
| CFL1-DMR-R | CTCAATAAAAAACTACAACTCAACC |
| CFL2-DMR-F | GGAAAATTTTTGAAAATGTTTTTT |
| CFL2-DMR-R | CTTTAAAATCATCCTAACCAATACC |

*2.7. Statistical Analysis*

All statistics are rendered as means $\pm$ SE (standard error) of three replications, each replication experiment in triplicate. The results were determined using the $2^{-\Delta\Delta Ct}$ method [18] and analyzed by SPSS software (version 18.0). $p$-values < 0.05 or <0.01 were considered as statistically significant differences.

## 3. Results

*3.1. Spatiotemporal Expression Analysis of CFL2 Gene Compared to CFL1 Gene*

In order to understand the expression regulation between CFL1and CFL2, we examined the expression patterns of CFL1 and CFL2 in fetal bovine, calf, and adult bovine tissue using RT-PCR experiments. We calculated the relative mRNA levels of CFL1 and CFL2 genes in different tissues of QC cattle at three developmental stages. The expression level of the CFL1 gene is a benchmark.

In fetal bovine tissue, the expression level of CFL2 in seven tissue samples, including heart, liver, spleen, lung, kidney, adipose, and muscle, was significantly lower than that of CFL1 ($p < 0.01$) (Figure 1A). In calf tissue, the expression level of CFL2 in the heart, liver, lung, and muscle was higher than that of CFL1. The expression level of CFL2 was the highest in heart tissue compared to CFL1, and the expression of CFL2 in liver and muscle tissues was also significantly higher than that of CFL1. There was no significant difference in expression in the spleen, lung, and adipose tissue ($p > 0.05$) (Figure 1B). In adult cattle tissue, the expression level of CFL2 in seven tissues, including heart, liver, spleen, lung, kidney, fat, and muscle, was significantly higher than that of CFL1 ($p < 0.01$). The expression level of CFL2 was extremely high in muscle and heart tissue compared with CFL1 expression ($p < 0.01$). The fold difference in relative expression level in heart tissue was nearly 40-fold and in muscle tissue was more than 40-fold. In addition, the expression levels of CFL2 and CFL1 in liver, spleen, lung, kidney, and adipose tissue were also significantly different, and the expression of CFL2 was significantly higher than that of CFL1 ($p < 0.01$) (Figure 1C).

The expression of CFL1 was significantly higher than that of CFL2 in embryonic cattle tissue, while the expression in calf and adult cattle tissue was significantly lower than that of CFL2, especially in the heart and muscle tissue of adult cattle. This may reveal that in muscle tissues, the expression level of the CFL1 gene is the highest in fetal bovines and lowest in adult bovines, and the expression level decreases significantly with age; while the expression level of the CFL2 gene increases with bovine age, showing a very significant upward trend.

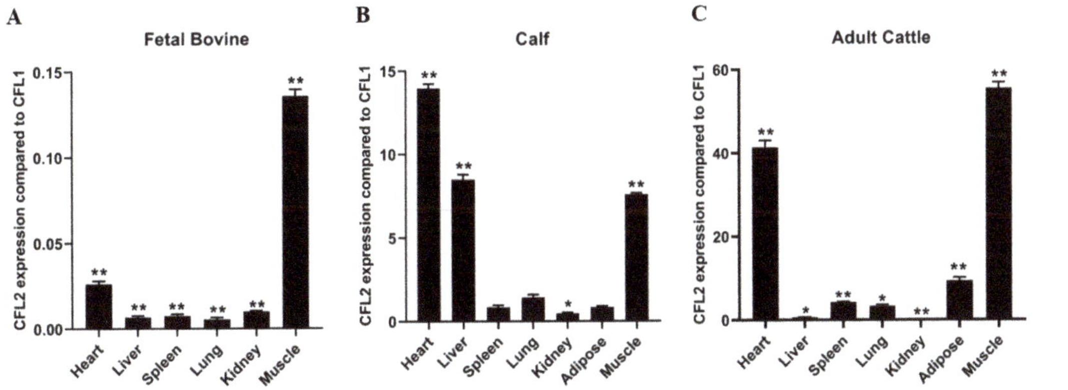


**Figure 1.** Spatial expression patterns of the CFL2 gene compared to the CFL1 gene in Qinchuan cattle. The relative spatial expression level of CFL2 was analyzed by qRT-PCR in different tissues of Qinchuan cattle at three growth and development stages: fetal bovine (**A**), calf (**B**), adult cattle (**C**). The expression level of the CFL1 gene was considered as 1. The mRNA expression level of CFL1 and CFL2 were normalized to GAPDH. Data are means ± SE of n = 3 independent experiments, each performed in triplicate. *, $p < 0.05$ and **, $p < 0.01$.

### 3.2. Spatiotemporal Expression of CFL1 and CFL2 in Cattle of Different Ages

In the heart and kidney, the expression of CFL1 in cattle of different ages showed an upward trend (Figure 2A–E). In the liver, the expression of CFL1 in cattle of different ages showed a trend of first decreasing and then increasing, and the overall expression level was highest in fetal cattle (Figure 2B). In the spleen and lung, the expression of CFL1 in cattle of different ages first increased and then decreased, and the expression level in calves was higher than that in fetal and adult cattle (Figure 2C,D).

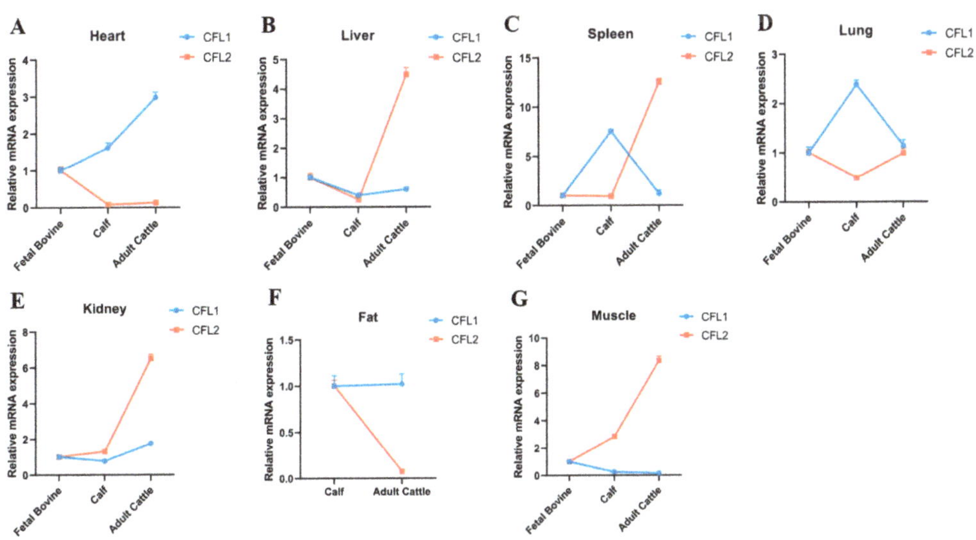


**Figure 2.** Temporal expression patterns of CFL1 and CFL2 genes in Qinchuan cattle. Expression trends of CFL1 and CFL2 genes in seven tissues (**A–G**). The expression level of all tissues in fetal bovine was considered as 1. The mRNA expression level of CFL1 and CFL2 were normalized to GAPDH. Data are means ± SE of n = 3 independent experiments, each performed in triplicate.

Due to the lack of fat deposition ability in fetal bovines, there was basically no CFL1 or CFL2 detected in fetal bovine fat, and there was no statistical change in the expression of CFL1 in calves and adult cattle. In muscle tissue, the expression of CFL1 showed a gradual downward trend with the increase in cattle age, and the expression in adult cattle was the lowest, close to zero (Figure 2G).

In heart and fat tissues, the expression of CFL2 in cattle of different ages showed a downward trend, and the presence of CFL2 was not detected in the adipose tissue of fetal cattle (Figure 2A–F). In lung tissues, the expression trend of CFL2 first decreased and then increased, and the expression level of CFL2 in the lung did not change significantly (Figure 2D). In the liver, spleen, kidney, and muscle, the expression of CFL2 showed an overall upward trend in cattle of different ages (Figure 2B,C), and the expression change in muscle tissue was especially obvious (Figure 2G).

In addition, it is worth noting that the expression level of CFL1 is higher than that of CFL2 in calf heart, spleen, and lung tissues; in adult cattle liver, spleen, kidney, and muscle, the expression level of CFL1 is lower than that of CFL2. Interestingly, we can clearly find that in muscle tissue, the expression level of CFL1 gradually decreases with the increase in cattle age, and the expression level of CFL2 gradually increases with the increase in cattle age and CFL2 expression level in calves and adults. The expression level in cattle was significantly higher than that of CFL1.

### 3.3. *Study on DNA Methylation Level of CFL1 and CFL2 Gene Promoters*

To further testify the effects of CFL1 and CFL2 genes on muscle tissues, we performed a DNA methylation study on myogenesis and muscle maturation stages.

CFL1 promoter DMR (185 bp) has thirteen CpGs, and CFL2 promoter DMR (177 bp) has eight CpGs; we sequenced fifteen clones in each muscle sample, then we obtained 195 CpGs of CFL1 DMR and 120 CpGs of CFL2 DMR. The methylated CpGs percentages of the CFL1 gene were 73.3%, 76.9%, 76.4%, and 75.9% in the FB group. There were 61.7%, 61.7%, 60.8%, and 60.0% of the CFL1 gene. Statistical results showed high DNA methylation levels of CFL1 promotor (mean 75.6%) and CFL2 promotor (mean 61.1%) in the FB group (Figure 3A,B).

In the AC group, the DNA methylation level values of CFL1 promotor were 87.7%, 86.2%, 88.7%, and 88.2%, CFL2 promotor values were 47.5%, 47.5%, 46.7%, and 48.3%. Statistical results presented significantly higher DNA methylation levels CFL1 promotor region (mean 87.7%) compared to CFL2 (mean 47.5%) (Figure 4A,B).

To further demonstrate whether the methylation levels of CFL1 and CFL2 affect their expression level in muscle tissues, we compared the mRNA level of CFL1 and CFL2 genes in the FB group and AC group. Contrasted with CFL1, the expression level of the CFL2 gene was significantly lower in the FB group ($p < 0.01$) (Figure 3C). In the AC group, the CFL2 gene had remarkably higher expression than CFL1, and the fold difference in relative expression level was about 55-fold (Figure 4C). Combined with the above results, there is a fairly straightforward correlation for tissue-specific CFL2 gene between expression and DNA methylation.

### 3.4. *Expression of CFL1 and CFL2 Genes during Differentiation of C2C12 Myoblasts*

The mouse myoblast C2C12 differentiates rapidly, forms retractable myotubes, and produces characteristic myoproteins. It is a widely used cell line to study myoblast proliferation and differentiation. During the growth and differentiation of C2C12, we detected the relative expression levels of myogenesis marker factors MYOD, MYOG, MYH3, and CFL1 and CFL2 (Figure 5) in order to initially understand the expression patterns of CFL1 and CFL2 in C2C12 muscle cells and the expression patterns of CFL1 and CFL2 in myoblasts, the regulatory relationship in the differentiation process, and its effect on the myogenic differentiation process.

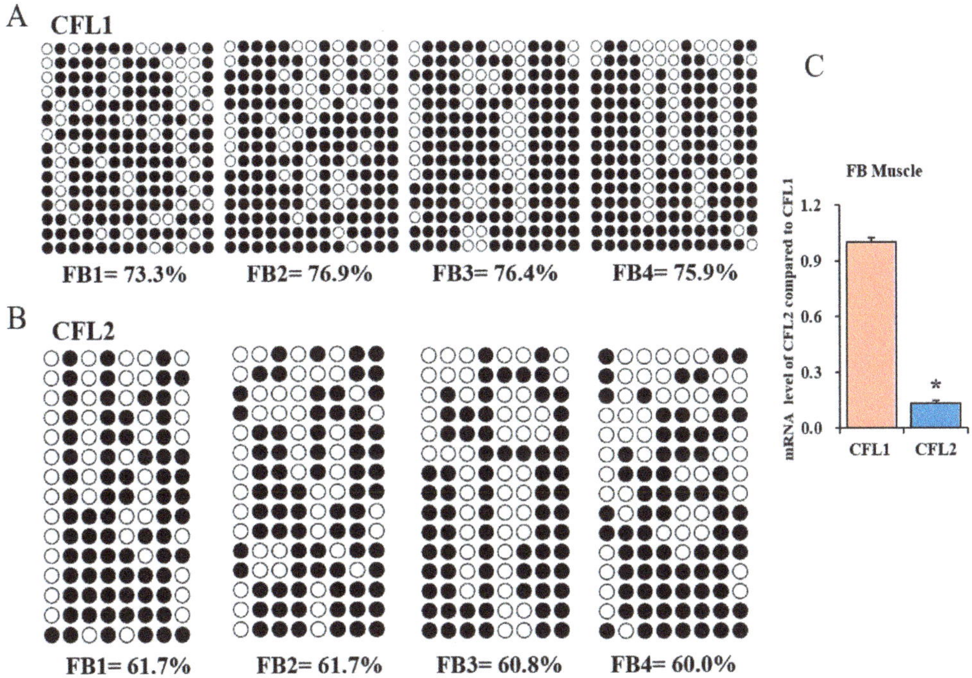


**Figure 3.** DNA methylation level of CFL1 and CFL2 gene promoters in fetal bovine muscle tissues. DNA methylation patterns of CFL1 (**A**) and CFL2 (**B**) gene promoters in fetal bovine muscle tissues (FB, n = 4) analyzed by BSP. Open circles show unmethylated CpGs and black circles methylated CpGs. mRNA expression level of the CFL2 gene compared to CFL1 was analyzed in muscle tissues (**C**). The mRNA expression level of CFL1 and CFL2 were normalized to GAPDH. The data represent the mean ± SE based on three independent experiments (*, $p < 0.05$).

During 1–8 days after induction of C2C12 differentiation (DM1-8), the overall expression trend of myoblast marker factors MYOD, MYOG, MYH3, and CFL2 gene was higher than that of C2C12 during growth and proliferation (GM), while the final expression level of CFL1 (DM8) was lower than that of C2C12 during growth (GM). The expression of MYOD increased from 0 to 6 days (GM-DM6) and then decreased from 6 to 8 days (GM6-8). The expression level of MYOD was higher than that of MYOG, MYH3, CFL1, and CFL2 in the late 6–8 days (GM6-8) of differentiation. The expression of MYOG increased from 0 to 2 days (GM-DM2), decreased from 2 to 4 days (DM2-4), increased from 4 to 6 days (DM4-6), and decreased from 6 to 8 days, and the expression level of MYOG was the highest on 2 days (DM2). MYH3 has an upward trend in 0–4 days (GM-DM4) and a larger upward trend in 2–4 days (DM2-4), and a downward trend in 4–8 days (DM4-8). However, the expression level of CFL1 showed an upward trend in the early 1–2 days (DM1-2) of differentiation and gradually decreased in the 4–8 days (DM4-8) after differentiation. On the eighth day of differentiation, the expression level of CFL1 was lower than that in the growth period of C2C12; CFL2 showed an upward trend in 0–6 days and a larger increase in 2–6 days (DM2-6), and CFL2 showed a downward trend in 6–8 days (DM6-8) in the late differentiation stage. The expression level of CFL1 was higher than that of CFL2 in the early stage of differentiation 0–2 days (GM-DM2), while the expression level of CFL2 was higher than that of CFL1 in the 2–6 days differentiation stage, and the expression level of CFL2 always increased in the 0–6 days (GM-DM6), and the expression level of CFL1 was decreased.

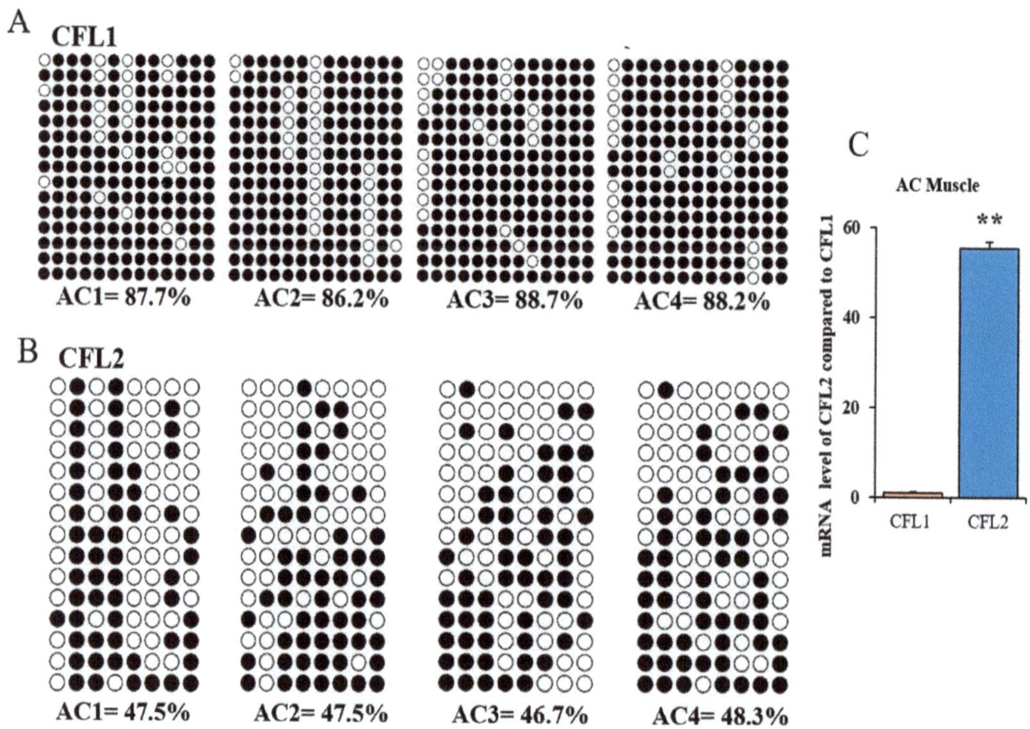


**Figure 4.** DNA methylation level of CFL1 and CFL2 gene promoters in adult cattle muscle tissues. DNA methylation patterns of CFL1 (**A**) and CFL2 (**B**) gene promoters in adult cattle muscle tissues (AC, n = 4) analyzed by BSP. Open circles show unmethylated CpGs and black circles methylated CpGs. mRNA expression level of the CFL2 gene compared to CFL1 was analyzed in muscle tissues (**C**). The mRNA expression level of CFL1 and CFL2 were normalized to GAPDH. The data represent the mean ± SE based on three independent experiments (**, $p < 0.01$).

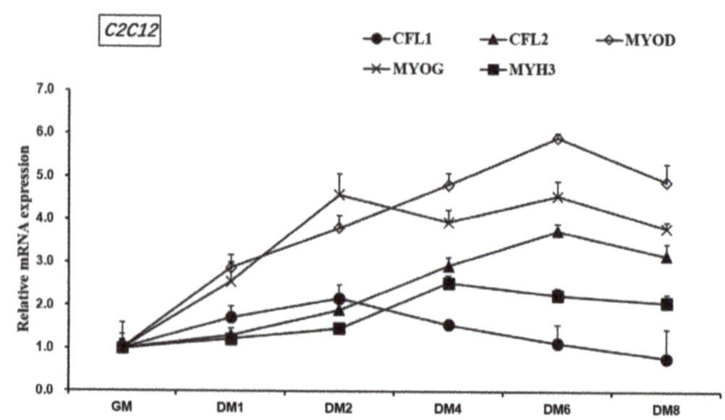


**Figure 5.** Expression trends of CFL1 and CFL2 gene and myogenesis factors were detected in C2C12 cells. C2C12 cells were induced for 1, 2, 4, 6, and 8 days by 2% horse serum, and the relative expression trends of CFL1gene, CFL2 gene, and myoblast differentiation marker genes MYOD, MYOG, and MYH3 were detected by qRT-PCR. The mRNA expression level was normalized to GAPDH. Data are mean ± SE of n = 3 independent experiments, each performed in triplicate. GM, growth medium. DM, differentiation medium.

### 3.5. Regulation of CFL1 and CFL2 Genes on Myoblast Differentiation

In order to determine the involvement of CFL1 and CFL2 in myogenic differentiation and to explore the roles of CFL1 and CFL2 in the process of myogenic differentiation and their relationship, we constructed adenovirus overexpression vectors and adenovirus interference vectors of CFL1 and CFL2. HEK293 cells packaged adenovirus and collected adenovirus venom, and then the adenovirus venom of CFL1, CFL2 interference, and overexpression vectors were transfected into PBMs and cell differentiation was induced. qRT-PCR detection showed that after transfection of pAd CMV-CFL1 (CFL1-OE), pAd CMV-CFL2 (CFL2-OE), pAd-shRNA-CFL1 (shCFL1), and pAd-shRNA-CFL2 (shCFL2), the expression of muscle marker factors MYOD, MYOG, and MYHC were significantly up-regulated or down-regulated ($p < 0.01$) (Figure 6).

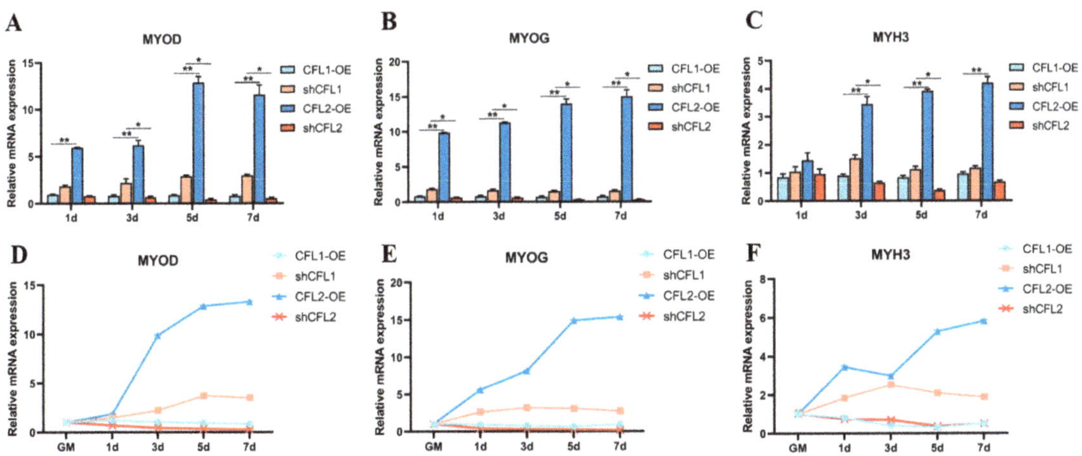


**Figure 6.** Functional comparison of CFL1 and CFL2 genes on the differentiation of PBMs. Overexpression vectors (CFL1-OE and CFL2-OE) and interference vectors (shCFL1-2 and shCFL2-1) of CFL1 and CFL2 genes were transfected into PBMs and induced for 1, 3, 5, 7 days by 2% horse serum. Then, the relative expression of marker genes MYOD, MYOG, and MYH3 were detected by qRT-PCR, transfected empty vector of overexpression or interference as negative control (**A–C**). Expression trends of myogenesis factors were detected by qRT-PCR, and the expression levels of four vectors in GM were considered as 1 (**D–F**). The mRNA expression levels of MYOD, MYOG, and MYH3 were normalized to GAPDH. Data are means ± SE of n = 3 independent experiments, each performed in triplicate. *, $p < 0.05$ and **, $p < 0.01$, two-tailed $t$-test. GM, growth medium.

After culturing PBMs and transfecting adenovirus overexpression vector CFL2-OE, the expression levels of MYOD and MYOG were significantly up-regulated at 1–7 d of differentiation ($p < 0.01$) (Figure 6A,B), and the expression level MYH3 at 3–7 d was significantly up-regulated (Figure 6C). Transfected with adenovirus overexpression vector CFL1-OE, the expression levels of MYOD, MYOG, and MYH3 did not change significantly ($p > 0.01$), while transfected adenovirus interference vector shCFL1, the expression levels of MYOD, MYOG, MYH3 at 1–7 d of differentiation were significantly trending upwards ($p < 0.01$) (Figure 6D–F).

When CFL2 was overexpressed, the expression levels of MYOD and MYOG at differentiation 1–7 d were significantly higher than those at overexpression CFL1 ($p < 0.01$) (Figure 6A,B), and the expression level of MYH3 at differentiation 3–7 d was significantly higher than that at overexpression CFL1 (Figure 6C). When CFL2 was interfered, the expression levels of MYOD and MYH3 at 3–7 d of differentiation were significantly lower than those when CFL1 was interfered ($p < 0.01$) (Figure 6A–C), and the expression levels of MYOG at 1–7 d of differentiation were significantly lower than those when CFL1 was interfered expression level ($p < 0.01$) (Figure 6B). Interestingly, the expression levels of MYOD, MYOG, and MYH3 were higher when CFL1 was overexpressed.

When CFL2 was overexpressed and CFL1 was interfered, the expression levels of MYOD, MYOG, and MYH3 were increased at the differentiation stage and had a large increase in the differentiation period. When CFL2 was overexpressed, the expression levels of MYOD, MYOG, and MYH3 were always higher than those when CFL1 was interfered during 1–7 d differentiation. When CFL2 was interfered and CFL1 was overexpressed, the expression levels of MYOD, MYOG, and MYH3 decreased at the differentiation stage. When CFL2 was interfered, the expression levels of MYOD and MYOG at 1–7 d differentiation were always lower than those when CFL1 was overexpressed (Figure 6D–F). In summary, during the differentiation stage of PBMs, a high expression level of CFL2 can promote differentiation, while a high expression level of CFL1 can inhibit differentiation.

## 4. Discussion

Muscle is an important factor affecting the growth and development of animals and one of the essential indicators to gauge meat quality. For cattle, the tenderness of beef refers to the freshness and tenderness of the meat when it is eaten and the quality of the meat. It is determined by the structural characteristics of the protein in the muscle. Meat quality is also a central quantitative trait regulated by a series of genes, and its phenotype has a special regulatory mechanism and a complicated genetic basis. At the transcription or translation level, gene expression regulation could affect the expression level of mRNA or protein [19]. Muscle fiber is the basic unit of muscle, muscle fiber type composition is an important factor of variation of meat quality, muscle fiber type composition has medium to upper heritability, and is tightly bound to meat quality traits [20]. Muscle cell differentiation plays a major role in accelerating muscle fiber formation. PBMs serve as an excellent model system for studying muscle cell differentiation in vitro, and serum supplements can be altered to stimulate myoblasts to differentiate into myocytes or to form myotubes. MRFs (muscle regulatory factors) modulate the occurrence and development of muscle, produce a marked effect on myoblast differentiation, and serve as marker genes during myoblast differentiation. MYOD, MYOG, and MYH3 are myogenic regulators of the MRF family that activate muscle gene transcription and drive non-muscle cells to transform into muscle cells. It is especially essential to expound on the genetic regulatory network of genes that is related to muscle growth and development from the viewpoint of genetics.

The actin-binding protein family, ADF/cofilin, is a major actin depolymerization factor and is ubiquitous in eukaryotes. It regulates actin dynamics by promoting the cleavage of F-actin and the depolymerization of actin monomers (G-actin) [1,21,22]. Cofilin family proteins include non-muscle adseverin cofilin-1 (n-cofilin, CFL1) and muscle adseverin cofilin-2 (m-cofilin, CFL2) [23]. The amino acid coincidence rate of the two proteins was as high as 80%. In addition, due to the high sequence conservation, antibodies often do not distinguish between subtype CFL1 and CFL2. In the chicken genome, only one cofilin protein (CFL2) was expressed, and xenopus could express two cofilin proteins (CFL1/CFL2) [24]. However, invertebrates (drosophila melanogaster, nematode, and the eggs of Echinococcus multilocularis) and eukaryotic microorganisms (acanthamoeba and Saccharomyces cerevisiae, etc.) only expressed one protein of the ADF/cofilin family. In toxoplasma gondii and coccidia, two ADF/cofilin family proteins could be expressed [25,26]. A previous study has revealed that cofilin genes regulate normal muscle function and regeneration. Therefore, as two subtypes of cofilin, CFL1 and CFL2 genes produce a marked effect on muscle growth and development. In healthy mice, the expression type of cofilin changes with the growth and development of muscle cells. In embryonic skeletal muscle, both non-muscle-type CFL1 and muscle type CFL2 were expressed simultaneously, and both of them can promote the recombination of actin filaments in the early stage of myofibrillary formation. However, with muscle growth and development, the expression of the CFL2 gene substitutes the CFL1 gene at the late stage of myocyte differentiation, and cofilin-2 is the only isoform in mature skeletal muscle [24]. The CFL2 gene is mainly expressed in the end stage of muscle development and may produce a marked effect on muscle physiology or development [27].

The cDNA sequence analysis of CFL1 and CFL2 genes showed that both CFL1 and CFL2 have a broad tissue expression profile. This observation implies that CFL1 and CFL2 genes may have different biological functions in different tissues of the organism. CFL1 not only participates in the regulation of cytoskeleton structure but also participates in the regulation of various cell functions. Studies have shown that CFL1 is involved in regulating the contractile response mechanism of vascular smooth muscle [28]. The phosphorylation and dephosphorylation of CFL1 can cause changes in tumor extracellular matrix adhesion, degree of metastasis, cell motility, and vascular invasion, and inhibition of CFL1 activity can reduce tumor cell motility and migration [29]. Phosphorylated CFL1 also directly activates phospholipase D1(PLD1) and regulates cell chemotaxis and phospholipid metabolism [30]. CFL1 is also involved in the regulation of mitochondrial function and structure [31]. The entry of CFL1 into mitochondria helps to open the mitochondrial permeability transition pore and promote the release of pigment C, thereby causing apoptosis [32]. CFL2 has the ability to bind G- and F-actin in a 1:1 ratio of cofilin to actin. It is the major component of intranuclear and cytoplasmic actin rods. CFL2 plays an important role in regulating the dynamics of striated muscle and fine muscle filaments [12,33,34]. CFL2 is the sixth gene found to be associated with NM disease in humans, and several cases of familial NM caused by mutations in the CFL2 gene have been reported [13,35,36]. Studies have shown that lack of CFL2 leads to cell cycle arrest and blocks the differentiation of satellite cells, leading to failure of skeletal muscle repair [34]. In mice, CFL2 gene knockout does not affect skeletal muscle development but shows progressive musculoskeletal muscle destruction and accumulation of F-actin [37,38]. It has also been shown that CFL2 regulates the length of filaments by participating in the regulation of actin dynamics of filaments in myocytes [12,39].

In conclusion, with the development of molecular biology techniques, many pieces of evidence show that CFL1 and CFL2 are related to a variety of effects, and the activities of CFL1 and CFL2 affect intracellular signal transduction and cytoskeletal actin recombination. The complex kinase signaling pathways of the actin family and cytoskeletal proteins are affected by CFL1. The inactivation and reactivation of CFL1 determine the formation and disassembly of F-actin, and CFL1 also allows the aggregation of actin (G-actin), resulting in gene mutations that determine the direction of cell migration. Currently, studies on the CFL2 gene are mainly focused on the proliferation and differentiation of muscle cells and the development, repair, and myopathy of skeletal muscle, with relatively few studies on the growth and development of domestic animals and its specific mechanism of action still needs to be further explored. Growth traits are important indicators of livestock production performance, and growth rate is one of the main target traits in livestock breeding. The bovine CFL1 gene (NC_037356.1) has 5 exons located on chromosome 29, with a full length of 3426bp; the bovine CFL2 gene (NC_037348.1) has four exons located on chromosome 21, with a full length 2653bp. Sun [40,41] et al. analyzed the correlation between the nucleotide polymorphisms (SNPs) of CFL1 and CFL2 and the body size traits of 488 QC cattle, and found that SNP T 2052C was significantly related to body length, chest width, chest depth, and body weight ($p < 0.05$), G 2107C was significantly related to hip length ($p < 0.05$), T 2169C was significantly related to chest width and chest depth ($p < 0.05$), G1500A was significantly correlated with hip width ($p < 0.05$), and T1694A was significantly correlated with chest circumference, chest depth, waist width, and body weight ($p < 0.05$). These results suggest that the CFL1 gene may be a strong candidate gene for influencing growth traits in QC cattle breeding programs. The CFL2 gene is not only closely related to muscle development and maintenance in mammals but also plays an important role in growth traits. Exploring and clarifying the effects of CFL1 and CFL2 genes on the growth and development of domestic animals will provide a new idea for molecular breeding and improvement of domestic animal economic traits.

Our study showed that the overexpression of the CFL1 gene had no significant effect on the expression of MYOD, MYOG, and MYH3, while the overexpression of the CFL2 gene significantly increased the expression of MYOD, MYOG, and MYH3. Down-regulated CFL2 expression significantly reduced the expression of MYOD, MYOG, and MYH3 during the differentiation of PBMs, indicating that increased CFL1 expression had no significant effect on the differentiation of PBMs, but increased CFL2 expression could promote the differentiation of PBMs. This fact may be related to the myogenic regulation of the CFL2 gene in myoblasts. We intend to search for a new mechanism of myogenic differentiation regulated by CFL2 from the viewpoint of genetics and epigenetics. CFL2 depletion inactivates the cellular differentiation process and suppresses the expression of myogenic transcription factors, ultimately resulting in impaired myogenic differentiation in myoblasts [33,42]. However, the knockdown CFL1 gene increased the expression of MYOD, MYOG, and MYH3 during PBMs differentiation, suggesting that CFL1 could suppress myoblast differentiation. This may be associated with the gradual disappearance of CFL1 at the end of myofiber differentiation.

## 5. Conclusions

Our study revealed that CFL1 was mainly expressed in embryonic muscle and CFL2 was predominantly expressed in adult muscle; there is a fairly straightforward correlation between muscle tissue-specific CFL2 gene and myoblasts differentiation. This study illustrated that CFL2 may be superior to CFL1 as a candidate gene for subsequent research on cattle genetics and breeding and will provide a theoretical research basis for improving the meat production and meat quality of cattle.

**Supplementary Materials:** The following are available online at https://www.mdpi.com/article/10.3390/agriculture12091420/s1, Figure S1: The efficiency detection of recombinant vectors by qRT-PCR and Western blot. Table S1: Primer information for vector construction. Table S2: Stability data expression of GAPDH in eight tissues of Qinchuan cattle at three growth stages.

**Author Contributions:** Conceptualization, Y.S.; validation, Y.M.; formal analysis, Y.S.; investigation, X.W. and T.Z.; data curation, T.Z.; writing—original draft preparation, Y.M.; writing—review and editing, Y.S.; supervision, Z.Y.; project administration, L.L.; funding acquisition, Y.S. and Z.Y. All authors have read and agreed to the published version of the manuscript.

**Funding:** This work was supported by the National Natural Science Foundation of China (No. 31902147, 31872324).

**Institutional Review Board Statement:** All animal experiments were carried out in accordance with the guidelines of the Institutional Administrative Committee and Ethics Committee of Laboratory Animals (license number: SYXK [Su] 2017-0044) and were approved by the Yangzhou University Institutional Animal Care and Use Committee.

**Informed Consent Statement:** Not applicable.

**Data Availability Statement:** Not applicable.

**Acknowledgments:** We are grateful to the Hong Chen team (Northwest A&F University) for their technical support.

**Conflicts of Interest:** The authors declare no conflict of interest.

## References

1. Bernstein, B.W.; Bamburg, J.R. ADF/cofilin: A functional node in cell biology. *Trends Cell Biol.* **2010**, *20*, 187–195. [CrossRef] [PubMed]
2. Yonezawa, N.; Nishida, E.; Koyasu, S.; Maekawa, S.; Sakai, H. Distribution among tissues and intracellular localization of cofilin, a 21kDa actin-binding protein. *Cell Struct. Funct.* **1987**, *12*, 443–452. [CrossRef] [PubMed]
3. Maekawa, S.; Nishida, E.; Ohta, Y.; Sakai, H. Isolation of low molecular weight actin-binding proteins from porcine brain. *J. Biochem.* **1984**, *95*, 377–385. [CrossRef] [PubMed]
4. Kanellos, G.; Zhou, J.; Patel, H.; Ridgway, R.; Huels, D.; Gurniak, C.; Sandilands, E.; Carragher, N.; Sansom, O.; Witke, W. ADF and Cofilin1 Control Actin Stress Fibers, Nuclear Integrity, and Cell Survival. *Cell Rep.* **2015**, *13*, 1949–1964. [CrossRef] [PubMed]

5. Tahtamouni, L.H.; Shaw, A.E.; Hasan, M.H.; Yasin, S.R.; Bamburg, J.R. Non-overlapping activities of ADF and cofilin-1 during the migration of metastatic breast tumor cells. *BMC Cell Biol.* **2013**, *14*, 45. [CrossRef]
6. Wang, A.Y.; Liu, H. The past, present, and future of CRM1/XPO1 inhibitors. *Stem Cell Investig.* **2019**, *6*, 6. [CrossRef]
7. Gressin, L.; Guillotin, A.; Guérin, C.; Blanchoin, L.; Michelot, A. Architecture dependence of actin filament network disassembly. *Curr. Biol. CB* **2015**, *25*, 1437–1447. [CrossRef]
8. Zhao, W.; Su, Y.H.; Su, R.J.; Ba, C.F.; Zeng, R.X.; Song, H.J. The full length cloning of a novel porcine gene CFL2b and its influence on the MyHC expression. *Mol. Biol. Rep.* **2009**, *36*, 2191–2199. [CrossRef]
9. Mohri, K.; Takano-Ohmuro, H.; Nakashima, H.; Hayakawa, K.; Obinata, T. Expression of cofilin isoforms during development of mouse striated muscles. *J. Muscle Res. Cell Motil.* **2000**, *21*, 49–57. [CrossRef]
10. Bellenchi, G.C.; Gurniak, C.B.; Perlas, E.; Middei, S.; Ammassari-Teule, M.; Witke, W. N-cofilin is associated with neuronal migration disorders and cell cycle control in the cerebral cortex. *Genes Dev.* **2007**, *21*, 2347–2357. [CrossRef]
11. Ikeda, S. Aberrant actin cytoskeleton leads to accelerated proliferation of corneal epithelial cells in mice deficient for destrin (actin depolymerizing factor). *Hum. Mol. Genet.* **2003**, *12*, 1029–1036. [CrossRef] [PubMed]
12. Kremneva, E.; Makkonen, M.H.; Skwarek-Maruszewska, A.; Gateva, G.; Michelot, A.; Dominguez, R.; Lappalainen, P. Cofilin-2 Controls Actin Filament Length in Muscle Sarcomeres. *Dev. Cell* **2014**, *31*, 215–226. [CrossRef] [PubMed]
13. Ong, R.W.; Alsaman, A.; Selcen, D.; Arabshahi, A.; Yau, K.S.; Ravenscroft, G.; Duff, R.M.; Atkinson, V.; Allcock, R.J.; Laing, N.G. Novel cofilin-2 (CFL2) four base pair deletion causing nemaline myopathy. *J. Neurol. Neurosurg. Psychiatry* **2014**, *85*, 1058–1060. [CrossRef] [PubMed]
14. Luo, J.; Deng, Z.L.; Luo, X.; Tang, N.; Song, W.X.; Chen, J.; Sharff, K.A.; Luu, H.H.; Haydon, R.C.; Kinzler, K.W. A protocol for rapid generation of recombinant adenoviruses using the AdEasy system. *Nat. Protoc.* **2007**, *2*, 1236. [CrossRef]
15. Rekawiecki, R.; Rutkowska, J.; Kotwica, J. Identification of optimal housekeeping genes for examination of gene expression in bovine corpus luteum. *Reprod. Biol.* **2012**, *14*, 362–367. [CrossRef]
16. Li, P.; Zhu, Y.; Kang, X.; Dan, X.; Ma, Y.; Shi, Y. An integrated approach in gene-expression landscape profiling to identify housekeeping and tissue-specific genes in cattle. *Anim. Prod. Sci.* **2021**, *61*, 1643–1651. [CrossRef]
17. Li, L.C.; Dahiya, R. Methprimer: Designing Primers for Methylation PCRs. *Bioinformatics* **2002**, *18*, 1427–1431. [CrossRef]
18. Schmittgen, T.D. Analyzing real-time PCR data by the comparative CT method. *Nat. Protoc.* **2008**, *3*, 1101–1108. [CrossRef]
19. Sun, Y.; Liu, K.; Huang, Y.; Lan, X.; Chen, H. Differential expression of FOXO1 during development and myoblast differentiation of Qinchuan cattle and its association analysis with growth traits. *Sci. China-Life Sci.* **2018**, *61*, 825–835. [CrossRef]
20. Choe, J. Overview of muscle metabolism, muscle fiber characteristics, and meat quality. *Korean J. Agric. Sci.* **2018**, *45*, 50–57.
21. Kanellos, G.; Frame, M.C. Cellular functions of the ADF/cofilin family at a glance. *J. Cell Sci.* **2016**, *129*, 3211. [CrossRef] [PubMed]
22. Ono, S.h.o.i.c.h.i.r.o. Regulation of Actin Filament Dynamics by Actin Depolymerizing Factor/Cofilin and Actin-Interacting Protein 1: New Blades for Twisted Filaments. *Biochemistry* **2003**, *42*, 13363–13370. [CrossRef] [PubMed]
23. Maciver, S.K.; Hussey, P.J. The ADF/cofilin family: Actin-remodeling proteins. *Genome Biol.* **2002**, *5*, 3007.1–3007.12.
24. Ono, S.; Minami, N.; Abe, H.; Obinata, T. Characterization of a novel cofilin isoform that is predominantly expressed in mammalian skeletal muscle. *J. Biol. Chem.* **1994**, *269*, 15280–15286. [CrossRef]
25. Makioka, A.; Kumagai, M.; Hiranuka, K.; Kobayashi, S.; Takeuchi, T. Entamoeba invadens: Identification of ADF/cofilin and their expression analysis in relation to encystation and excystation. *Exp. Parasitol.* **2011**, *127*, 195–201. [CrossRef]
26. Ueno, A.; Dautu, G.; Saiki, E.; Haga, K.; Igarashi, M. Toxoplasma gondii deoxyribose phosphate aldolase-like protein (TgDPA) interacts with actin depolymerizing factor (TgADF) to enhance the actin filament dynamics in the bradyzoite stage. *Mol. Biochem. Parasitol.* **2010**, *173*, 39–42. [CrossRef]
27. Obinata, T.; Nagaoka-Yasuda, R.; Ono, S.; Kusano, K.; Abe, H. Low Molecular-weight G-actin Binding Proteins Involved in the Regulation of Actin Assembly during Myofibrillogenesis. *Cell Struct. Funct.* **1997**, *22*, 181–189. [CrossRef]
28. Montenegro, M.F.; Valdivia, A.; Smolensky, A.; Verma, K.; Taylor, W.R.; Martín, A.S. Nox4-dependent activation of cofilin mediates VSMC reorientation in response to cyclic stretching. *Free. Radic. Biol. Med.* **2015**, *85*, 288–294. [CrossRef]
29. Oleinik, N.V.; Helke, K.L.; Kistner-Griffin, E.; Krupenko, N.I.; Krupenko, S.A. Rho GTPases RhoA and Rac1 Mediate Effects of Dietary Folate on Metastatic Potential of A549 Cancer Cells through the Control of Cofilin Phosphorylation. *J. Biol. Chem.* **2014**, *289*, 26383–26394. [CrossRef]
30. Sequential Regulation of DOCK2 Dynamics by Two Phospholipids During Neutrophil Chemotaxis. *Science* **2009**, *324*, 384. [CrossRef]
31. Hoffmann, L.; Rust, M.B.; Culmsee, C. Actin(g) on mitochondria—A role for cofilin1 in neuronal cell death pathways. *Biol. Chem.* **2019**, *400*, 1089–1097. [CrossRef] [PubMed]
32. Chua, B.T.; Volbracht, C.; Tan, K.O.; Li, R.; Yu, V.C.; Li, P. Mitochondrial translocation of cofilin is an early step in apoptosis induction. *Nat. Cell Biol.* **2003**, *5*, 1083–1089. [CrossRef]
33. Zhu, H.; Yang, H.; Zhao, S.; Zhang, J.; Liu, D.; Tian, Y.; Shen, Z.; Su, Y. Role of the cofilin 2 gene in regulating the myosin heavy chain genes in mouse myoblast C2C12 cells. *Int. J. Mol. Med.* **2018**, *41*, 1096–1102. [CrossRef] [PubMed]
34. Morton, S.U.; Mugdha, J.; Talia, S.; Beggs, A.H.; Agrawal, P.B.; Ashok, K. Skeletal Muscle MicroRNA and Messenger RNA Profiling in Cofilin-2 Deficient Mice Reveals Cell Cycle Dysregulation Hindering Muscle Regeneration. *PLoS ONE* **2015**, *10*, e0123829.

35. Agrawal, P.B.; Greenleaf, R.S.; Tomczak, K.K.; Lehtokari, V.L.; Wallgren-Pettersson, C.; Wallefeld, W.; Laing, N.G.; Darras, B.T.; Maciver, S.K.; Dormitzer, P.R. Nemaline myopathy with minicores caused by mutation of the CFL2 gene encoding the skeletal muscle actin-binding protein, cofilin-2. *Am. J. Hum. Genet.* **2007**, *80*, 162–167. [CrossRef]
36. Ockeloen, C.W.; Gilhuis, H.J.; Pfundt, R.; Kamsteeg, E.J.; Alfen, N.V. Congenital myopathy caused by a novel missense mutation in the CFL2 gene. *Neuromuscul. Disord. NMD* **2012**, *22*, 632–639. [CrossRef]
37. Agrawal, P.B.; Mugdha, J.; Talia, S.; Zoe, C.; Beggs, A.H. Normal myofibrillar development followed by progressive sarcomeric disruption with actin accumulations in a mouse Cfl2 knockout demonstrates requirement of cofilin-2 for muscle maintenance. *Hum. Mol. Genet.* **2012**, *21*, 2341–2356. [CrossRef]
38. Gurniak, C.B.; Chevessier, F.; Jokwitz, M.J.; Nsson, F.; Perlas, E.; Richter, H.; Matern, G.; Boyl, P.P.; Chaponnier, C.; Fürst, D. Severe protein aggregate myopathy in a knockout mouse model points to an essential role of cofilin2 in sarcomeric actin exchange and muscle maintenance. *Eur. J. Cell Biol.* **2014**, *93*, 252–266. [CrossRef]
39. Yamashiro, S.; Cox, E.A.; Baillie, D.L.; Hardin, J.D.; Ono, S. Sarcomeric actin organization is synergistically promoted by tropomodulin, ADF/cofilin, AIP1 and profilin in C. elegans. *J. Cell Sci.* **2008**, *121*, 3867. [CrossRef]
40. Sun, Y.; Lan, X.; Lei, C.; Zhang, C.; Chen, H. Haplotype combination of the bovine CFL2 gene sequence variants and association with growth traits in Qinchuan cattle. *Gene* **2015**, *563*, 136–141. [CrossRef]
41. Sun, Y.; Xu, C.; Yang, Z.; Li, M.; Chen, Z.; Xu, T.; Zhang, H.; Mao, Y. The polymorphism of bovine Cofilin-1 gene sequence variants and association analysis with growth traits in Qinchuan cattle. *Anim. Biotechnol.* **2022**, *33*, 63–69. [CrossRef] [PubMed]
42. Mai, T.N.; Min, K.H.; Kim, D.; Park, S.Y.; Wan, L. CFL2 is an essential mediator for myogenic differentiation in C2C12 myoblasts. *Biochem. Biophys. Res. Commun.* **2020**, *533*, 710–716.

agriculture

MDPI

Communication

# The Prevalence of *Escherichia coli* Derived from Bovine Clinical Mastitis and Distribution of Resistance to Antimicrobials in Part of Jiangsu Province, China

**Tianle Xu [1,2,*], Wendi Cao [3], Yicai Huang [3], Jingwen Zhao [3], Xinyue Wu [3] and Zhangping Yang [1,3]**

1 Joint International Research Laboratory of Agriculture and Agri-Product Safety, Ministry of Education of China, Yangzhou University, Yangzhou 225009, China
2 International Joint Research Laboratory in Universities of Jiangsu Province of China for Domestic Animal Germplasm Resources and Genetic Improvement, Yangzhou University, Yangzhou 225009, China
3 College of Animal Science and Technology, Yangzhou University, Yangzhou 225009, China
* Correspondence: tl-xu@yzu.edu.cn

**Abstract:** Bovine mastitis is often taken as one of the most common diseases in dairy farms, which its pathophysiology leads to a reduction of milk production and its quality. The penetration of pathogenic bacteria into the mammary gland, through either a contagious or environmental approach, has been determined the way of infection. The mastitis derived bacteria have become a challenge in practice, since the increasing exposure of antimicrobial. In order to identify characteristics of the epidemiological regulation and drug resistance of the pathogenic bacteria of bovine mastitis in northern Jiangsu, 156 clinical mastitis milk samples were collected from 3 large-scale farms for the epidemiological investigation and analysis of the drug resistance of *E. coli*. The bacteria were positively isolated in a total of 143 milk samples. The results showed that 78 strains of *E. coli* were detected, with a prevalence rate of 26.99%, followed by 67 strains of *K. pneumoniae*, with a prevalence of 23.19%, and 38 strains of *Staphylococcus*, with a prevalence of spp. 13.15%. In addition, 78 strains of *E. coli* isolated from bovine mastitis were tested for susceptibility to 8 kinds of antibiotics. It was shown that gentamicin and tetracycline were the most effective against *E. coli*, with the susceptibility rate of 83.3%, followed by streptomycin and ciprofloxacin, with 73.1% and 71.8% respectively. However, β-lactams including penicillin, cefothiophene, and amoxicillin showed serious resistance to *E. coli* isolates. There were 12 drug resistance genes detected by PCR, including β-lactam (blaTEM, blaCTX-M, and blaSHV), aminoglycoside (armA and armB), tetracycline (tetA, tetB, and tetC), and quinolone (qnrS, qepA, oqxA, and oqxB) related genes. Notably, all *E. coli* isolates carried blaTEM gene (100%). The detection rate of blaCTX-M was 53.8%, followed by the detection of blaSHV (20.5%), armA (9.0%), tetA (26.9%), tetB (2.6%), tetC (20.5%), qnrS (29.5%), oqxA (37.2%) and oqxB (1.3%). The present study provides crucial information on the distribution of bovine mastitis derived bacterial pathogens in Jiangsu province, as well as highlighting the antimicrobial resistance which might help to improve the efficiency of antibiotics treatment on bovine mastitis.

**Keywords:** *Escherichia coli*; drug resistance; bovine mastitis; β-lactams

**Citation:** Xu, T.; Cao, W.; Huang, Y.; Zhao, J.; Wu, X.; Yang, Z. The Prevalence of *Escherichia coli* Derived from Bovine Clinical Mastitis and Distribution of Resistance to Antimicrobials in Part of Jiangsu Province, China. *Agriculture* **2023**, *13*, 90. https://doi.org/10.3390/agriculture13010090

Academic Editor: Jiaqi Wang

Received: 19 December 2022
Revised: 25 December 2022
Accepted: 27 December 2022
Published: 29 December 2022



## 1. Introduction

Bovine mastitis is one of the diseases that cause great economic losses in the dairy industry [1,2]. It can lead to a decreased milk production, lowered milk quality, reproductive barriers, and a higher elimination rate [3]. The occurrence and development of mastitis over the past years appears to have changed, due to differences in the scales and management of the dairy farming industry, in China [4]. The pathogenic microorganisms that induce mastitis in dairy cows include bacteria, fungi, viruses, and mycoplasma, among which bacteria are the most important pathogens [4]. Mastitis derived pathogens can be divided into contagious pathogens and environmental pathogens [5]. The contagious pathogens include coagulase-negative *Staphylococci*, *Streptococcus agalactiae*, and *Streptococcus dysgalactiae*, which can be further spread among dairy cows through milking equipment and other means [6]. The environmental pathogens include *Escherichia coli* (*E. coli*), *Streptococcus uberis*, *Klebsiella pneumoniae*, and coagulase-negative *Staphylococcus*, which mainly exist in habitats such as soil, feces, and beds, and are transmitted mainly through udder contact [7].

Studies have shown that *E. coli* infection is usually characterized with persistence and a higher possibility of causing recurrent infections [5]. Moreover, the cases of cow mastitis caused by *E. coli* have shown an increasing trend in recent years [8]. As *E. coli* is an opportunistic pathogen widely present in the environment, it can cause human and animal infections with zoonotic diseases [9]. Currently, antibiotics are usually used to treat *E. coli* infected mastitis with better efficiency in animal husbandry and veterinary clinics. However, studies have shown that *E. coli* capable of developing broad-spectrum resistance to commonly used antibiotics, along with rapid variation [10]. For a long time, under selective pressure, the antibiotic resistance of *E. coli* continues to rise, which brings great challenges to the prevention and treatment of clinical upper colibacillosis [11]. Therefore, it is of great significance for the prevention and control of colibacillosis and public health to carry out the detection of pathogenic bacteria of bovine mastitis and the systematic detection of antibiotic resistance of *E. coli* from mastitis, as well as the resistance genes carried [12].

Although studies have recently focused on tackling antimicrobial resistance by designing alternative treatment regimens to antimicrobial, the discussion of issues and the emergence of resistance are still encouraged [13]. Among the factors involved in antimicrobial resistance, the bacterial status, including susceptibility and resistance, tolerance, persistence, and biofilm formation, are the main drivers of resistance outcome [14,15]. *E. coli* has many accessory genomes of plasmids and chromosome gene loci. In Beijing, China, 87.1% of the 70 *E. coli* isolates obtained from bovine mastitis cases were resistant to streptomycin, kanamycin, and ampicillin [16]. With the phenotypes of antibiotic resistance, the *E. coli* isolates are known as efficient reservoirs for resistant genes and are capable of transfering the genes among pathogenic bacteria [17]. The common harbored resistance genes for *E. coli* in America, mentioned in previous reports, are blaCTX-M, bla-TEM, ampC, tetA, tetB, sulI, and sulII [18,19]. Moreover, 5.3% of the isolates from bovine mastitis cases in Ningxia, China, were found to be carrying rmtB, bla-TEM1, and bla-CTX-M-15 which are related to β-lactams resistance [20]. More recently, we have characterized the phenotypic resistance of *Klebsiella pneumonia* isolated from clinical bovine mastitis in some parts of Jiangsu province [21]. As a result, isolates were found resistant to β-lactams including penicillin and amoxicillin, with the rates of 85.3% and 67.6%, respectively [21]. However, the phenotypic resistance of *E. coli* in northern Jiangsu province is rarely investigated. The objective of the current study is to underscore the prevalence of *E. coli* derived from bovine clinical mastitis, as well as investigate the phenotypes of the multi-drug resistant *E. coli* isolates, in Jiangsu province, China.

## 2. Materials and Methods

### 2.1. Isolation and Identification

A total of 156 samples were collected from the milk of dairy cows with clinical mastitis during July to October, in 2021. All strains isolated from the selected farms were in three regions (Xinyi, Sihong, and Huai'an) of Jiangsu Province in China. The milk samples were obtained from cows with clinical mastitis, diagnosed by using visible symptoms and testing with the California Mastitis Test (CMT), as officially recommended [22,23]. A volume of 100 μL of collected milk samples was plated onto a cultivation with blood agar (made with 5% freshly prepared whole blood, collected from sheep, and aerobically incubated at 37 °C for 24 h). By checking the morphology of the colonies, a single colony with identical morphology from each milk sample was then re-cultured by streaking on MacConkey (MC) agar, for purification. The sub-cultured bacteria were picked and amplified in Luria-Bertani (LB) broth at 37 °C for 24 h. The suspected isolates were further confirmed by PCR verification (Vazyme, Nanjing, China) with extracted genomic DNA from bacteria (Tiangen, Beijing, China), following 16S rDNA pyrosequencing (Tsingke., Ltd., Beijing, China), and sequences were proofread with SnapGene software (GSL Biotech, Chicago, IL, USA) [24]. Finally, the obtained sequences from the PCR products were analyzed using a BLAST sequence search, used for species identification, on the NCBI database. The confirmed isolates were stored in tubes containing 15% glycerol at −80 °C. The antimicrobial susceptibility of *E. coli* isolates was then determined using the Kirby–Bauer test and resistance gene detection.

### 2.2. Determination of Antimicrobial Susceptibility

The antimicrobial susceptibility was tested on a plate with Mueller–Hinton agar (MHA) by Kirby–Bauer (K-B) disk diffusion [25]. The 8 disks (Hangzhou Microbial Reagent Co., Hangzhou, China) used in the experiment were incubated with the following antibiotics: ciprofloxacin (CIP, 5 μg), streptomycin (STR, 10 μg), gentamycin (CN, 10 μg), tetracycline (TCY, 30 μg), cephalothin (KF, 30 μg), penicillin (PEN, 10 μg), amoxicillin (AML, 20 μg), and lincomycin (MY, 2 μg). The *E. coli* ATCC 25,922 strain was taken as the reference. The results of zone diameter for each disk were recorded as susceptible, intermedium, or resistant, on the basis of the interpretative standards of the Clinical and Laboratory Standards Institute [26]. When the recommended values for the antimicrobial disks were not available, following the manufacturer's instructions was advised.

### 2.3. Resistance Gene Detection

The detection of 12 resistance genes positively carried in *E. coli* was determined using PCR assays on 78 isolated *E. coli* isolates used in this study, including blaTEM, blaCTX-M, blaSHV, armA, armB, tetA, tetB, tetC, qnrS, qepA, oqxA, and oqxB [27,28]. Primers used in the current study has been listed in Table 1. The PCR was performed as follows: denaturing at 94 °C for 5 min; then 35 cycles, at 94 °C for 30 s, corresponding to extension and annealing, for 30 s and 72 °C for 1 min, along with an additional elongation at 72 °C for 10 min. The reference strains of ATCC 25,922 were taken for PCRs as well. PCR amplicons were then verified by running gels with 1% agar and performing DNA sequencing (Tsingke, Beijing, China). Visualization of bands on the gel images was used to identify the presence of each targeted gene.

**Table 1.** Primers designed for the PCR assay.

| Genes | Primers (5′-3′) | Annealing (°C) | Products (bp) |
|---|---|---|---|
| blaTEM | F: TCGCCGCATACACTATTCTCAGAATGA<br>R: ACGCTCACCGGCTCCAGATTTAT | 50 | 445 |
| blaCTX-M | F: ATGTGCAGACCAGTAAGTATGGC<br>R: TGGGTAATAGTACCAGAACAG | 57 | 593 |
| blaSHV | F: ATGCGTTATATTCGCCTGTG<br>R: TGCTTTGTTATTCGGGCCAA | 57 | 747 |
| armA | F: GGGTCTTACTATTCTGCCTAT<br>R: ATTCCCTTCTCCTTTCCAG | 50 | 503 |
| armB | F: TTTCTGCGGGCGATGTAA<br>R: AGTTCTGTTCCGATGGTCTTT | 57 | 523 |
| tetA | F: GCTACATCCTGCTTGCCTTC<br>R: CATAGATCGCCGTGAAGAGG | 60 | 210 |
| tetB | F: TTGGTTAGGGGCAAGTTTTG<br>R: GTAATGGGCCAATAACACCG | 60 | 659 |
| tetC | F: CTTGAGACCTTCAACCCAG<br>R: ATGGTCGTCATCTACCTGCC | 60 | 418 |
| qnrS | F: ACATAAAGACTTAAGTGATC<br>R: CAATTAGTCAGGATAAAC | 52 | 619 |
| qepA | F: CCAGCTCGGCAACTTGATAC<br>R: ATGCTCGCCTTCCAGAAAA | 60 | 570 |
| oqxA | F: CTCGGCGCGATGATGCT<br>R: CCACTCTTCACGGGAGACGA | 57 | 392 |

**Table 1.** *Cont.*

| Genes | Primers (5′-3′) | Annealing (°C) | Products (bp) |
|---|---|---|---|
| oqxB | F: TTCTCCCCCGGCGGGAAGTAC<br>R: CTCGGCCATTTTGGCGCGTA | 64 | 512 |
| 16SrDNA | TTCGGACCTCACGCTATCA<br>GAAGGCACCAA TCCATCTC | 62 | 824 |

## 3. Results

### 3.1. Isolation of Bacterial from Bovine Clinical Mastitis

Among the 156 milk samples, 12 species of bacteria were identified. However, 13 milk samples showed no bacterial detection, due to the non-visible growth of colonies on the cultured plates. The detectable rate of bacterial pathogens from clinical mastitis was 91.67% (Table 2). The infected quarters of bovine mammary glands are usually contaminated with various pathogens and the incidence is becoming more complicated to manage. As shown in Table 3, out of 143 samples that were culture-positive, the infected patterns from those samples showed 1 species of bacteria in 44 samples (30.77%), 2 species of bacteria in 52 samples (36.36%), and 3 species of bacteria in 47 samples (32.87%). Of the 289 strains of bacteria isolated, 8 species with 220 strains were closely related to mastitis, with a total detection rate of 76.12%. Among 289 isolated strains, *E. coli* was the most frequently isolated (78 strains, 26.99%), followed by 67 strains of *Klebsiella* spp. (23.19%), 38 strains of *Staphylococcus* spp. (13.15%), 8 strains of *Streptococcus agalactis* (2.77%), 4 strains of *Streptococcus uberis* (1.38%), 3 strains of *Streptococcus dysgalactiae* (1.04%), and 2 strains of *Pseudomonas aeruginosa* (0.69%) (Figure 1).

**Table 2.** Isolation of 156 clinical mastitis milk samples.

| Items | Positive Isolates | Negative Isolates | Total Isolates |
|---|---|---|---|
| Samples (No.) | 143 | 13 | 156 |
| Ratio (%) | 91.67 | 8.33 | 100 |

**Table 3.** Pathogen infection pattern of 143 positive isolation.

| Items | Single Isolates | Double Isolates | Triple Isolates | Total Isolates |
|---|---|---|---|---|
| Samples (No.) | 44 | 52 | 47 | 143 |
| Ratio (%) | 30.77 | 36.36 | 32.87 | 100 |

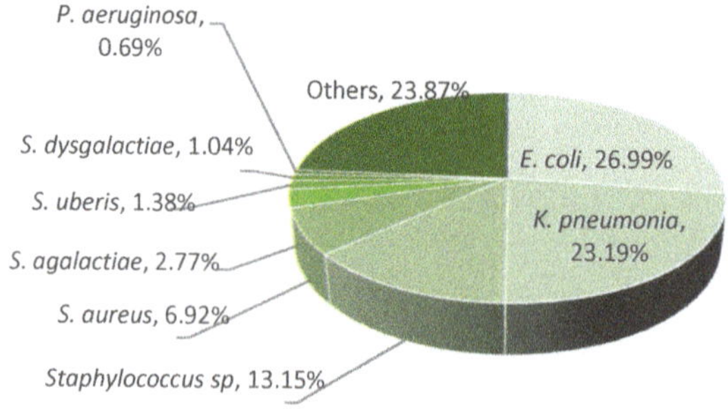


**Figure 1.** Analysis of 289 strains of pathogenic bacteria of bovine mastitis. All isolates were obtained and identified through bacterial isolation, purification, and 16S rDNA identification.

### 3.2. Phenotypic Resistance of E. coli Isolated from Bovine Clinical Mastitis

As shown in Figure 2, *E. coli* strains isolated from cow mastitis in northern Jiangsu have developed resistance to the eight antibiotics used, to varying degrees, with the resistance rate ranging from 11.6–66.7%. The resistance of *E. coli* isolates to lincomycin was the most serious, with the resistance rate of 66.7%, followed by the resistance to β-lactams including penicillin, cefothiophene, and amoxicillin, with rates of 60.2%, 57.7% and 55.1%, respectively. In contrast, *E. coli* isolates were more sensitive to gentamicin, tetracycline, streptomycin, and ciprofloxacin than others tested, and the resistance rates were 16.7%, 11.6%, 26.9% and 26.9%, respectively.

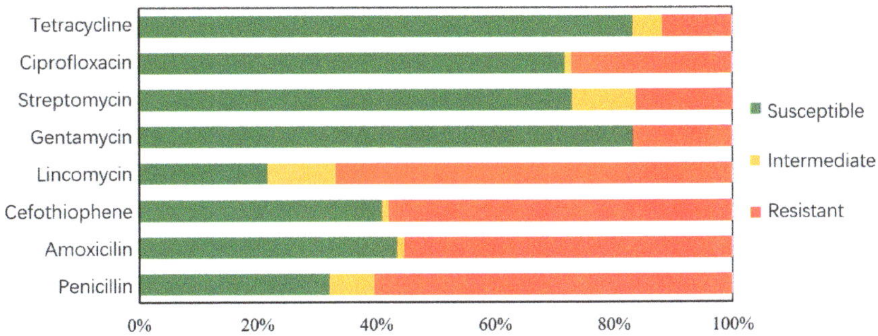


**Figure 2.** Phenotypic drug-resistance proportions of *E. coli.* All the 78 isolates of *E. coli* were analyzed with 8 antimicrobial agents. The bars in green, yellow, and red indicate the antimicrobial's susceptibility with susceptible, intermediate, and resistant, respectively.

*E. coli* isolates were found to be multi-drug resistant in farms of northern Jiangsu province. As shown in Figure 3, 22 strains were the most resistant to 5 antibiotics, accounting for 28.2%. Similarly, 16 strains resistant to an amount of 6 antibiotics, followed by 11 and 7 strains resistant to 1 and 2 antibiotics, respectively. However, only 2 strains appeared to be resistant to 4 or 7 antibiotics. Notably, major resistance to β-lactam drugs indicate that the isolated *E. coli* strains from bovine mastitis in northern Jiangsu are dominated by lactam drugs and are becoming relatively serious.

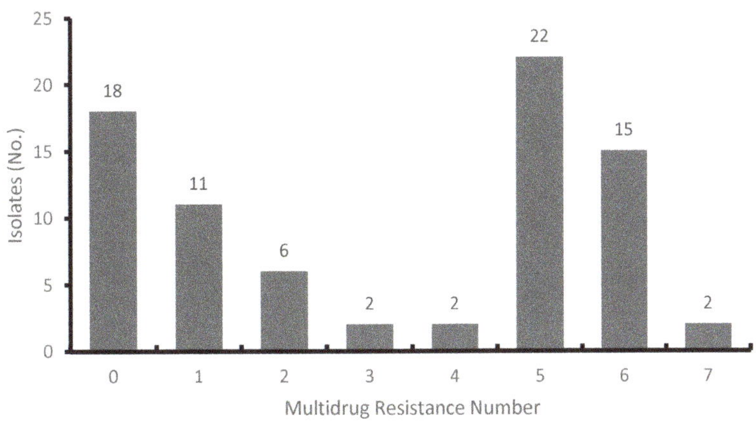


**Figure 3.** Multidrug Resistance of 78 *E. coli* strains to 8 Antibiotics. All 78 *E. coli* isolates were collected and analyzed for multi-drug resistance with tetracycline, ciprofloxacin, streptomycin, gentamycin, lincomycin, cefothiophene, amoxicilin, and penicillin.

### *3.3. Genotypic Analysis of Antimicrobial-Resistance in Isolated E. coli Strains*

According to the resistance phenotype of isolated *E. coli* to various antibiotics in this study, and the clinical use of antibiotics in the farms, 12 drug resistance genes related to four types of antibiotics were detected. The results are shown in Figure 4. For β-Lactams: 78 strains of *E. coli* have the detected blaTEM gene (100%). Moreover, nearly half of the

*E. coli* carried the blaCTX-M gene, and had a detection rate of 53.8%. A total of 16 strains of *E. coli* had the detectable blaSHV gene, accounting for 20.5% of the total. For aminoglycoside: only 7 strains had the detectable armA gene: however, no armB gene was detected. For tetracyclines: The detection rates of tetA and tetC genes were 26.9% and 20.5% of the total isolates, respectively, and only two strains were detected with tetB gene. For quinolone: The detection rates of oqxA and qnrS genes were 37.2% and 29.5%, respectively. Only one *E. coli* strain was detected with the oqxB gene, but no strain carried the qepA gene. The detection rate of each drug resistance gene was not significantly different from the corresponding antibiotic resistance rate.

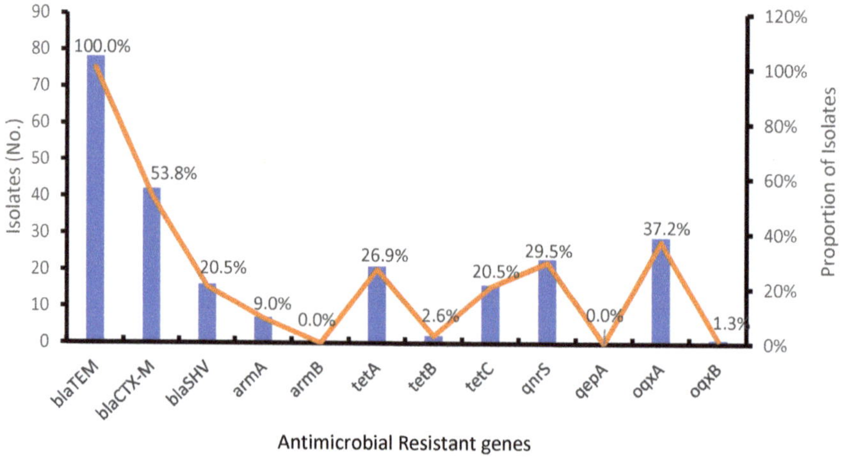


**Figure 4.** Detection of drug-resistant genes in *E. coli*. PCR was performed to identify the carriage of β-lactam (blaTEM, blaCTX-M, and blaSHV), aminoglycoside (armA and armB), tetracycline (tetA, tetB, and tetC), and quinolone (qnrS, qepA, oqxA, and oqxB) related genes in 78 *E. coli* isolates.

## 4. Discussion

In this study, 156 clinical mastitis milk samples collected from three dairy farms in Xinyi, Suqian, and Xuyi in northern Jiangsu were isolated and identified. The results showed that the pathogenic bacteria were mainly dominated by *E. coli*, *Klebsiella* spp., *Staphylococcus* spp., *Staphylococcus aureus*, *Streptococcus agalactis*, *Streptococcus dysgalactiae*, *Streptococcus uberis*, and *Pseudomonas aeruginosa*, and the detection rate of gram-negative bacteria was high. However, it has been reported in an earlier publication that the isolated bacteria from farms in Ningxia, China displayed different distributions of pathogens as compared to the current study [29]. The reason for the difference may be attributed to the sampling time and region, since the patterns of the infection for bovine mastitis are region specific. The sampling time in this study was from July to October, which was hot and humid. Cows are prone to heat stress, which will cause the decline of cow resistance, accelerate the growth and reproduction of bacteria, and impact mammary tissue. Therefore, this period is associated with a high incidence of clinical mastitis in cows. During the sampling and investigation, it was also found that although these farms are large-scale between each other, the environmental sanitation, milking methods, and other feeding management aspects of these farms in northern Jiangsu still need to be improved. This is indicative that external factors such as farm management have an important impact on the induction of cow mastitis and its infection types. In addition, of 143 milk samples shown to be bacteria positive, 99 had mixed infection, accounting for 69.23% of the total samples. The variety of bacterial pathogens isolated further indicates that the environmental sanitation and feeding management of these dairy farms need to be improved. Attention should be paid to the timely cleaning of urine and feces during the rainy season, and the regular disinfection of the environment, to prohibit the growth of environmentally derived pathogens such as *E. coli*.

*E. coli* is one of the main pathogens that cause bovine mastitis. However, due to the increased incidence of antibiotic resistant bacterial infections, the effect of antibiotic treatment is becoming limited. By comparing and analyzing the drug susceptibility of 78 *E. coli* strains from bovine mastitis, we found that at present, coliform bacteria from bovine mastitis in northern Jiangsu Province is highly resistant to β- Lactam antibiotics. However, the susceptibility of the coliforms to tetracycline and gentamicin are still maintained. By referring to relevant data, we found that *E. coli* from bovine mastitis in some regions of China, maintained high resistance to β-Lactam antibiotics, which is consistent to the current research. As reported in a previous study, the main pathogenic bacteria isolated from three farms in Jiangsu Province was tested for drug susceptibility, where almost all isolates of *E. coli* were shown to be resistant to penicillin, ampicillin, and lincomycin, but sensitive to gentamycin, ciprofloxacin, and streptomycin [30]. It has been reported that *E. coli* from dairy cows in Henan Province showed the highest drug resistance rate of β-lactam antibiotics, followed by the resistance of ampicillin sodium, cefotaxime, and ceftiofur, which exceeded 60%. In contrast, a more comprehensive study revealed that 75 (22.9%) out of 328 *E. coli* isolates were confirmed as quinolone-resistant, in a nationwide sampling investigation [8]. However, *E. coli* from bovine mastitis in Tianjin and Ningxia areas has low sensitivity to streptomycin, which is inconsistent with the current study [31]. During the sampling, we noticed that lincomycin, penicillin, and cephalosporin are routine drugs used in farms, which also explains why *E. coli* from bovine mastitis in northern Jiangsu province is highly resistant to these antibiotics. In addition, 26.9% of the 78 strains of *E. coli* isolated in the present study were resistant to quinolone (ciprofloxacin), which was much higher than the data collected from developed countries. The prevalence of quinolone resistant *E. coli* in bovine mastitis cases reported by Finland, Canada, and the United States were 0.7% and 0.0%, respectively [32,33]. It is speculated that the widespread addition of quinolones in the animal feeding diet is related to the high drug resistance rate of quinolones in China [34].

The resistance genes can be horizontally transferred by inserting sequence, plasmids, transposons, integrons, bacteriophages, and other movable genetic elements [35]. Extended-Spectrum β-Lactamases (ESBLs) producing enzymes are important factors for the resistance of *E. coli* to cephalosporins, of which blaCTX-M has the highest prevalent gene in ESBL pathogens. It has been evidenced that more than half of the 70 *E. coli* strains were found to be blaCTX-M positive in four regions of China's farms, which was consistent with the results of this study [10]. The efflux pump is determined as the main drug resistance mechanism of tetracyclines [36]. The predominant genes tetA and tetB are in charge of the resistance of *E. coli* to tetracyclines, in which the current study detected tetA (26.9%) and tetB (2.6%) in tetracycline-resistant strains, which further proved the conclusion. Quinolones are effective antibacterial agents for the treatment of various infections caused by *E. coli*. Unfortunately, due to the widespread use of quinolone drugs, the number of quinolone resistant *E. coli*, derived from animals produced for food in China has increased [36–38]. Multiple mutations that cause amino acid changes in the targeted enzyme related to the quinolone resistance determining region (QRDR) is the most common mechanism of high-level resistance of clinical strains to quinolones [38]. In addition, the common resistance of β-lactam and quinolone drugs is a common phenomenon, due to the presence of β-lactamases and the substitution mutations in QRDR, regardless of whether the plasmid mediated quinolone resistance (PMQR) gene is present [39]. In this study, 21 ciprofloxacin resistant strains were all resistant to β-lactam, including penicillin, amoxicillin, and cefothiophene, which further confirmed this theory.

Collectively, the dominant pathogens causing bovine mastitis in 3 dairy farms in northern Jiangsu are *E. coli* and *K. pneumoniae*, which provides an experimental basis for the prevention and treatment of bovine mastitis and clinical medication. Among them, the resistance of isolated *E. coli* to β-lactams was the most serious in the selected farms.

The detection rate of resistance genes blaTEM (100%) and blaCTX-M (53.8%) was high, which was consistent with the β-lactam drug resistance phenotype. Our study highlights the distribution and characteristics of pathogenic *E. coli* derived from bovine mastitis, and provides reference of antimicrobial selection for the treatment.

**Author Contributions:** Z.Y. and T.X. performed study concept and design; W.C., Y.H. and T.X. performed experiment and writing, review of the paper; T.X., J.Z. and X.W. provided acquisition and interpretation of data; Z.Y. and W.C. provided material. All authors have read and agreed to the published version of the manuscript.

**Funding:** This study was supported by the National Natural Science Foundation of China (32102731). the Open Project Program of Joint International Research Laboratory of Agriculture and Agri-Product Safety, the Ministry of Education of China, Yangzhou University (JILAR-KF202204). The Open Project Program of International Joint Research Laboratory in Universities of Jiangsu Province Of China for Domestic Animal Germplasm Resources and Genetic Improvement: (IJRLD-KF202214).

**Institutional Review Board Statement:** Not applicable.

**Informed Consent Statement:** Not applicable.

**Data Availability Statement:** The datasets used and/or analyzed during the current study are available from the corresponding author on reasonable request.

**Acknowledgments:** We appreciate the support of the milk sample provisions from the dairy farms located in the 3 correlated farms.

**Conflicts of Interest:** The authors declare no conflict of interest.

## References

1. Bar, D.; Tauer, L.W.; Bennett, G.; González, R.; Hertl, J.A.; Schukken, Y.H.; Schulte, H.F.; Welcome, F.L.; Gröhn, Y.T. The cost of generic clinical mastitis in dairy cows as estimated by using dynamic programming. *J. Dairy Sci.* **2008**, *91*, 2205–2214. [CrossRef] [PubMed]
2. Hertl, J.A.; Groehn, Y.T.; Leach, J.; Bar, D.; Bennett, G.J.; Gonzalez, R.N.; Rauch, B.J.; Welcome, F.L.; Tauer, L.W.; Schukken, Y.H. Effects of clinical mastitis caused by gram-positive and gram-negative bacteria and other organisms on the probability of conception in New York State Holstein dairy cows. *J. Dairy Sci.* **2010**, *93*, 1551–1560. [CrossRef] [PubMed]
3. Gao, J.; Barkema, H.W.; Zhang, L.; Liu, G.; Deng, Z.; Cai, L.; Shan, R.; Zhang, S.; Zou, J.; Kastelic, J.P. Incidence of clinical mastitis and distribution of pathogens on large Chinese dairy farms. *J. Dairy Sci.* **2017**, *100*, 4797–4806. [CrossRef] [PubMed]
4. He, W.; Ma, S.; Lei, L.; He, J.; Li, X.; Tao, J.; Wang, X.; Song, S.; Wang, Y.; Wu, C.; et al. Prevalence, etiology, and economic impact of clinical mastitis on large dairy farms in China—ScienceDirect. *Vet. Microbiol.* **2020**, *242*, 108570. [CrossRef]
5. Zhao, X.; Lacasse, P. Mammary tissue damage during mastitis: Causes and controls. *J. Dairy Sci.* **2005**, *88*, 2.
6. Esener, N.; Green, M.J.; Emes, R.D.; Jowett, B.; Davies, P.L.; Bradley, A.J.; Dottorini, T. Discrimination of contagious and environmental strains of Streptococcus uberis in dairy herds by means of mass spectrometry and machine-learning. *Sci. Rep.* **2018**, *8*, 17517. [CrossRef]
7. White, D.G.; Zhao, S.; Simjee, S.; Wagner, D.D.; McDermott, P.F. Antimicrobial resistance of foodborne pathogens. *Microbes Infect.* **2002**, *4*, 405–412. [CrossRef]
8. Yang, F.; Zhang, S.; Shang, X.; Wang, L.; Li, H.; Wang, X. Characteristics of quinolone-resistant *Escherichia coli* isolated from bovine mastitis in China. *J. Dairy Sci.* **2018**, *101*, 6244–6252. [CrossRef]
9. Blanco, M.; Blanco, J.E.; Mora, A.; Dahbi, G.; Bernárdez, M. Serotypes, virulence genes, and intimin types of Shiga toxin (verotoxin)-producing *Escherichia coli* isolates from cattle in Spain and identification of a new intimin variant gene (eae-xi). *J. Clin. Microbiol.* **2004**, *42*, 645. [CrossRef]
10. Yu, Z.N.; Wang, J.; Ho, H.; Wang, Y.T.; Huang, S.N.; Han, R.W. Prevalence and antimicrobial-resistance phenotypes and genotypes of *Escherichia coli* isolated from raw milk samples from mastitis cases in four regions of China. *J. Glob. Antimicrob. Resist.* **2020**, *22*, 94–101. [CrossRef]
11. Bradley, A.J.; Green, M.J. Adaptation of *Escherichia coli* to the Bovine Mammary Gland. *J. Clin. Microbiol.* **2001**, *39*, 1845–1849. [CrossRef] [PubMed]
12. Li, J.; Xie, S.; Ahmed, S.; Wang, F.; Gu, Y.; Zhang, C.; Chai, X.; Wu, Y.; Cai, J.; Cheng, G. Antimicrobial Activity and Resistance: Influencing Factors. *Front. Pharmacol.* **2017**, *8*, 364. [CrossRef] [PubMed]
13. Xu, T.; Zhu, H.; Liu, R.; Wu, X.; Chang, G.; Yang, Y.; Yang, Z. The protective role of caffeic acid on bovine mammary epithelial cells and the inhibition of growth and biofilm formation of Gram-negative bacteria isolated from clinical mastitis milk. *Front. Immunol.* **2022**, *13*, 1005430. [CrossRef] [PubMed]

14. Tyerman, J.G.; Ponciano, J.M.; Joyce, P.; Forney, L.J.; Harmon, L.J. The evolution of antibiotic susceptibility and resistance during the formation of Escherichia colibiofilms in the absence of antibiotics. *BMC Evol. Biol.* **2013**, *13*, 22. [CrossRef] [PubMed]
15. Gędas, A.; Olszewska, M.A. Chapter 1—Biofilm formation and resistance. In *Recent Trends in Biofilm Science and Technology*; Simoes, M., Borges, A., Chaves Simoes, L., Eds.; Academic Press: Cambridge, MA, USA, 2020; pp. 1–21.
16. Liu, Y.; Liu, G.; Liu, W.; Liu, Y.; Ali, T.; Chen, W.; Yin, J.; Han, B. Phylogenetic group, virulence factors and antimicrobial resistance of *Escherichia coli* associated with bovine mastitis. *Res. Microbiol.* **2014**, *165*, 273–277. [CrossRef]
17. Hayer, S.S.; Lim, S.; Hong, S.L.; Elnekave, E.; Alvarez, J. Genetic determinants of extended spectrum cephalosporin and fluoroquinolone resistance in *Escherichia coli* isolated from diseased pigs in the U.S.A. *mSphere* **2020**, *5*, 920–990. [CrossRef]
18. Srinivasan, V.; Gillespie, B.E.; Lewis, M.J.; Nguyen, L.T.; Headrick, S.I.; Schukken, Y.H.; Oliver, S.P. Phenotypic and genotypic antimicrobial resistance patterns of *Escherichia coli* isolated from dairy cows with mastitis. *Vet. Microbiol.* **2007**, *124*, 319–328. [CrossRef]
19. López Meza, J.E.; Higuera Ramos, J.E.; Ochoa Zarzosa, A.; Chassin Noria, O.; Valdez Alarcón, J.J.; Bravo Patiño, A.; Baizabal Aguirre, V.M. Molecular characterization of *Staphylococcus* spp. isolates associated with bovine mastitis in Tarímbaro, Michoacán, Mexico. *Técnica Pecu. México* **2006**, *44*, 91–106.
20. Yu, T.; He, T.; Yao, H.; Zhang, J.B.; Wang, G.Q. Prevalence of 16S rRNA Methylase Gene rmtB Among *Escherichia coli* Isolated from Bovine Mastitis in Ningxia, China. *Foodborne Pathog. Dis.* **2015**, *12*, 770–777. [CrossRef]
21. Xu, T.; Wu, X.; Cao, H.; Pei, T.; Zhou, Y.; Yang, Y.; Yang, Z. The Characteristics of Multilocus Sequence Typing, Virulence Genes and Drug Resistance of Klebsiella pneumoniae Isolated from Cattle in Northern Jiangsu, China. *Animals* **2022**, *12*, 2627. [CrossRef]
22. Hogan, J.S.; Gonzalez, R.N.; Harmon, R.J.; Nickerson, S.C.; Smith, K.L. *Laboratory Handbook on Bovine Mastitis*; National Mastitis Council: Madison, WI, USA, 1999.
23. Blum, S.; Heller, E.D.; Krifucks, O.; Sela, S.; Hammer-Muntz, O.; Leitner, G. Identification of a bovine mastitis *Escherichia coli* subset. *Vet. Microbiol.* **2008**, *132*, 135–148. [CrossRef]
24. Frank, J.A.; Reich, C.I.; Sharma, S.; Weisbaum, J.S.; Wilson, B.A.; Olsen, G.J. Critical evaluation of two primers commonly used for amplification of bacterial 16S rRNA genes. *Appl. Environ. Microbiol.* **2008**, *74*, 2461–2470. [CrossRef]
25. Ceriotti, F.; Zakowski, J.; Sine, H.; Altaie, S.; Horowitz, G.; Pesce, A.J.; Boyd, J.; Horn, P.; Gard, U.; Horowitz, G. Clinical and Laboratory Standards Institute (CLSI). 2012. Available online: https://www.scienceopen.com/document?vid=1d021ccc-8583-4e6b-a47c-8735734df154 (accessed on 1 December 2022).
26. CLSI. *Performance Standards for Antimicrobial Susceptibility Testing, M100*, 31st ed.; Clinical and Laboratory Standards Institute: Wayne, PA, USA, 2021.
27. Turton, J.F.; Perry, C.; Elgohari, S.; Hampton, C.V. PCR characterization and typing of Klebsiella pneumoniae using capsular type-specific, variable number tandem repeat and virulence gene targets. *J. Med. Microbiol.* **2010**, *59*, 541. [CrossRef] [PubMed]
28. Yu, W.L.; Ko, W.C.; Cheng, K.C.; Lee, C.C.; Lai, C.C.; Chuang, Y.C. Comparison of prevalence of virulence factors for Klebsiella pneumoniae liver abscesses between isolates with capsular K1/K2 and non-K1/K2 serotypes. *Diagn. Microbiol. Infect. Dis.* **2008**, *62*, 1–6. [CrossRef]
29. Dong, W.; Fawan, L.; Xuqing, W.; Jianbo, W.; Jianming, P.; Linjun, D.; Junhong, L.; Dengqiang, H. Isolation and identification of mastitis pathogens in dairy cows from some areas of Ningxia. *Prograss Vet. Med. Chin.* **2012**, *33*, 4. [CrossRef]
30. Dan, W.; Feng, Y.; Xinpu, L.; Jinyin, L.; Longhai, L.; Zhe, Z.; Yaru, Z.; Hongsheng, L. Isolation, identification and drug resistance detection of the pathogenic bacteria causing dairy cow mastitis in Suzhou and capsular polysaccharide typing test of the pathogens. *Chin. J. Prev. Vet. Med.* **2018**, *40*, 7. [CrossRef]
31. Qian, S.; Juyun, D.; Yina, L.; Jinlian, S.; Zongshuai, L.; Xingxu, Z.; Yong, Z. Isolation, Identification and Drug Resistance Analysis of Main Pathogens ofDairy Cow Clinical Mastitis in Wuzhong Area of Ningxia. *Prog. Veterianry Med. Chin.* **2022**, *43*, 6.
32. Saini, V.; Mcclure, J.T.; Léger, D.; Keefe, G.P.; Scholl, D.T.; Morck, D.W.; Barkema, H.W. Antimicrobial resistance profiles of common mastitis pathogens on Canadian dairy farms. *J. Dairy Sci.* **2012**, *95*, 4319–4332. [CrossRef]
33. Metzger, S.A.; Hogan, J.S. Short communication: Antimicrobial susceptibility and frequency of resistance genes in *Escherichia coli* isolated from bovine mastitis. *J. Dairy Sci.* **2013**, *96*, 3044–3049. [CrossRef] [PubMed]
34. Xiao, Y.H.; Giske, C.G.; Wei, Z.Q.; Shen, P.; Heddini, A.; Li, L.J. Epidemiology and characteristics of antimicrobial resistance in China. *Drug Resist. Updates* **2011**, *14*, 236–250. [CrossRef]
35. Zhang, S.; Abbas, M.; Rehman, M.U.; Wang, M.; Jia, R.; Chen, S.; Liu, M.; Zhu, D.; Zhao, X.; Gao, Q.; et al. Updates on the global dissemination of colistin-resistant Escherichia coli: An emerging threat to public health. *Sci. Total Environ.* **2021**, *799*, 149280. [CrossRef] [PubMed]
36. Neela, F.A.; Nonaka, L.; Rahman, M.H.; Suzuki, S. Transfer of the chromosomally encoded tetracycline resistance gene tet(M) from marine bacteria to *Escherichia coli* and Enterococcus faecalis. *World J. Microbiol. Biotechnol.* **2009**, *25*, 1095–1101. [CrossRef]
37. Wang, G.Q.; Wu, C.M.; Du, X.D.; Shen, Z.Q.; Song, L.H.; Chen, X.; Shen, J.Z. Characterization of integrons-mediated antimicrobial resistance among *Escherichia coli* strains isolated from bovine mastitis. *Vet. Microbiol.* **2008**, *127*, 73–78. [CrossRef] [PubMed]

38. Zhao, H.X.; Zhao, J.L.; Shen, J.Z.; Fan, H.L.; Guan, H.; An, X.P.; Li, P.F. Prevalence and molecular characterization of fluoroquinolone resistance in *Escherichia coli* isolates from dairy cattle with endometritis in China. *Microb. Drug Resist.* **2014**, *20*, 162–169. [CrossRef] [PubMed]
39. Bajaj, P.; Kanaujia, P.K.; Singh, N.S.; Sharma, S.; Kumar, S.; Virdi, J.S. Quinolone co-resistance in ESBL- or AmpC-producing *Escherichia coli* from an Indian urban aquatic environment and their public health implications. *Environ. Sci. Pollut. Res. Int.* **2016**, *23*, 1954–1959. [CrossRef]

agriculture

MDPI

Article

# *DDIT3* Governs Milk Production Traits by Targeting IL-6 to Induce Apoptosis in Dairy Cattle

**Xiaogang Cui [1,2], Changqing Li [1], Zhangqi Wei [1], Hangting Meng [2], Fengfeng Zhang [2], Yue Liu [2], Changxin Wu [2] and Shaohua Yang [1,3,*]**

1 College of Food and Biological Engineering, Hefei University of Technology, 193 Tunxi Road, Hefei 230009, China
2 Key Lab of Medical Molecular Cell Biology of Shanxi Province, Key Laboratory of Chemical Biology and Molecular Engineering of Ministry of Education, Shanxi Provincial Key Laboratory for Major Infectious Disease Response, Institutes of Biomedical Sciences, Shanxi University, Taiyuan 030006, China
3 College of Animal Science and Technology, China Agricultural University, Beijing 100193, China
* Correspondence: yangshaohua309@hfut.edu.cn

**Abstract:** The mechanisms of modulating milk production traits remain largely unknown. Based on our previous RNA-seq, *DDIT3* was presumed as a novel, promising candidate gene for regulating milk protein and fat traits in dairy cattle. To further detect the genetic effect of *DDIT3* and its potential molecular mechanisms in regulating milk production traits in dairy cattle, here, we performed a genotype-phenotype association study. Two SNPs, g.-1194 C>T and g.-128 C>T, were significantly associated with MY ($p$ = 0.0063), FY ($p$ = 0.0001) and PY ($p$ = 0.0216), respectively. A luciferase assay demonstrated that the allele T of g.-128 C>T increased the promoter activity by binding the HSF2, while allele C did not. To further reveal the molecular regulatory mechanisms, the *DDIT3*-knockdown MAC-T cells were established. It was observed that *DDIT3* silencing could induce apoptosis and increase the number of PI-positive cells. Meanwhile, *DDIT3* silencing led to increased expression of inflammatory markers, such as *IL-6*, *IL6R*, *IL1B*, *IL7R*, *IL1RL2*, *IL1A*, *STAT1-5*, *MYC*, *IGFBP4*, and *IGFBP5*, and especially for *IL-6* ($\log_2$FC = 4.22; $p = 3.49 \times 10^{-112}$). Additionally, compared with the control group, increased lipid accumulation was found in the *DDIT3*-knockdown MAC-T cells. Thus, our results proved that lower expression of *DDIT3* could result in increased lipid accumulation and apoptosis via up-regulating the expression of *IL-6*. These findings provided clues about the regulatory mechanisms of milk production traits in dairy cattle.

**Keywords:** *DDIT3*; genetic effect; milk production traits; apoptosis; dairy cattle

**Citation:** Cui, X.; Li, C.; Wei, Z.; Meng, H.; Zhang, F.; Liu, Y.; Wu, C.; Yang, S. *DDIT3* Governs Milk Production Traits by Targeting IL-6 to Induce Apoptosis in Dairy Cattle. *Agriculture* **2023**, *13*, 117. https://doi.org/10.3390/agriculture13010117

Academic Editors: Zhi Chen and Cong Li

Received: 4 December 2022
Revised: 22 December 2022
Accepted: 27 December 2022
Published: 31 December 2022



## 1. Introduction

The milk production traits of dairy cattle, including milk yield (MY), fat yield (FY), protein yield (PY), fat percentage (FP), and protein percentage (PP), are the most important economical traits, which are influenced by many genes [1]. Therefore, the most profitable breeding goal for milk-producing traits remains improvement [2]. As is known, RNA sequencing (RNA-seq) has enabled gene discovery and expedited the genetic improvement of dairy cattle [3,4]. Our previous RNA-seq study identified 31 differentially expressed genes associated with milk production traits in the mammary glands of lactating Holstein cows [5]. Of these, DNA damage-inducible transcript 3 (*DDIT3*) was significantly down-regulated ($p = 4.01 \times 10^{-5}$) in the Holstein cows [5]. Moreover, it was found that *DDIT3* located only 0.46 Mb, from the SNP (ARS-BFGL-NGS-14781), which was significantly associated with fat percentage [6]. In addition, *DDIT3* was also adjacent to a previously reported QTL for PP and FP [7]. These findings indicated that the *DDIT3* gene may be a promising key gene for milk production traits in dairy cattle.

*DDIT3*, known as the CAAT/enhancer binding protein homologous transcription factor (*CHOP*) or growth retardation and DNA damage-inducible gene 153 (*GADD153*), is involved in various biological processes, including differentiation, proliferation, expression, and energy metabolism [8]. As a transcription factor, *DDIT3* has specifically been shown to regulate DNA damage and ER stress-induced apoptosis by regulating the transcription of antiapoptotic and proapoptotic genes [9,10]. Recent publications have indicated that an overexpression of *DDIT3* promotes apoptosis in several cell lines [11–13]. Overexpression of *DDIT3* in human cells led to cell death and growth arrest at the G1/S phase in murine 3T3 cells [14]. *DDIT3*$^{-/-}$ mouse experiments revealed that *DDIT3*-mediated apoptosis contributes to the pathogenesis of ER stress-related diseases in murine 3T3 cells [10]. Current evidence suggests that *DDIT3* is an important mediator of cellular functions. However, the effects of *DDIT3* on cellular functions have not been tested and clarified in dairy cattle.

To explore the potential causal genetic variants associated with milk production traits, we detected the genetic effects with an independent dairy cattle population. We also generated *DDIT3* knock-out bovine MAC-T cells using RNA interference (RNAi) and applied RNA sequencing technology to investigate the biological function of *DDIT3*. Subsequently, the differentiation model was established to understand the molecular mechanisms of *DDIT3*. Our results may lead to a better understanding of the biochemical mechanisms by which *DDIT3* regulates milk production ability and can be used via marker-assisted breeding.

## 2. Materials and Methods

### 2.1. Ethics Statement

The experimental programs and protocols were approved by the Institutional Animal Care and Use Committee of Hefei University of Technology (approval number HFUT20200518005).

### 2.2. Animals and Phenotypic Data

A daughter design was used in this study. A total of 717 Chinese Holstein cows from 12 corresponding sires were collected to construct the study population. The dairy cattle were from 16 dairy farms in Sanyuanlvhe Dairy Farming Center (Beijing, China). The estimated breeding values (EBVs) for milk yield (MY), protein yield (PY), fat yield (FY), protein percentage (PP), and fat percentage (FP) were predicted by the Dairy Data Center of the Dairy Association of China.

### 2.3. DNA Extraction and SNP Identification

Blood samples were collected from the dairy cows and stored at −20 °C. Genomic DNA was extracted using the Tiangen DP (318) Blood DNA Kit (Beijing, China). A DNA pool was constructed with equal DNA concentration for each. The results of the isolated DNA were detected by a NanoDrop™ ND-2000c Spectrophotometer (Thermo Scientific, Waltham, MA, USA). Based on the genomic sequence of the bovine *DD*IT3 gene (ENSBTAT00000044712.2), 8 pairs of primers were designed by Primer Premier 5.0 to amplify all four exons and 1500 bp of 5′ flanking sequences (Table S1).

### 2.4. Genotyping and Haplotype Analysis

The SNPs of all individuals were genotyped using the SEQUENOM MassArray MALDI-TOF mass spectrometer (Sequenom, San Diego, CA, USA). To further explore the LD extent, haplotypes of the *DDIT3* gene were inferred using the software BEAGLE 3.2 program, and, where necessary, sporadic missing genotype data were also imputed. The measure of pairwise LD for all three SNPs was performed using the software Haploview 4.2 (Broad Institute of MIT and Harvard, Cambridge, MA, USA). Accordingly, the LD block was generated using the linkage disequilibrium coefficient D′. Haplotypes within these blocks were applied to test their associations with the milk production traits.

### 2.5. Association Analyses

The pedigrees of the dairy cows detected in the present study were traced back for three generations to create a genetic matrix. Based on the estimated matrices, associations between genotypes and/or haplotype combinations and the five milk production traits were evaluated by the mixed model in SAS 9.1.2 (University of Notre Dame, Notre Dame, IN, USA). A linear mixed regression model was shown as follows:

$$\mathbf{y} = \mathbf{1}\mu + b\mathbf{x} + \mathbf{Za} + \mathbf{e}$$

where **y** is the EBVs value of the five milk production traits, and $\mu$ is the overall mean for the values. *B* is the regression coefficient, **x** is a design vector of the SNP genotype or haplotype combination, **a** is the vector of residual polygenetic effects, **Z** is a diagonal incidence (0 or 1) matrix that links each animal additive genetic value to its "phenotype", and **e** is the vector of residual errors.

The additive (a), dominance (d), and allele substitution ($\alpha$) effects were further evaluated as follows: a = (AA − BB)/2, d = AB − (AA + BB)/2, and $\alpha$ = a + d (q − p), where AB represents heterozygous genotype, AA and BB indicate the two homozygous genotypes, and p and q are the A and B frequencies, respectively (Falconer and Mackay, 1996). For both analyses above, the Bonferroni method was used according to the number of SNP or haplotype blocks tested. A $p$ value < 0.05/N was considered significant; here, N is the SNP loci numbers or haplotype blocks tested in the analyses.

### 2.6. Construction of Recombinant Plasmid and Luciferase Assay

To directly detect the allele-specific effects of the SNP g.-128 C>T on the promoter activity, three luciferase fragments, which correspond to the DDIT3 promoter (−281 to +52, +1 represents transcription start site) were synthesized. Two restriction enzyme recognition sites, KpnI and BglII, were used at the 5′ and 3′ termini, respectively. The fragments were then cloned into the Luciferase Assay Vector pGL4.14 (Promega, Madison, WI, USA).

Then, these recombinant plasmids were co-transfected to HEK-293 cells. Transfection reactions were performed using 490 ng of plasmid DNAs and 10 ng of pRL-TK as an internal control, containing Lipofectamine 2000 (Invitrogen, CA, USA), with pGL4.14 vector as a negative control. After 48 h transfection, the luciferase activities were measured using a Dual-Luciferase Reporter Assay System (Promega).

### 2.7. siRNA Design, Synthesis, and Assessment

Three siRNAs' sequences targeting *DDIT3* mRNA and a non-specific siRNA were designed and synthesized by GenePharma (GenePharma, Suzhou, China) (Table 1). MAC-T bovine mammary epithelial cells were cultured in DMEM/F12 basal medium (Gibco Life Technologies, Grand Island, NY, USA) supplied with 10% FBS (Gibco) and maintained with 5% $CO_2$ at 37 °C. When the cells were 60–80% confluent, 100 pmol of siRNA was transfected into cells in one well using 5 μL of Lipofectamine$^{TM}$ 2000 (Invitrogen), and the cells were then incubated (37 °C, 5% $CO_2$) for 48 h. Next, the total RNA of the MAC-T cells was extracted with a TRIzol reagent (Invitrogen), and a qRT-PCR analysis was explored to detect the *DDIT3* expression in the cells transfected by siRNAs. Each of the siRNAs was detected in triplicate.

**Table 1.** Associations of the two SNPs in the *DDIT3* gene with the five traits (LSM ± SE).

| SNPs | Genotype (No. Individuals) | MY | FY | FP | PY | PP |
|---|---|---|---|---|---|---|
| g.-1194 C>T | CC(313) | 426.07 ± 63.03 $^{a}$ | 9.29 ± 2.81 $^{A}$ | −0.04 ± 0.029 | 11.34 ± 1.97 $^{a}$ | −0.014 ± 0.009 |
| | TC(321) | 302.21 ± 62.33 $^{b}$ | 2.90 ± 2.78 $^{B}$ | −0.06 ± 0.029 | 9.16 ± 1.95 $^{ab}$ | 0.001 ± 0.009 |
| | TT(83) | 232.53 ± 82.51 $^{b}$ | −0.96 ± 3.05 $^{B}$ | −0.07 ± 0.036 | 6.20 ± 2.45 $^{b}$ | −0.004 ± 0.012 |
| | $p$ value | 0.0063 $^{**}$ | 0.0001 $^{**}$ | 0.5409 | 0.0216 $^{*}$ | 0.1536 |
| g.-128 C>T | CC(313) | 426.07 ± 63.03 $^{a}$ | 9.29 ± 2.81 $^{A}$ | −0.04 ± 0.029 | 11.34 ± 1.97 $^{a}$ | −0.014 ± 0.009 |
| | TC(321) | 302.21 ± 62.33 $^{b}$ | 2.90 ± 2.78 $^{B}$ | −0.06 ± 0.029 | 9.16 ± 1.95 $^{ab}$ | 0.001 ± 0.009 |
| | TT(83) | 232.53 ± 82.51 $^{b}$ | −0.96 ± 3.05 $^{B}$ | −0.07 ± 0.036 | 6.20 ± 2.45 $^{b}$ | −0.004 ± 0.012 |
| | $p$ value | 0.0063 $^{**}$ | 0.0001 $^{**}$ | 0.5409 | 0.0216 $^{*}$ | 0.1536 |

$^{A,B}$ means $p < 0.01$; $^{a,b}$ means $p < 0.05$. * $p$ indicates the significant association after Bonferroni correction for multiple testing at the significance level of 0.05.** $p$ indicates the extremely significant association after Bonferroni correction for multiple testing at the significance level of 0.01.

### 2.8. Construction of shRNA Vectors and Transfection in MAC-T cells

Based on the qRT-PCR results, the siRNA *DDIT3*-287 was found to reduce the *DDIT3* expression by at least 60%. Hence, the DNA sense and antisense sequences for this shRNA were synthesized, and restriction sites were used for connecting synthetic DNA fragments into the LV3 shRNA lentiviral expression vectors (GenePharma)—namely LV3-287. With the appropriate lentiviral expression vectors, the HEK-293T cells (GenePharma) were transiently transfected, then following packaging vectors pGag/Pol, pRev, and pVSV-G to obtain lentiviral particles.

The MAC-T cells were propagated and incubated in DMEM/F12, with 10% FBS (Gibco), and 1×Penicillin-*Streptomycin* (Gibco) at 37 °C and 5% $CO_2$. One day before infection, the cells were seeded in T25 culture flasks and infected when reaching approximately 80% confluence. The 500 μL mixed solution was as follows: 50 μL of lentiviral particles ($1 \times 10^9$ TU/ml) carrying shRNA for *DDIT3*-287, 25 μg of Polybrene (Sigma), and 450 μL of Opti-MEM I Reduced-Serum Medium (Gibco). After one day, the cells were cultured in fresh growth media without lentiviral particles for another two days.

### 2.9. RNA Sequencing

Three parallel samples (cells from 3 culture flasks) of the experimental group (*DDIT3*-287 shRNA) and control group (NC shRNA) were prepared, respectively. The integrity of the obtained RNA was confirmed using Agilent BioAnalyzer 2100 (Agilent Technologies, Santa Clara, CA, USA). A total of 5 μg of RNA with RIN value >= 7.0 was used to standardize between the RNA samples. Using the Kit NEBNext® Ultra ™ RNA Library Prep Kit (NEB), a sequencing library was prepared. Then, clustering of the index-coded samples was carried out by the TruSeq PE Cluster Kit v3-cBot-HS (Illumina). After cluster generation, these paired-end reads were generated, and RNA-seq libraries were sequenced using an Illumina HiSeq 2000 platform. Meanwhile, FASTQ sequence files were generated by CASAVA ver.1.8.2 (Illumina), accompanied by the removal of failed reads.

### 2.10. Differential Expression Analysis and Pathway Enrichment Analysis

The mapping of reads to the reference genome (Bos Taurus ARS-UCD1.2) was performed, and a differential expression analysis of the data was then performed using the package Bowtie v0.12.8 and paired-end and TopHat v2.0.0. To make estimates and determine differentially expressed genes, DESeq2 relies on the negative binomial distribution by using a generalized linear regression model. The Log2 fold-change (L2FC) values, *p*-values, and q-values (p-adjusted) of the differentially expressed genes were obtained in the output data. A q-value of 0.05 was considered significant.

Gene Ontology (GO) enrichment was analyzed by the GOseq R package. The enriched Kyoto Encyclopedia of Genes and Genomes (KEGG) pathways were evaluated by KOBAS software. The GO term and KEGG pathway analysis results with corrected $p$ values < 0.05 were considered to be significantly enriched.

### 2.11. Validation of RNA-Seq Results by qPCR

Eight DEGs were randomly selected to validate the RNA sequencing data (Table S1 shows the primers that were used in sequencing). A qRT-PCR was carried out using a SYBR Green I Master Kit using a LightCycler 96 real-time PCR system (Roche Applied Science, Penzberg, Germany), and the program was 95 °C for 5 min, 40 cycles of 95 °C for 12 s, 58 °C for 12 s, 72 °C for 12 s, and 72 °C for 6 min. The relative expression level was normalized to the *GAPDH* and *ACTB* by the $2^{-\Delta\Delta Ct}$ method.

### 2.12. Oil Red O Staining and TG Content Measurement

After washing with PBS, the cells in the chamber slides were then fixed with 4% paraformaldehyde for 30 min, incubated for 5 min with Oil Red O at room temperature, and then visualized using ECLIPSE Ti-S light microscopy (Nikon Corporation, Tokyo,

Japan). Additionally, the TG content was determined by a triglyceride detection kit (Sangon Biotech Co., Ltd., Shanghai, China).

### *2.13. Calcein-AM/Propidium Iodide (PI) Staining*

After different stimulation, the MAC-T cells were then collected by trypsinization, centrifuged at 1500× *g* for 2 min at room temperature to collect the cell pellet, and washed once with 1× Assay Buffer. Then, the cells were mixed with 1× assay buffer and stained with the mixture by 2 μM of calcein-AM and 4.5 μM of PI per well for 15 min at room temperature. The images of the cells were collected immediately and analyzed using a Nikon eclipse Ti-S fluorescence microscope.

## 3. Results

### *3.1. SNP Identification and Its Associations with the Five Milk Traits*

Through resequencing of the whole coding and 5′ regulatory regions, two SNPs were discovered. These two variants, g.-1194 C>T and g.-128 C>T, were in the 5′ regulatory regions (Figure 1a,b). Chi-squared testing showed that the SNPs of the DDIT3 gene were in Hardy–Weinberg equilibrium ($p > 0.05$). The allele frequencies and locations of the SNPs are shown in Table S2.

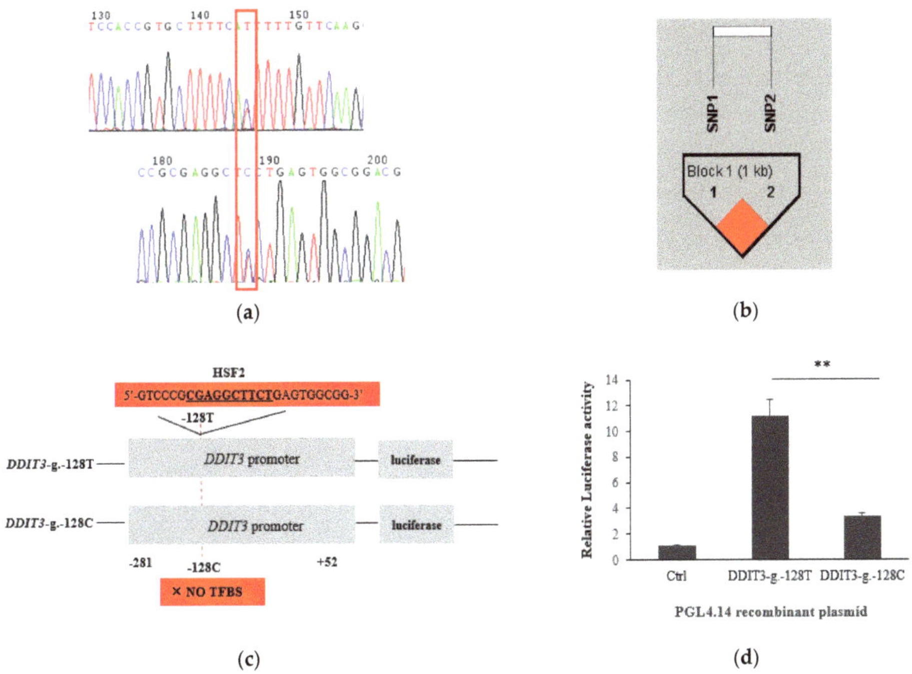


**Figure 1.** The variants of DDIT3 and luciferase activity analysis in HEK293 cells. Note: (**a**) The variants g.-1194 C>T and g.-128 C>T in the 5′ regulatory regions. (**b**) The haplotype block and LD pattern formed by g.-1194 C>T and g.-128 C>T in the *DDIT3*; the darker shading indicates higher LD. The length of the block is provided in kilobases (kb), and pairwise linkage disequilibrium (D′) is given for each SNP combination. (**c**) TFBS modified from TFSEARCH software output; underlined nucleotides denote the TFBS sequences. (**d**) Luciferase activity analysis; the value was shown as the mean ± SEM. *p* values are from a *t*-test (two-tailed). ** $p < 0.01$.

The results of association between the EBVs of the five milk production traits and the two identified SNPs are shown in Table 1. After applying the Bonferroni correction for multiple t-testing, both the SNPs, g.-1194 C>T and g.-128 C>T, were significantly associated with MY ($p = 0.0063$), FY ($p = 0.0001$) and PY ($p = 0.0216$), and these two SNPs showed the same results. Correspondingly, significant dominant effects, additive effects, and allele substitution effects were observed (Table S3). On the other hand, one haplotype block from the HAPLOVIEW program, composed of g.-1194 C>T and g.-128 C>T, was observed in Figure 1a.

### 3.2. Prediction of TFBSs and Its Effect on Transcriptional Activity

The online software TFSEARCH was conducted to predict the putative differential TFBSs of the two regulatory SNPs, g.-1194 C>T and g.-128 C>T. It was shown that the allele of T in g.-128 C>T generated a putative TFBS for the transcription factors HSF2 (Heat shock factor protein 2), while that was abolished upon its substitution by the C allele (Figure 1c). However, there were no significant differences between the allele of T or C for the SNP g.-1194 C>T. Meanwhile, as shown in Figure 1d, the allele of T showed 68% higher luciferase activity compared with the allele of C ($p < 0.01$). Therefore, g.-128 C>T mutation can be considered as a functional SNP for gene expression via the alteration of transcription factor binding sites.

### 3.3. Interference Efficiency of Designed siRNAs in Bovine MAC-T Cells

A total of three siRNAs were designed against *DDIT3* mRNA (Table 2; Figure S1). With the qRT-PCR assay, only siRNA D287 was found to decrease the mRNA expression of the *DDIT3* gene in the MAC-T cells compared with the controls ($p < 0.05$), while D82 and D394 siRNAs did not have an obvious effect ($p > 0.05$) (Figure 2A). Of them, D287 was chosen for further RNAi experiments, which reduced the *DDIT3* expression by about 50%. The shRNA lentiviral vector for D287, LV3-287, was constructed and transfected into the MAC-T cells. Green fluorescence signals of GFP were seen after 72 h of culturing by transient transfection of shRNA virus particles. As shown in Figure 2B, the surviving cells simultaneously expressed GFP and shRNA. Correspondingly, the mRNA levels of the *DDIT3* gene in the knockdown lines were found to decrease by about 50% compared with the NC group using the qRT-PCR ($p < 0.01$).

**Table 2.** Sequences and positions of 3 siRNAs against *DDIT3* gene.

| siRNA Name | Sense (5′-3′) | Target Sites in *DDIT3* | Target Sites in *DDIT3* Gene |
|---|---|---|---|
| D394 | GCTGGCTGAAGAGAATGAACG | 394–415 bp | Exon 4 |
| D287 | GACCAAGGAAGAACCAGAAAA | 287–308 bp | Exon 4 |
| D80 | GTGCTGTCCTCAGATGAAAAT | 80–101 bp | Exon 3 |
| NC | GATCCGTTCTCCGAACGTGTCACGT | | |

**A**

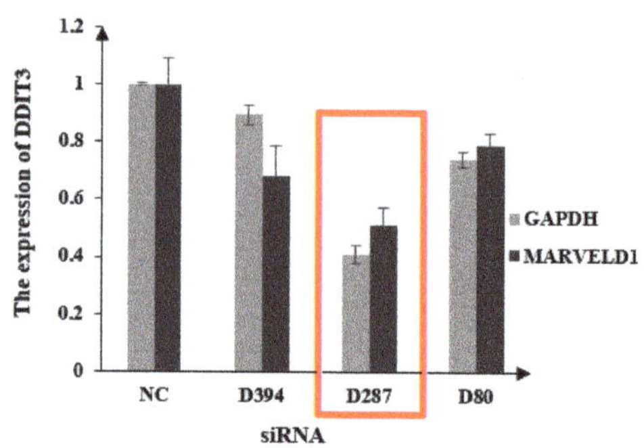


**Figure 2.** *Cont.*

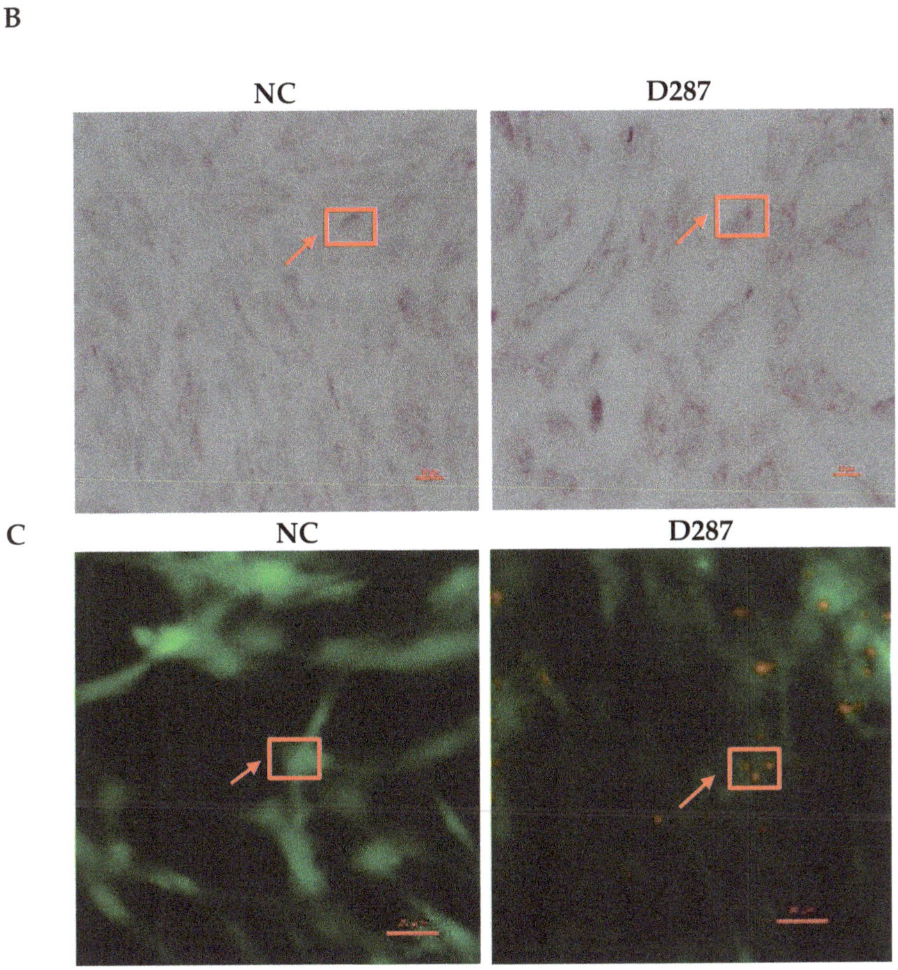


**Figure 2.** shRNA interference of *DDIT3* and its effect on MAC-T cells. **Note**: (**A**) The shRNA interference effects of LV3-DDIT3-D287 in MAC-T cells. (**B**) Lipid accumulation of NC and LV3-DDIT3-D287 groups in MAC-T cells by Oil Red O staining; NC-versus Ctrl group and D287-LV3-DDIT3-D287 group. (**C**) Calcein-AM/PI staining assays.

### 3.4. DEGs and Functional Annotation

In total, 3838 differentially expressed genes were identified (Table S4). Of these, 1940 genes were down-regulated and 1898 were genes up-regulated. The volcano plot of the DEGs is displayed in Figure S2. Moreover, the top 10 genes associated with the regulation of fat metabolism are shown in Table 3. Of them, IL6 was the most significantly up-regulated gene ($Log_2$fold change = 4.22). The 3838 common DEGs were then processed by GO and KEGG enrichment analysis. We found that the expression pattern significantly changed the genes involved in cellular senescence, cell cycle and death, DNA replication and lysosome, etc. (Table S5).

**Table 3.** Top 10 differentially expressed genes between knockdown and control groups.

| Symbol | Gene Name | Regulation | $Log_2$ Fold Change | padj | Gene Function |
|---|---|---|---|---|---|
| *IL6* | interleukin 6 | up | 4.22 | $3.49 \times 10^{-112}$ | plays important roles in fatty acid biosynthesis |
| *CXCL8* | C-X-C motif chemokine ligand 8 | up | 3.01 | $2.04 \times 10^{-90}$ | catalyzes the first step in the hydrolysis of triglycerides |
| *GRO1* | chemokine (C-X-C motif) ligand 1 | up | 2.68 | $1.21 \times 10^{-87}$ | involved in the biosynthesis of unsaturated fatty acids |
| *TNFAIP3* | TNF alpha-induced protein 3 | up | 2.08 | $6.41 \times 10^{-80}$ | linoleic acid metabolism, n-3, and n-6 fatty acid metabolism |
| *EGR1* | early growth response 1 | up | 3.03 | $7.80 \times 10^{-78}$ | involved in sterol biosynthesis |
| *PHLDA1* | pleckstrin homology-like domain family A1 | up | 1.72 | $7.19 \times 10^{-53}$ | involved in adipocyte differentiation and glucose homeostasis |
| *SERPINE1* | serpin family E member 1 | up | 1.48 | $7.74 \times 10^{-49}$ | plays a key role in lipid biosynthesis and fatty acid degradation |
| *CYR61* | cysteine rich angiogenic inducer 61 | up | 1.66 | $8.90 \times 10^{-49}$ | catalyzes the carboxylation of acetyl-CoA to malonyl-CoA |
| *CCL20* | C-C motif chemokine ligand 20 | up | 4.27 | $2.00 \times 10^{-46}$ | involved in lipid synthesis and energy generation |
| *IFI27* | putative ISG12(a) protein | down | −3.01 | $1.06 \times 10^{-46}$ | plays critical roles in lipid metabolism |

Meanwhile, these results indicated that not only the regulation of autophagy, cellular senescence, and cell cycle, but also those involved in lipid metabolic signaling pathways could be regulated by *DDIT3*, for their role in cell growth, proliferation, and differentiation.

### 3.5. *DDIT3 Knocking down Caused Lipid Accumulation and Apoptosis in Bovine MAC-T Cells*

The differentiation of MAC-T cells was detected using Oil Red O staining and triglycerides content assay. It showed that the number and size of the lipid droplets in the *DDIT3* knock-down group were significantly increased compared with the NC group (Figure 2B). Furthermore, the concentration of TG was significantly lower in the *DDIT3* knock-down group than that of the control group. We then determined the role of *DDIT3* in cell viability and apoptosis. Calcein-AM/propidium iodide (PI) staining showed that the transfection of *DDIT3* siRNA could induce apoptosis and increase the number of PI-positive cells (Figure 2C). *DDIT3* silencing by D287 led to increased expression of inflammatory markers.

### 3.6. *The Expression Alteration of the Key Genes*

Here, we first detected the expression of these genes (*IL-6, IL6R, IL1B, IL7R, IL1RL2, IL1A, STAT1-5, MYC, IGFBP4,* and *IGFBP5*), which were acquired by our RNA-seq and pathway analysis. Compared with the control group, the mRNA expression levels of these genes in the *DDIT3* knock-down group altered significantly. As shown in Figure 3A,C,D, *DDIT3* knocking down can markedly increase the mRNA expressions of *IL-6, IL6R, IL1B, IL7R, MYC,* and *IGFBP4* ($p < 0.01$), especially for *IL-6,* which was known as pro-inflammatory cytokine. Moreover, the expression levels of *STAT1-5, IL1RL2, IL1A,* and *IGFBP5* mRNA were lower in the *DDIT3* knock-down group than those in the control group ($p < 0.01$) (Figure 3A,B,E). All the expressions of these detected genes were consistent with our RNA-seq data. Altogether, these results indicated that these key genes may play a critical role in regulating milk composition traits, which were induced by knocking down *DDIT3*. Then, based on the mRNA expression, we proposed that *DDIT3* might modulate the lipid metabolism and apoptosis in dairy cattle through regulating the expression of IL6. Thus, these results demonstrated that *DDIT3* knocking down caused lipid accumulation and apoptosis in bovine MAC-T cells.

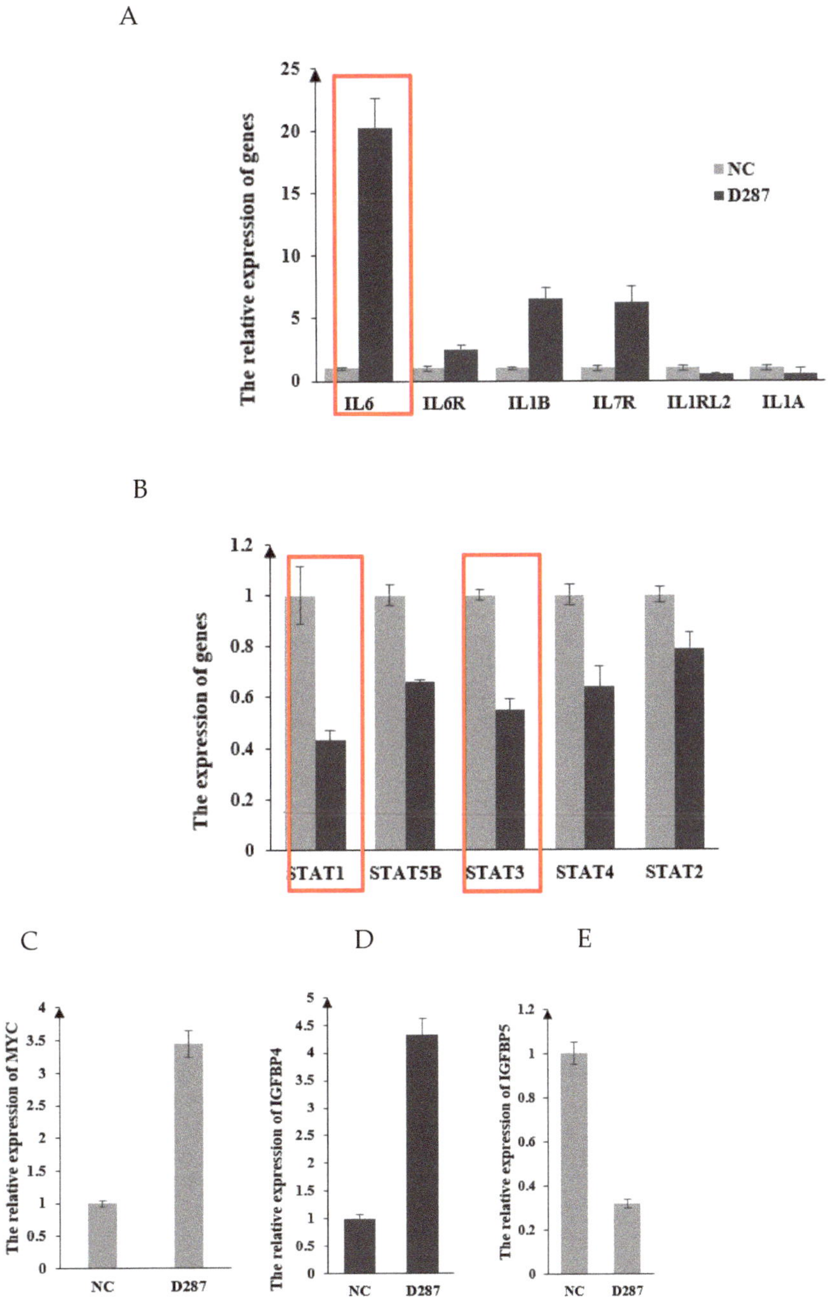


**Figure 3.** Effects of *DDIT3* knocking down on the mRNA expression of genes regulating apoptosis in MAC-T cells. **Note:** (**A**). The mRNA expression of genes regulating inflammatory response. (**B**). The expression of STAT1-5. (**C**–**E**). The mRNA expression of MYC, IGFBP4, and IGFBP5, respectively.

## 4. Discussion

In the present study, we confirmed that the DDIT3 gene was significantly associated with five milk production traits. *DDIT3* was initially identified in our previous RNA-seq, where it was found to be significantly differentially expressed in the high and low PP and FP of lactating Holstein cows [5]. Genetic variations in key promoter regions may cause significant potential phenotype diversity [15]. In the present study, we found that the SNPs g.-1194 C>T and g.-128 C>T, which are located in the promoter region, were significantly associated with MY, FY, and PY. Cows with the TT genotype had lower MY, FY, and PY than those with the TT genotype. In the association analyses, the EBVs of daughters were used as phenotypic observations. By using phenotypic values for association analyses, Li et al. (2019) found that, in the first lactation, the SNP g.56284880C>T (which was g.-128 C>T in our study) was significantly associated with milk MY, FY, and PY; likewise, in the second lactation, it was significantly associated with milk MY and PY. These results were consistent with our current study. Hence, the findings for the association analyses based on the two different variables (phenotypic values and EBVs) basically overlapped. Thus, *DDIT3* was considered to be a major gene influencing milk production traits.

Meanwhile, the promoter activity analysis clearly demonstrated that the allele T of g.-128 C>T of the *DDIT3* gene was influenced by the increased promoter activity by binding the HSF2 factor, while allele C did not, indicating that the HSF2 (heat shock factor) transcription factor may up-regulate the expression of *DDIT3*. HSF2, as a transcription factor, participated in the expression of heat shock genes by interacting directly with HSF1 or HSF4 [16–18]. Additionally, HSF2 plays important roles in regulating the HSF1-mediated stress response [19]. Numerous studies have shown that the SNPs in the promoter region are associated with the changes of gene expression by altering putative transcription factor-binding sites [20]. Considering the significant association effects of g.-128 C>T on the milk fat and protein production traits, it is possible that this SNP regulates the *DDIT3* expression by changing the binding status of the transcription factor, HSF2, to affect the formation of milk and protein traits. Additionally, there was no TFBS for the loci g.-1194 C>T; the significant association of g.-1194 C>T with MY, FY, or PY may be due to the very strong LD between g.-1194 C>T and g.-128 C>T (D′ = 1.00). Therefore, the locus g.-128 C>T is directly responsible for the *DDIT3* expression. Based on these findings, HSF2 can be used for marker-assisted selection to modulate the expression of the *DDIT3* gene in dairy cattle to improve milk production.

It has been widely reported that *DDIT3* affects ER stress and induces cell apoptosis or autophagy [21–23], but its relationship with milk production traits remains unclear in dairy cattle. In the present study, for the loci g.-128 C>T, cows with the TT genotype had lower MY, FY, and PY than those with the TT genotype. Therefore, *DDIT3* negatively regulated milk production traits, which were all consistent with our previous studies [5,24]. Furthermore, the concentration of TG in the *DDIT3* knock-down group showed significantly higher content than that of the NC group. In addition, *DDIT3*-knockdown induced an increased number of PI-positive cells and could induce apoptosis by calcein-AM/ PI staining in the MAC-T cell line. This is the first report on *DDIT3* inhibiting lipid accumulation and inducing apoptosis in the bovine MAC-T cell line.

To further exploit the effects of DDIT3 on the apoptosis of the bovine MAC-T cell line, we knocked down *DDIT3* in bovine MAC-T cells and found that interleukin-6 (IL6) was regulated by *DDIT3*. It has been reported that IL6 regulates inflammatory responses, can be produced for a short period of time after the body is stimulated by inflammation and independently of stimulation by tumor necrosis factor-alpha (TNFα) and accompanied by the increase in other inflammatory substances [25,26]. In addition, IL6 mediates cellular hypertrophy through the activation of the JAK/STAT pathway in different types of cells [27,28]. In our present study, *DDIT3* knockdown promoted the expression of IL6, and induced apoptosis of the bovine MAC-T cell line, indicating that the regulatory effects of DDIT3 on the vitality of MAC-T cells is mediated by IL6. As is known, the JAK/STAT pathway is stimulated in the early stage of diabetic nephropathy [26,29,30]. Our present results only showed that IL6 could further stimulate STAT3-related pathways, resulting in an aggravating effect on inflammatory responses, whereas JAK2 was not induced significantly. The different results probably stemmed from the differences between the in vivo and in vitro studies, as well as the heterogeneity of the cells used in the study. Collectively, these findings and our present results suggest that the *DDIT3*/IL6/STAT3 signaling pathway may be involved in affecting milk production traits. Thus, further studies of animal models with IL6/STAT3 double deficiency will be required to provide additional evidence for this possibility.

**Supplementary Materials:** The following are available online at https://www.mdpi.com/article/10.3390/agriculture13010117/s1, Table S1. Primers used for pooled DNA sequencing for the DDIT3 gene; Table S2. Genotypic and allelic frequencies of the two SNPs in the DDIT3 gene; Table S3. Genetic effects of g.-1194 C>T and g.-128 C>T on milk production traits in Chinese Holsteins; Table S4. DEGs within two different comparison groups; Table S5. Significantly enriched KEGG terms of DEGs; Figure S1. Sequencing of recombinant plasmid; Figure S2. DEGs among the DDIT3-KDvsNC groups Note: FC presented in this figure is fold change.

**Author Contributions:** Conceptualization, X.C. and S.Y.; methodology, C.L.and F.Z.; software, X.C., C.L., Z.W. and H.M.; validation, X.C., C.L., Z.W. and H.M.; formal analysis, X.C., and C.L.; investigation, F.Z.; resources, C.W. and S.Y.; data curation, F.Z and Y.L..; writing—original draft preparation, X.C. and S.Y.; writing—review and editing, X.C., Y.L., C.W. and S.Y.; visualization, S.Y.; supervision, C.W. and S.Y.; project administration, C.W. and S.Y.; funding acquisition, C.W. and S.Y. All authors have read and agreed to the published version of the manuscript.

**Funding:** This work was supported by the National Science Foundation Project of Anhui (1908085MC63), the Beijing Research and Technology Program (D121100003312001), and the Program of Introducing Talents of Discipline to Universities under grant (D21004).

**Acknowledgments:** The authors also appreciate the kind help from the official Dairy Data Center of China in providing the official EBVs data.

**Conflicts of Interest:** The authors declare that they have no known competing financial interests or personal relationships that could have appeared to influence the work reported in this paper.

## References

1. Caroli, A.M.; Chessa, S.; Erhardt, G.J. Invited review: Milk protein polymorphisms in cattle: Effect on animal breeding and human nutrition. *J. Dairy Sci.* **2009**, *92*, 5335–5352. [CrossRef] [PubMed]
2. Bovenhuis, H.; Van Arendonk, J.A.M.; Korver, S. Associations between milk protein polymorphisms and milk production traits. *J. Dairy Sci.* **1992**, *75*, 2549–2559. [CrossRef] [PubMed]
3. McCabe, M.; Waters, S.; Morris, D.; Kenny, D.; Lynn, D.; Creevey, C. RNA-seq analysis of differential gene expression in liver from lactating dairy cows divergent in negative energy balance. *BMC Genom.* **2012**, *13*, 193. [CrossRef] [PubMed]
4. Wickramasinghe, S.; Rincon, G.; Islas-Trejo, A.; Medrano, J.F. Transcriptional profiling of bovine milk using RNA sequencing. *BMC Genom.* **2012**, *13*, 45. [CrossRef] [PubMed]
5. Cui, X.; Hou, Y.; Yang, S.; Xie, Y.; Zhang, S.; Zhang, Y.; Zhang, Q.; Lu, X.; Liu, G.E.; Sun, D. Transcriptional profiling of mammary gland in Holstein cows with extremely different milk protein and fat percentage using RNA sequencing. *BMC Genom.* **2014**, *15*, 226. [CrossRef] [PubMed]
6. Cole, J.B.; Wiggans, G.R.; Ma, L.; Sonstegard, T.S.; Lawlor, T.J., Jr.; Crooker, B.A.; Van Tassell, C.P.; Yang, J.; Wang, S.; Matukumalli, L.K.; et al. Genome-wide association analysis of thirty one production, health, reproduction and body conformation traits in contemporary U.S. Holstein cows. *BMC Genom.* **2011**, *12*, 408. [CrossRef]
7. Bagnato, A.; Schiavini, F.; Rossoni, A.; Maltecca, C.; Dolezal, M.; Medugorac, I.; Sölkner, J.; Russo, V.; Fontanesi, L.; Friedmann, A.; et al. Quantitative trait loci affecting milk yield and protein percentage in a three-country Brown Swiss population. *J. Dairy Sci.* **2008**, *91*, 767–783. [CrossRef]
8. Marola, O.J.; Syc-Mazurek, S.B.; Libby, R.T. DDIT3 (CHOP) contributes to retinal ganglion cell somal loss but not axonal degeneration in DBA/2J mice. *Cell Death Discov.* **2019**, *5*, 140. [CrossRef]
9. Nashine, S.; Liu, Y.; Kim, B.J.; Clark, A.F.; Pang, I.H. Role of C/EBP homologous protein in retinal ganglion cell death after ischemia/reperfusion injury. *Investig. Ophthalmol. Vis. Sci.* **2014**, *56*, 221–231. [CrossRef]
10. Silva, R.M.; Ries, V.; Oo, T.F.; Yarygina, O.; Jackson-Lewis, V.; Ryu, E.J.; Lu, P.D.; Marciniak, S.J.; Ron, D.; Przedborski, S.; et al. CHOP/GADD153 is a mediator of apoptotic death in substantia nigra dopamine neurons in an in vivo neurotoxin model of parkinsonism. *J. Neurochem.* **2005**, *95*, 974–986. [CrossRef]

11. Liao, Y.; Fung, T.S.; Huang, M.; Fang, S.G.; Zhong, Y.; Liu, D.X. Upregulation of CHOP/GADD153 during coronavirus infectious bronchitis virus infection modulates apoptosis by restricting activation of the extracellular signal-regulated kinase pathway. *J. Virol.* **2013**, *87*, 8124–8134. [CrossRef] [PubMed]
12. Abreu, M.M.; Sealy, L. The C/EBPβ isoform, liver-inhibitory protein (LIP), induces autophagy in breast cancer cell lines. *Exp. Cell Res.* **2010**, *316*, 3227–3238. [CrossRef] [PubMed]
13. Wang, J.; Wang, L.; Ho, C.T.; Zhang, K.; Liu, Q.; Zhao, H. Garcinol from Garcinia indica Downregulates Cancer Stem-like Cell Biomarker ALDH1A1 in Nonsmall Cell Lung Cancer A549 Cells through DDIT3 Activation. *J. Agric. Food Chem.* **2017**, *65*, 3675–3683. [CrossRef] [PubMed]
14. Barone, M.V.; Crozat, A.; Tabaee, A.; Philipson, L.; Ron, D. CHOP (GADD153) and its oncogenic variant, TLS-CHOP, have opposing effects on the induction of G1/S arrest. *Genes Dev.* **1994**, *8*, 453–464. [CrossRef]
15. Wang, X.; Tomso, D.J.; Liu, X.; Bell, D.A. Single nucleotide polymorphism in transcriptional regulatory regions and expression of environmentally responsive genes. *Toxicol. Appl. Pharmacol.* **2005**, *207*, 84–90. [CrossRef]
16. He, H.; Soncin, F.; Grammatikakis, N.; Li, Y.; Siganou, A.; Gong, J.; Brown, S.A.; Kingston, R.E.; Calderwood, S.K. Elevated expression of heat shock factor (HSF) 2A stimulates HSF1-induced transcription during stress. *J. Biol. Chem.* **2003**, *278*, 35465–35475. [CrossRef]
17. Ostling, P.; Björk, J.K.; Roos-Mattjus, P.; Mezger, V.; Sistonen, L. Heat shock factor 2 (HSF2) contributes to inducible expression of hsp genes through interplay with HSF1. *J. Biol. Chem.* **2007**, *282*, 7077–7086. [CrossRef]
18. Sandqvist, A.; Björk, J.K.; Akerfelt, M.; Chitikova, Z.; Grichine, A.; Vourc'h, C.; Jolly, C.; Salminen, T.A.; Nymalm, Y.; Sistonen, L. Heterotrimerization of heat-shock factors 1 and 2 provides a transcriptional switch in response to distinct stimuli. *Mol. Biol. Cell* **2009**, *20*, 1340–1347. [CrossRef]
19. Yang, L.N.; Ning, Z.Y.; Wang, L.; Yan, X.; Meng, Z.Q. HSF2 regulates aerobic glycolysis by suppression of FBP1 in hepatocellular carcinoma. *Am. J. Cancer Res.* **2019**, *9*, 1607–1621.
20. Zi, C.; Wu, Z.; Wang, J.; Huo, Y.; Zhu, G.; Wu, S.; Bao, W. Transcriptional activity of the FUT1 gene promoter region in pigs. *Int. J. Mol. Sci.* **2013**, *14*, 24126–24134. [CrossRef]
21. Posey, K.L.; Coustry, F.; Veerisetty, A.C.; Liu, P.; Alcorn, J.L.; Hecht, J.T. Chop (Ddit3) is essential for D469del-COMP retention and cell death in chondrocytes in an inducible transgenic mouse model of pseudoachondroplasia. *Am. J. Pathol.* **2012**, *180*, 727–737. [CrossRef] [PubMed]
22. Wu, Y.; Sun, H.; Song, F.; Fu, D.; Wang, J. DDIT3 overexpression increases odontoblastic potential of human dental pulp cells. *Cell Prolif.* **2014**, *47*, 249–257. [CrossRef] [PubMed]
23. Wang, S.; Hou, P.; Pan, W.; He, W.; He, D.C.; Wang, H.; He, H. DDIT3 Targets Innate Immunity via the DDIT3-OTUD1-MAVS Pathway To Promote Bovine Viral Diarrhea Virus Replication. *J. Virol.* **2021**, *95*, e02351-20. [CrossRef] [PubMed]
24. Li, Y.; Han, B.; Liu, L.; Zhao, F.; Liang, W.; Jiang, J.; Yang, Y.; Ma, Z.; Sun, D. Genetic association of DDIT3, RPL23A, SESN2 and NR4A1 genes with milk yield and composition in dairy cattle. *Anim. Genet.* **2019**, *50*, 123–135. [CrossRef] [PubMed]
25. Bobbo, V.C.; Engel, D.F.; Jara, C.P.; Mendes, N.F.; Haddad-Tovolli, R.; Prado, T.P.; Sidarta-Oliveira, D.; Morari, J.; Velloso, L.A.; Araujo, E.P. Interleukin-6 actions in the hypothalamus protects against obesity and is involved in the regulation of neurogenesis. *J. Neuroinflam.* **2021**, *18*, 192. [CrossRef]
26. Jo, H.A.; Kim, J.Y.; Yang, S.H.; Han, S.S.; Joo, K.W.; Kim, Y.S.; Kim, D.K. The role of local IL6/JAK2/STAT3 signaling in high glucose-induced podocyte hypertrophy. *Kidney Res. Clin. Pract.* **2016**, *35*, 212–218. [CrossRef]
27. Serrano, A.L.; Baeza-Raja, B.; Perdiguero, E.; Jardí, M.; Muñoz-Cánoves, P. Interleukin-6 is an essential regulator of satellite cell-mediated skeletal muscle hypertrophy. *Cell Metab.* **2008**, *7*, 33–44. [CrossRef]
28. Begue, G.; Douillard, A.; Galbes, O.; Rossano, B.; Vernus, B.; Candau, R.; Py, G. Early activation of rat skeletal muscle IL-6/STAT1/STAT3 dependent gene expression in resistance exercise linked to hypertrophy. *PloS ONE* **2013**, *8*, e57141. [CrossRef]
29. Mir, S.A.; Chatterjee, A.; Mitra, A.; Pathak, K.; Mahata, S.K.; Sarkar, S. Inhibition of signal transducer and activator of transcription 3 (STAT3) attenuates interleukin-6 (IL-6)-induced collagen synthesis and resultant hypertrophy in rat heart. *J. Biol. Chem.* **2012**, *287*, 2666–2677. [CrossRef]
30. Ropelle, E.R.; Flores, M.B.; Cintra, D.E.; Rocha, G.Z.; Pauli, J.R.; Morari, J.; de Souza, C.T.; Moraes, J.C.; Prada, P.O.; Guadagnini, D.; et al. IL-6 and IL-10 anti-inflammatory activity links exercise to hypothalamic insulin and leptin sensitivity through IKKβ and ER stress inhibition. *PLoS Biol.* **2010**, *8*, e1000465. [CrossRef]

agriculture

Article

# Genetic Polymorphism and mRNA Expression Studies Reveal *IL6R* and *LEPR* Gene Associations with Reproductive Traits in Chinese Holsteins

**Hailiang Zhang [1,†], Abdul Sammad [1,†], Rui Shi [1], Yixin Dong [1], Shanjiang Zhao [2], Lin Liu [3], Gang Guo [4], Qing Xu [5], Aoxing Liu [6,*] and Yachun Wang [1,*]**

1 Key Laboratory of Animal Genetics, Breeding and Reproduction, MARA, National Engineering Laboratory for Animal Breeding, College of Animal Science and Technology, China Agricultural University, Beijing 100193, China
2 Embryo Biotechnology and Reproduction Laboratory, Institute of Animal Sciences, Chinese Academy of Agricultural Sciences, Beijing 100193, China
3 Beijing Dairy Cattle Center, Beijing 100192, China
4 Beijing Sunlon Livestock Development Co., Ltd., Beijing 100176, China
5 Institute of Life Sciences and Bio-Engineering, Beijing Jiaotong University, Beijing 100044, China
6 Center for Quantitative Genetics and Genomics, Aarhus University, 8830 Tjele, Denmark
* Correspondence: aoxing.liu@helsinki.fi (A.L.); wangyachun@cau.edu.cn (Y.W.)
† These authors contributed equally to this work.

**Abstract:** Genetic selection of milk yield traits alters the energy distribution of high producing cows, resulting in gene-induced negative energy balance, and consequently, poor body condition scores and reduced reproductive performances. Here, we investigated two metabolic-syndrome pathway genes, *IL6R* (Interleukin 6 receptor) and *LEPR* (Leptin receptor), for their polymorphism effects on reproductive performance in dairy cows, by applying polymorphism association analyses in 1588 Chinese Holstein cows (at population level) and gene expression analyses in granulosa cells isolated from eight cows (at cell level). Among the six single nucleotide polymorphisms we examined (two SNPs for *IL6R* and four SNPs for *LEPR*), five were significantly associated with at least one reproductive trait, including female fertility traits covering both the ability to recycle after calving and the ability to conceive and keep pregnancy when inseminated properly, as well as calving traits. Notably, the identified variant SNP g.80143337A/C in *LEPR* is a missense variant. The role of *IL6R* and *LEPR* in cattle reproduction were further confirmed by observed differences in relative gene expression levels amongst granulosa cells with different developmental stages. Collectively, the functional validation of *IL6R* and *LEPR* performed in this study improved our understanding of cattle reproduction while providing important molecular markers for genetic selection of reproductive traits in high-yielding dairy cattle.

**Keywords:** Holstein cattle; fertility; gene expression; ovarian follicle; haplotype

**Citation:** Zhang, H.; Sammad, A.; Shi, R.; Dong, Y.; Zhao, S.; Liu, L.; Guo, G.; Xu, Q.; Liu, A.; Wang, Y. Genetic Polymorphism and mRNA Expression Studies Reveal *IL6R* and *LEPR* Gene Associations with Reproductive Traits in Chinese Holsteins. *Agriculture* **2023**, *13*, 321. https://doi.org/10.3390/agriculture13020321

Academic Editor: Nikola Puvača

Received: 4 December 2022
Revised: 18 January 2023
Accepted: 26 January 2023
Published: 28 January 2023



## 1. Introduction

Over the past few decades, the decline of reproductive performance has been widely observed in dairy cattle, partially resulting from unfavorable genetic correlations with intensively selected milk production traits [1,2]. Poor reproductive performance and increasing involuntary culling rate increase the cost of herd management and veterinary care, placing a significant burden on the dairy industry [3]. It has been noticed by breeders globally that genomic selection of female fertility, such as integrating potential causative mutations into prediction models [4], needs to be carefully strategized [5]. One foreseen challenge is that, given the complexity and the highly polygenic characteristic of cattle reproduction, candidate genes identified for reproduction traits might also exert pleiotropic effects for other economically important traits.

Genetic selection of milk yield traits alters the energy distribution of high producing cows, resulting in gene-induced negative energy balance, and consequently, lower body condition scores [6,7]. Coupled with challenging environmental restraints, some of the lactating cows may suffer from reduced reproductive performance [8,9]. We hypothesized that some of the genes highly involved in energy balance procedure might exert shared effects on female reproductive traits as well. Here, we used genes *IL6R* (interleukin 6 receptor, in BTA3) and *LEPR* (Leptin receptor, in BTA3) as examples, both of which are involved in metabolic syndrome pathways [10] and showed evidences of associations with reproductive traits from large scale association studies [11]. The gene *IL6R* encodes a protein containing antibody folding [12], which is a specific binding receptor of Interleukin 6 (IL6). IL6 binds with IL6R to form complexes, and then with trans-membrane protein gp130 (signal transducer of IL6 family cytokines), the formed high-affinity complexes can lead to signal transduction [13]. The gene *LEPR* is the controller of leptin, one type of hormone mainly produced by fat cells, playing an important role in controlling body weight, fat deposition, food intake, immune function, and reproduction. It has been reported that LEPR exists in the ovaries, at which leptin binds to LEPR to modulate steroid production of the ovaries [14].

Previous studies in human have shown that the high expression of IL-6 can change the characteristic of cytokines pattern [15], which can further induce infertility [16,17], habitual abortion [18], and preeclampsia [19] in women. Additionally, multiple polymorphisms in *IL6R* are reported to be associated with either function or pathophysiology of reproduction [20,21]. However, there was a lack of reporting in the literature regarding its role in reproduction in cattle. For *LEPR*, a few studies reported the associations between the polymorphism of *LEPR* and fertility consequences [22,23] and reproductive traits [24]. There is a lack of focused analyses on reproduction traits covering different perspectives of female reproduction. Furthermore, validation in expression profiles remains important for understanding the physiological and molecular process of *LEPR* gene regulating reproduction.

Bovine ovarian follicle, consisting of an oocyte that ovulates, fertilizes, and forms an embryo, is surrounded by granulosa and theca cells, which can produce molecular signals and hormones to cause the oocyte's developmental competence [25]. During follicle growth, GCs replicate, secrete various hormones, and ensure an essential micro-environment for follicle growth [26]. The proliferating and differentiating GCs are critical for normal follicle growth and are involved in oocyte development, ovulation, and the process of luteinization [27,28]. Due to its indispensable roles in reproduction process, ovarian follicles can be more suitable tissues for studying reproductive traits. Gene expressions in GCs may change at different stages of follicular development, which in turn regulate follicular development and, hence, the reproductive outcomes.

Collectively, the objectives of this study were to (1) examine the effects of genes related to metabolic syndrome pathways, including *IL6R* and *LEPR* on dairy cattle reproduction, employing both SNP- and haplotype-based association analyses; (2) investigate their roles in reproduction process by employing expression analyses in GCs. The findings from this study could improve our understanding of genes with pleiotropic effects for complex traits and positively contribute to the genetic improvement of reproduction performances in dairy cattle.

## 2. Materials and Methods

### 2.1. Genotype and Haplotype-Based Association Analyses

#### 2.1.1. SNP Identification in 68 Bulls

Premier 5.0 and Oligo 7.0 (Sangon Biotech Co., Ltd., Shanghai, China) were used to design primes of all exons and flanking regions (within 2 kb distance of 5′ and 3′ end) in two candidate genes. In total, 38 pairs of primers (Table S1) were designed, including 14 for *IL6R* and 24 for *LEPR*. These primers were then synthesized at Sangon Biotech (Shanghai, China).

The DNA samples were extracted from frozen semen samples of 68 Holstein bulls (Beijing Dairy Cattle Center, Beijing, China) using a standard phenol–chloroform method, and then, DNA concentrations were diluted to 200 ng/μL with TB buffer. We randomly mixed DNA samples into three pools (one pool contained samples from 28 bulls and two pools contained 20 samples each) and then used pooled samples for the subsequent PCR amplification (ABI3730XL DNA analyzer (Applied Biosystems, Foster, CA, USA)). Sequencing data were aligned to the reference genome (UMD 3.1) using Ensemble (https://asia.ensembl.org/, accessed on 1 March 2017) and examined for potential polymorphisms using Chromas (http://technelysium.com.au/wp/chromas/, accessed on 1 March 2017) and DNAman (https://www.lynnon.com/dnaman.html/, accessed on 1 March 2017). In total, 35 variants were detected for gene *IL6R* and *LEPR*.

#### 2.1.2. Genotyping and Phenotyping of 1588 Cows

A total of 1588 Holstein cows from eight dairy farms (Beijing Sunlon Livestock Development Co., Ltd., Beijing, China) were genotyped, of which complete pedigree and reproductive information have been recorded. For each cow, DNA of whole blood samples were extracted using TIANamp Blood DNA Kit (TIANgen, Beijing, China). To shortlist variants included for KASP (Kompetitive allele-specific PCR) genotyping [29], a pilot analysis was performed in 46 (out of 1588) randomly selected samples. Only one variant was kept if the distance between two variants was less than 200 bp. Ultimately, six variants (Table 1) remained and used for genotyping in all 1588 cows with the KASP method [29]. A chi-square test was used to determine whether allelic frequencies of any variant deviated from the Hardy-Weinberg equilibrium. For candidate gene *IL6R* and *LEPR*, haplotype blocks were constructed based on LD structures of identified SNPs by using the Haploview4.0 software [30,31], and the default parameters of Haploview4.0 software are used in haplotype block construction.

**Table 1.** The detail information of six SNPs within the gene *IL6R* and *LEPR* in 1588 Holstein cows.

| Gene | SNP | Genotype | Number of Individuals | Genotypic Frequency | Allele | Allelic Frequency | H-W-E Test, $\chi^2$ Value (df = 1) |
|---|---|---|---|---|---|---|---|
| | | G:G | 155 | 0.125 | G | 0.373 | |
| *IL6R* | g.16210680C/A | T:G | 614 | 0.496 | T | 0.627 | 0.034 |
| | | T:T | 469 | 0.379 | | | |
| | | A:A | 455 | 0.325 | A | 0.570 | |
| *IL6R* | g.16177843C/T | A:G | 684 | 0.489 | G | 0.430 | 0.945 |
| | | G:G | 259 | 0.185 | | | |
| | | G:G | 439 | 0.299 | G | 0.550 | |
| *LEPR* | g.80143337A/C | G:T | 736 | 0.502 | T | 0.450 | 0.581 |
| | | T:T | 291 | 0.198 | | | |
| | | -:- | 125 | 0.085 | - | 0.297 | |
| *LEPR* | g.80101398A/- | A:- | 621 | 0.423 | A | 0.703 | 0.600 |
| | | A:A | 722 | 0.492 | | | |
| | | A:A | 53 | 0.037 | A | 0.177 | |
| *LEPR* | g.80071722T/G | C:A | 403 | 0.280 | C | 0.823 | 0.145 |
| | | C:C | 985 | 0.684 | | | |
| | | A:A | 721 | 0.494 | A | 0.706 | |
| *LEPR* | g.80101081A/T | A:T | 619 | 0.424 | T | 0.294 | 0.387 |
| | | T:T | 119 | 0.082 | | | |

In this study, we detected the associations of six variants with nine reproductive traits covering the broad mechanisms of cattle reproduction [32,33], including age at the first service (AFS), age at the first calving (AFC), the interval from calving to the first insemination (ICF), the interval from first to last insemination in heifers (IFL_H) and cows (IFL_C), stillbirth in heifers (SB_H) and cows (SB_C), and calving ease score in heifers (CE_H) and cows (CE_C). Estimated breeding values (EBVs) for above nine reproduction traits were estimated using single-trait animal models [34] in routine genetic evaluation (Independent Innovation League of Dairy Breeding, Beijing, China), which were used as the response variable of association analyses. The effects in evaluation models for various reproductive traits included fixed effects (age at first insemination, herd-year of calving or herd-year of first insemination within parity, year-month of calving or year-month of first insemination within parity, AI technician, sexed semen, and parity) and random effects (additive genetic effect, permanent environmental effect, and random residual effect). Furthermore, a full pedigree dataset was provided by the Dairy Association of China (Beijing, China), and each animal was traced back as many generations as possible. In total, the pedigree included 109,676 females and 3609 males, with birth years from 1930 to 2020. Furthermore, the sub-pedigree linked to the genotyped population (1558 cows) included 7757 females and 1478 males. Descriptive statistics of EBVs for each reproductive trait were presented in Table S2.

#### 2.1.3. Association Analysis

The SNP- and haplotype block-based association analyses for each reproductive trait were performed by employing the GLM procedure of SAS 9.2 (www.sas.com, accessed on 1 March 2017). False positives resulting from multiple testing in association analyses was controlled using Bonferroni *t*-test. The model used for association analyses of each reproductive trait is as follows:

$$Y_{ij} = \mu + G_i + e_{ij}$$

where $Y_{ij}$ was an individual's EBV; μ was the overall mean; $G_i$ was the fixed effect corresponding to the genotype or haplotype combination; and $e_{ij}$ was the random residual effect. The proportion of the phenotypic variance ($\sigma_P^2$) explained by each identified SNP was calculated as $(\sigma_g^2)/(\sigma_P^2)$, where ($\sigma_g^2$) was the variance explained by SNP, and it was calculated as the following formula: $\sigma_g^2 = 2pq[a + d(q - p)]^2$ [35]; and ($\sigma_P^2$) was the phenotypic variance (Table S3).

### 2.2. *Gene Expression Assay of IL6R and LEPR*

#### 2.2.1. Ovary Collection, Follicle Selection, and Granulosa Cell Isolation

Bovine ovaries of eight dairy cows were collected at an abattoir and put into thermally insulated bottles (28–30 °C) containing sterile physiological saline (with 100 U/mL Penicillin and 0.1 mg/mL Streptomycin) immediately. Within the time of two hours, ovaries were transported to laboratory. After washing three times with warm (37.5 °C) 0.9% NaCl solution and rinsing in 70% warm ethanol for 30 s, ovaries were washed with Dulbecco's phosphate-buffered saline (DPBS) three times. Healthy ovarian follicles with amber-colored fluid were kept for the further operation. According to their diameters (d), ovarian follicles were classified into four developmental stages, including stage 1 ($d \leq 3$ mm), 2 (3 mm $< d \leq 7$ mm), 3 (7 mm $< d \leq 10$ mm), and 4 ($d > 10$ mm). The follicular fluid contained both cumulus–oocyte complexes (COCs) and GCs. A filter (diameter = 70 μm) was used to filter out COCs and then was used to centrifuge the filtrates (contained GCs) at 1500× *g* for 5 min. The supernatant of the follicular fluid was discarded by aspiration. The GC cells were washed three times in phosphate-buffered saline (pH 7.4) and placed in DMEM/F-12 (Gibco, Life Technologies Inc., Grand Island, NY, USA), which contained 1% penicillin-streptomycin and 10% fetal bovine serum (FBS, Gibco, Life Technologies Inc., Grand Island, NY, USA).

#### 2.2.2. Quantitative Reverse Transcription PCR (RT-qPCR)

The RT-qPCR was conducted to detect the relative expression level of *IL6R* and *LEPR* in GCs collected from ovarian follicles at different developmental stages. Total RNA was extracted from three biological replicates of GCs. Reverse transcription was carried out through first-strand cDNA synthesis kit (Thermo Fisher Scientific, Waltham, USA) with oligo (dT) 18 primers according to the manufacturer protocols. The gene-specific primers spanning the exons were designed using Primer 3 (version 4.0, https://bioinfo.ut.ee/primer3-0.4.0/, accessed on 1 March 2017) and Primer blast (www.ncbi.nlm.nih.gov/tools/primer-blast/, accessed on 1 March 2017) (Table S4).

The RT-qPCR was carried out through iTaq™ Universal SYBR® Green Super-mix (Bio-Rad Laboratories GmbH, Munich, Germany) in Applied Biosystem® Step-OnePlus™ (Ap-plied Biosystems, CA, USA). A reaction volume of 20 μL was used, including 7.4 μL of double-distilled $H_2O$, 0.3 μL of forwarding primer, 0.3 μL of reverse primer, 10 μL of 1× SYBR Green master mix (Bio-Rad Laboratories GmbH, Munich, Germany), and 2 μL of cDNA template. Light Cycler 480 instrument (Roche, Mannheim, Germany) was used to perform RT-qPCR. The second derivative maximum method was employed for data acquiring. In this study, the gene *GAPDH* was used as the reference gene, and the $2^{-\Delta\Delta CT}$ method was used to calculate relative expression levels of gene [36].

## 3. Results

### 3.1. Screening of Polymorphisms in Association Popualtion

In this study, a total of 35 variants (Table S5) were identified from polymorphisms screening using mixed pooled DNA of 68 bulls, including eight in *IL6R* and 27 in *LEPR*. The detailed information of the 35 variants are presented in Table S5. By the pilot analysis using 46 randomly selected samples, a total of six SNPs were confirmed using initial Kompetitive allele-specific PCR (KASP) genotyping. As shown in Figure 1, the clustering distributions of three genotypes for each SNP were obviously separated, suggesting that the reliability of KASP genotyping was guaranteed.

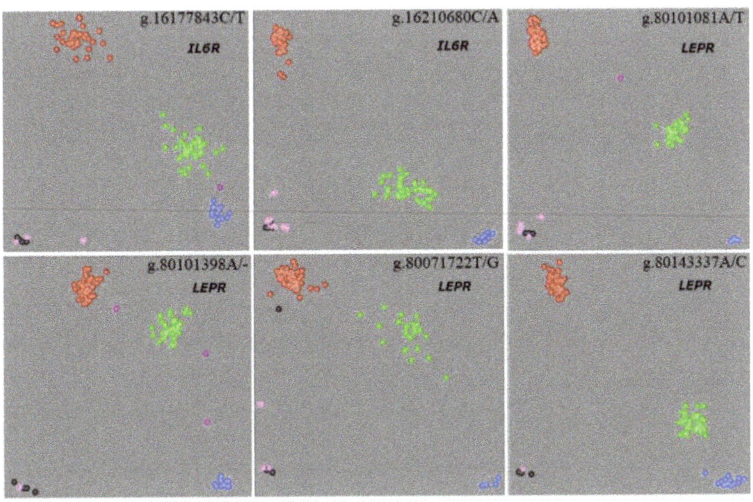


**Figure 1.** Distributions of SNP clustering for the identified six SNPs *IL6R* and *LEPR* using KASP genotyping method. The red and blue represents two homozygous genotypes; the green represents heterozygous genotype; the pink represents no or weak signal; the purple represents signal for no genotyping; and the black represents blank control.

Subsequently, these six SNPs in gene *IL6R* (two) and *LEPR* (four) were genotyped by KASP genotyping in 1588 cows, and the allele frequencies and the corresponding *p*-value derived from Hardy-Weinberg equilibrium test for six SNPs are presented in Table 1. We detected three SNPs (g.16210680C/A, gene *IL6R*; g.80071722T/G and g.80071722T/G, gene *LEPR*) in the exon region, one in the downstream regulatory region (g.16177843C/T, gene *IL6R*), and two in the intronic region (g.80101398A/- and g.80101081A/T, gene *LEPR*).

*3.2. Association Analyses of SNPs with Reproductive Traits*

In total, nine significant associations ($p < 0.05$ in F-test) for gene *IL6R* and *LEPR* were observed with AFC, CE_C, CE_H and ICF. For gene *LEPR*, all SNPs had the significant association with ICF. Furthermore, the SNP g.80143337A/C had the significant association with AFC, CE_H and CE_C. For gene *IL6R*, only one significant association was found between SNP g.16177843C/T and CE_H. The least square means of various genotypes for each SNP are presented in Table 2.

The proportion of phenotypic variance explained by each SNP for every reproductive trait are presented in Table S6. Among all associations between six SNPs and nine reproductive traits, the SNP g.80143337A/C in gene *LERP* explained the largest phenotypic variance for ICF (0.009%). For gene *IL6R*, the SNP g.16177843C/T and g.16210680C/A explained a large phenotypic variance for AFC (0.003%) and SB_C (0.003%), respectively.

*3.3. Association Analyses of Haplotype Blocks with Reproductive Traits*

In this study, the LD between every two SNPs were estimated for gene *IL6R* and *LEPR*. Only one haplotype block was constructed for *LEPR* consisting all screened SNPs, which included 10 haplotypes (Figure 2). As a low LD between SNP g.16210680C/A and g.16177843C/T, two SNPs in *IL6R* were not considered as haplotype blocks.

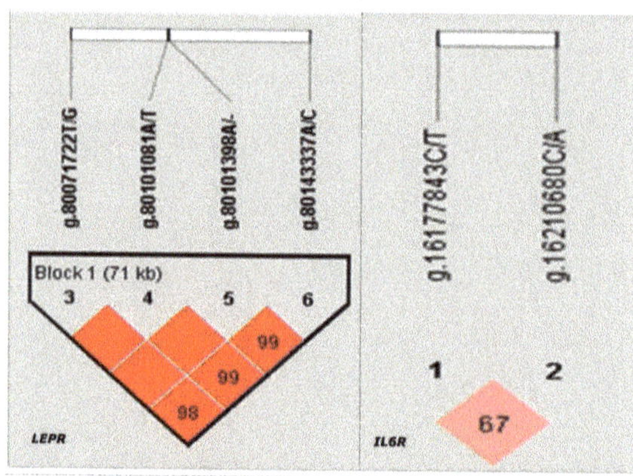


**Figure 2.** Haplotype blocks constructed based on linkage disequilibrium (LD) for gene *IL6R* and *LEPR* in Chinese Holstein cows.

The significant associations detected between haplotype block and nine reproductive traits are presented in Table 3. Consistent with the results from SNP-based analyses, haplotype block in *LEPR* had significant associations with AFC, ICF, and CE_C. Furthermore, the significant associations were also found with AFS and IFL_C. For traits ICF and CE_C, H1H1 was the most beneficial haplotype.

**Table 2.** Association of SNPs within gene *IL6R* and *LEPR* with reproductive traits in Holstein cows (least square mean ± standard error).

| SNP | Genotype | AFS | AFC | IFL_H | IFL_C | ICF | SB_H | SB_C | CE_H | CE_C |
|---|---|---|---|---|---|---|---|---|---|---|
| *IL6R* | GG (155) | −16.718 ± 0.631 | −10.085 ± 0.506 | 5.983 ± 0.499 | 2.365 ± 0.410 | 0.544 ± 0.090 | 0.025 ± 0.003 | 0.016 ± 0.001 | 0.009 ± 0.001 | 0.001 ± 0.000 |
| g.16210680C/A | TG (614) | −17.431 ± 0.317 | −9.317 ± 0.254 | 6.609 ± 0.251 | 2.585 ± 0.206 | 0.425 ± 0.045 | 0.029 ± 0.001 | 0.015 ± 0.001 | 0.009 ± 0.001 | 0.001 ± 0.000 |
| | TT (469) | −17.359 ± 0.363 | −9.020 ± 0.291 | 6.673 ± 0.287 | 2.475 ± 0.236 | 0.356 ± 0.052 | 0.028 ± 0.001 | 0.014 ± 0.001 | 0.008 ± 0.001 | 0.001 ± 0.000 |
| | *p* value | 0.593 | 0.189 | 0.466 | 0.870 | 0.182 | 0.450 | 0.089 | 0.331 | 0.162 |
| *IL6R* | AA (456) | −17.750 ± 0.354 | −9.726 ± 0.299 | 6.464 ± 0.295 | 2.505 ± 0.237 | 0.403 ± 0.052 | 0.027 ± 0.001 | 0.014 ± 0.001 | 0.007 ± 0.001 b | 0.001 ± 0.000 |
| g.16177843C/T | AG (684) | −17.392 ± 0.289 | −9.129 ± 0.244 | 6.921 ± 0.241 | 2.465 ± 0.193 | 0.355 ± 0.043 | 0.029 ± 0.001 | 0.015 ± 0.000 | 0.009 ± 0.000 a | 0.001 ± 0.000 |
| | GG (260) | −17.199 ± 0.469 | −8.731 ± 0.397 | 6.739 ± 0.391 | 2.344 ± 0.314 | 0.469 ± 0.070 | 0.027 ± 0.002 | 0.014 ± 0.001 | 0.009 ± 0.001 ab | 0.000 ± 0.000 |
| | *p* value | 0.598 | 0.107 | 0.486 | 0.917 | 0.367 | 0.495 | 0.690 | 0.018 | 0.310 |
| *LEPR* | GG (440) | −16.878 ± 0.370 | −8.753 ± 0.306a | 7.114 ± 0.299 | 2.666 ± 0.243 | 0.571 ± 0.053a | 0.028 ± 0.001 | 0.015 ± 0.001 | 0.007 ± 0.001 b | 0.001 ± 0.000 a |
| g.80143337A/C | GT (737) | −17.978 ± 0.285 | −9.694 ± 0.236b | 6.414 ± 0.231 | 2.605 ± 0.188 | 0.394 ± 0.041b | 0.029 ± 0.001 | 0.015 ± 0.000 | 0.009 ± 0.000 a | 0.001 ± 0.000 a |
| | TT (291) | −17.232 ± 0.454 | −8.942 ± 0.375ab | 6.609 ± 0.367 | 2.260 ± 0.299 | 0.076 ± 0.065c | 0.027 ± 0.002 | 0.015 ± 0.001 | 0.010 ± 0.001 a | 0.000 ± 0.000 b |
| | *p* value | 0.050 | 0.033 | 0.179 | 0.534 | <0.0001 | 0.794 | 0.897 | 0.002 | <0.0001 |
| *LEPR* | 119 | −16.770 ± 0.707 | −8.300 ± 0.590 | 7.485 ± 0.575 | 2.950 ± 0.469 | 0.641 ± 0.103a | 0.030 ± 0.003 | 0.014 ± 0.001 | 0.007 ± 0.001 | 0.001 ± 0.001 |
| g.80101398A/- | A- (621) | −17.610 ± 0.308 | −9.344 ± 0.257 | 6.690 ± 0.250 | 2.701 ± 0.204 | 0.437 ± 0.045 ab | 0.028 ± 0.001 | 0.014 ± 0.001 | 0.008 ± 0.001 | 0.001 ± 0.000 |
| | AA (723) | −17.570 ± 0.286 | −9.305 ± 0.239 | 6.547 ± 0.232 | 2.281 ± 0.189 | 0.294 ± 0.041 b | 0.028 ± 0.001 | 0.015 ± 0.000 | 0.009 ± 0.000 | 0.001 ± 0.000 |
| | *p* value | 0.538 | 0.250 | 0.318 | 0.201 | 0.002 | 0.850 | 0.354 | 0.388 | 0.510 |
| *LEPR* | AA (722) | −17.649 ± 0.286 | −9.332 ± 0.239 | 6.522 ± 0.233 | 2.277 ± 0.190 | 0.290 ± 0.042 b | 0.028 ± 0.001 | 0.015 ± 0.000 | 0.009 ± 0.000 | 0.001 ± 0.000 |
| g.80101081A/T | AT (619) | −17.598 ± 0.308 | −9.328 ± 0.258 | 6.723 ± 0.251 | 2.739 ± 0.205 | 0.434 ± 0.045 ab | 0.028 ± 0.001 | 0.014 ± 0.001 | 0.008 ± 0.001 | 0.001 ± 0.000 |
| | TT (120) | −16.613 ± 0.703 | −8.270 ± 0.588 | 7.291 ± 0.574 | 2.854 ± 0.467 | 0.633 ± 0.103 a | 0.030 ± 0.003 | 0.015 ± 0.001 | 0.007 ± 0.001 | 0.001 ± 0.001 |
| | *p* value | 0.383 | 0.225 | 0.445 | 0.192 | 0.002 | 0.741 | 0.334 | 0.366 | 0.470 |
| *LEPR* | AA (53) | −18.391 ± 1.060 | −9.591 ± 0.872 | 5.843 ± 0.852 | 2.571 ± 0.702 | 0.520 ± 0.154 ab | 0.029 ± 0.004 | 0.017 ± 0.002 | 0.006 ± 0.002 | 0.002 ± 0.001 |
| g.80071722T/G | CA (404) | −16.865 ± 0.385 | −8.618 ± 0.316 | 6.833 ± 0.309 | 2.911 ± 0.254 | 0.517 ± 0.056 a | 0.028 ± 0.002 | 0.015 ± 0.001 | 0.008 ± 0.001 | 0.001 ± 0.000 |
| | CC (986) | −17.799 ± 0.246 | −9.515 ± 0.202 | 6.641 ± 0.198 | 2.375 ± 0.163 | 0.335 ± 0.036 b | 0.028 ± 0.001 | 0.015 ± 0.000 | 0.009 ± 0.000 | 0.001 ± 0.000 |
| | *p* value | 0.090 | 0.054 | 0.537 | 0.207 | 0.017 | 0.969 | 0.448 | 0.168 | 0.336 |

AFS, age at the first calving; AFC, age at the first service; IFL_H, the interval from the first to last insemination in heifers; IFL_C, the interval from the first to last insemination in cows; ICF, the interval from calving to the first insemination; SB_H, stillbirth in heifers; SB_C, stillbirth in cows; CE_H, calving ease in heifers; CE_C, calving ease in cows. In multiple comparisons by employing Bonferroni *t*-test, different genotypes containing same letter superscripts presents no significant difference ($p < 0.05$), and whereas no same letter superscripts mean significant difference ($p > 0.05$).

**Table 3.** Association of haplotype block within gene *LEPR* with reproductive traits in Holstein cows (least square mean ± standard error).

| Haplotype | AFS | AFC | IFL_H | IFL_C | ICF | SB_H | SB_C | CE_H | CE_C |
|---|---|---|---|---|---|---|---|---|---|
| H1H1 (274) | −17.304 ± 0.462 | −9.194 ± 0.382 | 6.493 ± 0.376 | 2.178 ± 0.308 | 0.091 ± 0.067 b | 0.027 ± 0.002 | 0.015 ± 0.001 | 0.010 ± 0.001 | 0.000 ± 0.000 b |
| H1H2 (379) | −18.234 ± 0.393 | −9.720 ± 0.325 | 6.522 ± 0.320 | 2.886 ± 0.262 | 0.396 ± 0.057 a | 0.028 ± 0.002 | 0.014 ± 0.001 | 0.009 ± 0.001 | 0.000 ± 0.000 ab |
| H1H3 (215) | −17.282 ± 0.522 | −8.786 ± 0.431 | 6.588 ± 0.425 | 2.773 ± 0.348 | 0.449 ± 0.076 a | 0.029 ± 0.002 | 0.015 ± 0.001 | 0.008 ± 0.001 | 0.000 ± 0.000 ab |
| H1H4 (108) | −18.748 ± 0.736 | −10.791 ± 0.609 | 6.223 ± 0.600 | 1.189 ± 0.491 | 0.270 ± 0.107 ab | 0.029 ± 0.003 | 0.016 ± 0.001 | 0.010 ± 0.001 | 0.002 ± 0.001 a |
| H2H2 (115) | −16.655 ± 0.713 | −8.191 ± 0.590 | 7.465 ± 0.581 | 2.923 ± 0.475 | 0.652 ± 0.103 a | 0.030 ± 0.003 | 0.015 ± 0.001 | 0.007 ± 0.001 | 0.001 ± 0.001 ab |
| H2H3 (144) | −16.422 ± 0.637 | −8.239 ± 0.527 | 7.306 ± 0.519 | 3.264 ± 0.425 | 0.578 ± 0.092 a | 0.028 ± 0.003 | 0.014 ± 0.001 | 0.007 ± 0.001 | 0.001 ± 0.000 ab |
| H2H4 (62) | −16.877 ± 0.971 | −9.634 ± 0.803 | 7.058 ± 0.791 | 1.188 ± 0.648 | 0.474 ± 0.141 ab | 0.027 ± 0.004 | 0.014 ± 0.002 | 0.006 ± 0.002 | 0.002 ± 0.001 ab |
| H3H3 (52) | −18.382 ± 1.061 | −9.591 ± 0.877 | 5.872 ± 0.864 | 2.644 ± 0.707 | 0.526 ± 0.154 ab | 0.029 ± 0.004 | 0.017 ± 0.002 | 0.006 ± 0.002 | 0.002 ± 0.001 ab |
| H3H4 (29) | −17.430 ± 1.420 | −9.808 ± 1.175 | 6.159 ± 1.157 | 3.277 ± 0.947 | 0.457 ± 0.206 ab | 0.028 ± 0.006 | 0.015 ± 0.002 | 0.008 ± 0.002 | 0.002 ± 0.001 ab |
| H4H4 (9) | −23.454 ± 2.550 | −12.237 ± 2.108 | 9.755 ± 2.077 | 0.389 ± 1.700 | 0.568 ± 0.370 ab | 0.013 ± 0.010 | 0.010 ± 0.004 | 0.001 ± 0.004 | 0.001 ± 0.002 ab |
| *p* value | 0.044 | 0.025 | 0.545 | 0.012 | <0.001 | 0.942 | 0.850 | 0.087 | 0.001 |

AFS, age at the first calving; AFC, age at the first service; IFL_H, the interval from the first to last insemination in heifers; IFL_C, the interval from the first to last insemination in cows; ICF, the interval from calving to the first insemination; SB_H, stillbirth in heifers; SB_C, stillbirth in cows; CE_H, calving ease in heifers; CE_C, calving ease in cows. In multiple comparisons by employing Bonferroni *t*-test, different genotypes containing same letter superscripts presents no significant difference ($p < 0.05$), whereas no same letter superscripts mean significant differences ($p > 0.05$).

### 3.4. Gene Expression Analysis in GCs Isolated from Follicle of Different Developmental Stages

By using RT-qPCR, the relative expression level of gene *IL6R* and *LEPR* was measured in GCs, which were isolated from bovine ovarian follicular presenting different development stages. The relative mRNA expression of *IL6R* showed an increase trend along the development of follicles and reached the highest level in stage 3 (Figure 3A). The significant differences of the relative mRNA expression were observed between stages 1, 2, and 3. In contrast, its relative mRNA expression significantly decreased along the development of follicles for gene *LEPR* (Figure 3B). The significant differences of the relative mRNA expression for *LEPR* were found across different development stages, except for stages 3 and 4.

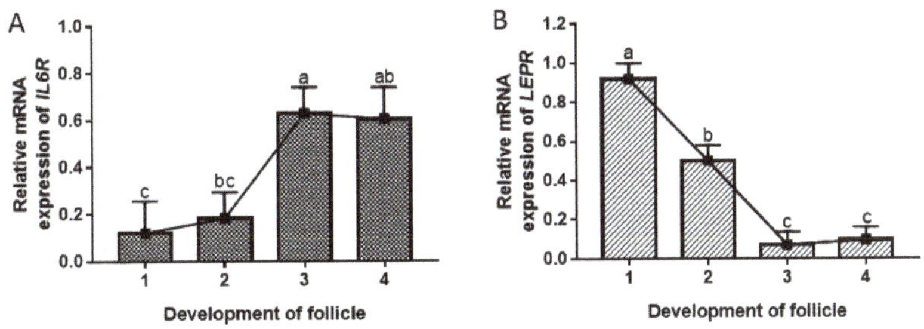


**Figure 3.** The relative mRNA expression level of the gene *IL6R* (**A**) and *LEPR* (**B**) in isolated granulosa. Based on the diameter (d), follicles were divided into four different development stages, including stage 1 ($d \leq 3$ mm), stage 2 ($3 \text{ mm} < d \leq 7$ mm), stage 3 ($7 \text{ mm} < d \leq 10$ mm), and stage 4 ($d > 10$ mm). Different letters are used when the expression levels are significantly different between two stages ($p < 0.05$).

## 4. Discussion

In this study, we examined two metabolic syndrome pathway genes, *IL6R* and *LEPR*, for their effects on cattle female reproduction by both association analyses at population level and the mRNA expression analyses at the cell level. Among six SNPs of gene *IL6R* and *LEPR* genotyped in resource population, five SNPs were significantly associated with five reproduction traits covering both the ability to recycle after calving (ICF) and the ability to conceive and keep pregnancy when inseminated properly (AFS and AFC), and calving ease (CE_H and CE_C), with the greatest number of significant SNPs in gene *LEPR* for ICF.

In our previous GWAS in Chinese and Nordic Holsteins, the gene *IL6R* was identified as the candidate gene of cattle reproduction [11]. In this study, the SNP g.16177843C/T located in the downstream region of *IL6R* was significantly associated with trait CE_H, which further confirmed our previous findings. In transition dairy cows, gene *IL6R* was differently expressed in subcutaneous adipose tissue at the close-up dry period [37]. Considering the relationship between body condition score (BCS) and incidence of calving ease [38], the *IL6R* may play roles in reproductive traits by regulating the body condition of both cows and calf at calving.

The effects of gene *LEPR* on economically important traits of cattle have been widely investigated, including reproduction [39–44], milk production [39,41–44], BCS [45], somatic cell score [40,46], and energy output and energy storage traits [47] in Holstein, Slovak Spotted, and Pinzgau cows, as well as growth [48] and fat deposition in Chinese indigenous cattle [49]. In this study, we firstly confirmed its genetic associations with reproductive traits in a large population from Chinese Holsteins. Amongst the four SNPs, the SNP g.80143337A/C located in exon 2 of *LEPR* had significant associations with five reproductive traits, including AFS, AFC, ICF, CE_H, and CE_C. The SNP g.80143337A/C is a missense mutation, which can cause changes in protein structure, thereby possibly altering protein function. Previous studies investigated the genetic effect of polymorphisms of *LEPR* gene on AFS in Polish Holstein [46] and AFC in Slovak Spotted and Pinzgau cows [39] and obtained similar results as our study. In Holstein primiparous cows, *LEPR* was identified to be significantly associated with days open (sum of ICF and IFL), but not for the services per conception [43]. Furthermore, there exists a weak association between the *LEPR* polymorphism and ICF in US Holstein cows ($p$ = 0.079) [24]. In the present study, all those four SNPs were significantly associated with ICF, which is similar to the findings of above studies. Although the association between *LEPR* polymorphism and reproductive traits of cattle has been widely reported in various population, less has been conducted in terms of mRNA expression analyses, especially in GC tissue with different development stages.

We observed that the relative expression of *IL6R* in GCs increased with increasing follicle size, which was consistent with the observations from the previous study [50]. Baskind et al. [51] also reported that the expression level of *IL6* gene was high before ovulation. IL6 can inhibit the secretion of progesterone-induced by LH [50] and further inhibits the secretion of estrogen in GCs [52], highlighting the role of *IL6R* in the follicular development process. This trend characteristic of *IL6R* may also be related to follicular atresia at later stages of development [53]. With the increase in follicular size, we observed that the relative expression of *LEPR* in GCs decreased, which is similar to the finding reported by Sarkar et al. (2010) [54]. Additionally, several studies have shown that leptin binds to LEPR to modulate steroid production in GCs in the ovary [14] and improve the developmental competence of oocytes at later stages [55], in which the mechanism of LEPR promoting steroid production remains less studied. Differential expression in varying follicular developmental stages indicates its potential role in follicular development.

## 5. Conclusions

In this study, we investigated two metabolic-syndrome pathway genes, *IL6R* and *LEPR*, for their effects on reproductive performance in dairy cows. Among the six SNPs we examined, five were significantly associated with at least one reproductive trait, including age at the first service, age at the first calving, the interval from calving to first insemination, and calving ease in heifers and cows. The roles of *IL6R* and *LEPR* in cattle reproduction were further confirmed by observed differences in relative gene expression levels amongst granulosa cells with different developmental stages. Collectively, the functional validation of *IL6R* and *LEPR* performed in this study provided important molecular markers for genetic selection of reproductive traits in high-yielding dairy cattle.

**Supplementary Materials:** The following supporting information can be downloaded at the figshare repository, accession number https://doi.org/10.6084/m9.figshare.21671198 (accessed on 12 December 2022), Table S1: Primers (38) of the gene *IL6R* and *LEPR* for pooled DNA sequencing; Table S2: Descriptive statistics of EBV for reproductive traits in Chinese Holstein cattle; Table S3: Genetic parameters and variance components for reproductive traits in Chinese Holstein cattle; Table S4: List of primers pairs of employed for RT-qPCR; Table S5: Polymorphic loci information of candidate gene *IL6R* and *LEPR* in Chinese Holstein cattle; Table S6: The percentage of phenotypic variance explained by six SNPs in the Holstein population.

**Author Contributions:** Conceptualization, Y.W. and Q.X.; methodology, Y.D. and S.Z.; formal analysis, Y.D. and H.Z.; resources, G.G. and L.L.; data curation, Y.D. and H.Z.; writing—original draft preparation, A.S. and R.S.; writing—review and editing, H.Z., A.L. and Y.W.; visualization, R.S., A.L. and H.Z.; supervision, Y.W. and A.L.; project administration, Y.W. All authors have read and agreed to the published version of the manuscript.

**Funding:** This research was funded by the earmarked fund for CARS-36, CARS-36; the Program for Changjiang Scholar and Innovation Research Team in University, IRT-15R62; Ningxia Agricultural Breeding Program, 2019NYYZ05; the Beijing Sanyuan Breeding Technology Ltd. Co. funded project, SYZYZ20190005; Ningxia Key Research and Development Projects, 2022BBF02017; and National Agriculture Key Science & Technology Project, NK20221201.

**Institutional Review Board Statement:** Ethical review and approval was not required for the animal study because ethical review and approval were waived for this study. Healthy ovaries were collected from the slaughterhouse. The blood samples and frozen semen were collected along with the regular quarantine inspection of the farms and breeding stations, so no ethical approval was required for this study. Written informed consent was obtained from the owners for the participation of their animals in this study.

**Data Availability Statement:** The data presented in the study are deposited in the figshare repository, accession number https://doi.org/10.6084/m9.figshare.21971585 (accessed on 28 January2022).

**Acknowledgments:** We thank the Dairy Association of China (Beijing, China) for providing the pedigree datasets.

**Conflicts of Interest:** Author G.G. is employed by Beijing Sunlon Livestock Development Co., Ltd. and author L.L. is employed by the Beijing Dairy Cattle Center. Both authors are involved in custody and analysis of the data of dairy herd improvement program. The remaining authors declare that the research was conducted in the absence of any commercial or financial relationships that could be construed as a potential conflict of interest. The funders had no role in the design of the study; in the collection, analyses, or interpretation of data; in the writing of the manuscript; or in the decision to publish the results.

## References

1. Walsh, S.W.; Williams, E.J.; Evans, A.C.O. A review of the causes of poor fertility in high milk producing dairy cows. *Anim. Reprod. Sci.* **2011**, *123*, 127–138. [CrossRef] [PubMed]
2. Jamrozik, J.; Fatehi, J.; Kistemaker, G.J.; Schaeffer, L.R. Estimates of genetic parameters for Canadian Holstein female reproduction traits. *J. Dairy Sci.* **2005**, *88*, 2199–2208. [CrossRef] [PubMed]
3. Guo, G.; Guo, X.; Wang, Y.; Zhang, X.; Zhang, S.; Li, X.; Liu, L.; Shi, W.; Usman, T.; Wang, X.; et al. Estimation of genetic parameters of fertility traits in Chinese Holstein cattle. *Can. J. Anim. Sci.* **2014**, *94*, 281–285. [CrossRef]
4. Brondum, R.F.; Su, G.; Janss, L.; Sahana, G.; Guldbrandtsen, B.; Boichard, D.; Lund, M.S. Quantitative trait loci markers derived from whole genome sequence data increases the reliability of genomic prediction. *J. Dairy Sci.* **2015**, *98*, 4107–4116. [CrossRef]
5. García-Ruiz, A.; Cole, J.B.; VanRaden, P.M.; Wiggans, G.R.; Ruiz-López, F.J.; Van Tassell, C.P. Changes in genetic selection differentials and generation intervals in US Holstein dairy cattle as a result of genomic selection. *Proc. Natl. Acad. Sci. USA* **2016**, *113*, E3995–E4004. [CrossRef] [PubMed]
6. Gutierrez, C.G.; Gong, J.G.; Bramley, T.A.; Webb, R. Selection on predicted breeding value for milk production delays ovulation independently of changes in follicular development, milk production and body weight. *Anim. Reprod. Sci.* **2006**, *95*, 193–205. [CrossRef] [PubMed]
7. Sejian, V.; Bhatta, R.; Soren, N.M.; Malik, P.K.; Ravindra, J.P.; Prasad, C.S.; Lal, R. *Introduction to Concepts of Climate Change Impact on Livestock and its Adaptation and Mitigation*; Springer: New Delhi, India, 2015; pp. 1–23.
8. Wathes, D.C.; Fenwick, M.; Cheng, Z.; Bourne, N.; Llewellyn, S.; Morris, D.G.; Kenny, D.; Murphy, J.; Fitzpatrick, R. Influence of negative energy balance on cyclicity and fertility in the high producing dairy cow. *Theriogenology* **2007**, *68*, S232–S241. [CrossRef] [PubMed]
9. Sammad, A.; Umer, S.; Shi, R.; Zhu, H.; Zhao, X.; Wang, Y. Dairy cow reproduction under the influence of heat stress. *J. Anim. Physiol. Anim. Nutr.* **2020**, *104*, 978–986. [CrossRef]
10. Ridker, P.M.; Pare, G.; Parker, A.; Zee, R.Y.; Danik, J.S.; Buring, J.E.; Kwiatkowski, D.; Cook, N.R.; Miletich, J.P.; Chasman, D.I. Loci related to metabolic-syndrome pathways including LEPR, HNF1A, IL6R, and GCKR associate with plasma C-reactive protein: The Women's Genome Health Study. *Am. J. Hum. Genet.* **2008**, *82*, 1185–1192. [CrossRef]
11. Liu, A.; Wang, Y.; Sahana, G.; Zhang, Q.; Liu, L.; Lund, M.S.; Su, G. Genome-wide association studies for female fertility traits in Chinese and Nordic Holsteins. *Sci. Rep.* **2017**, *7*, 8487. [CrossRef]
12. Van Snick, J. Interleukin-6: An overview. *Annu. Rev. Immunol.* **1990**, *8*, 253–278. [CrossRef]
13. Rose-John, S. IL-6 trans-signaling via the soluble IL-6 receptor: Importance for the pro-inflammatory activities of IL-6. *Int. J. Biol. Sci.* **2012**, *8*, 1237–1247. [CrossRef]
14. Spicer, L.J. Leptin: A possible metabolic signal affecting reproduction. *Domest. Anim. Endocrinol.* **2001**, *21*, 251–270. [CrossRef] [PubMed]
15. Markert, U.R.; Morales-Prieto, D.M.; Fitzgerald, J.S. Understanding the link between the IL-6 cytokine family and pregnancy: Implications for future therapeutics. *Expert Rev. Clin. Immunol.* **2011**, *7*, 603–609. [CrossRef]
16. Demir, B.; Guven, S.; Guven, E.S.; Atamer, Y.; Gul, T. Serum IL-6 level may have role in the pathophysiology of unexplained infertility. *Am. J. Reprod. Immunol.* **2009**, *62*, 261–267. [CrossRef] [PubMed]
17. Sherwin, J.R.; Smith, S.K.; Wilson, A.; Sharkey, A.M. Soluble gp130 is up-regulated in the implantation window and shows altered secretion in patients with primary unexplained infertility. *J. Clin. Endocrinol. Metab.* **2002**, *87*, 3953–3960. [CrossRef] [PubMed]
18. Zhu, L.; Chen, H.; Liu, M.; Yuan, Y.; Wang, Z.; Chen, Y.; Wei, J.; Su, F.; Zhang, J. Treg/Th17 cell imbalance and IL-6 profile in patients with unexplained recurrent spontaneous abortion. *Reprod. Sci.* **2017**, *24*, 882–890. [CrossRef] [PubMed]
19. Benyo, D.F.; Smarason, A.; Redman, C.W.; Sims, C.; Conrad, K.P. Expression of inflammatory cytokines in placentas from women with preeclampsia. *J. Clin. Endocrinol. Metab.* **2001**, *86*, 2505–2512. [CrossRef] [PubMed]

20. Nasiri, M.; Rasti, Z. CTLA-4 and IL-6 gene polymorphisms: Risk factors for recurrent pregnancy loss. *Hum. Immunol.* **2016**, *77*, 1271–1274. [CrossRef]
21. Medica, I.; Ostojic, S.; Pereza, N.; Kastrin, A.; Peterlin, B. Association between genetic polymorphisms in cytokine genes and recurrent miscarriage—A meta-analysis. *Reprod. Biomed. Online* **2009**, *19*, 406–414. [CrossRef] [PubMed]
22. Liefers, S.C.; Veerkamp, R.F.; Te Pas, M.F.W.; Chilliard, Y.; Van der Lende, T. Genetics and physiology of leptin in periparturient dairy cows. *Domest. Anim. Endocrinol.* **2005**, *29*, 227–238. [CrossRef]
23. Chin, J.R.; Heuser, C.C.; Eller, A.G.; Branch, D.W.; Nelson, L.T.; Silver, R.M. Leptin and leptin receptor polymorphisms and recurrent pregnancy loss. *J. Perinatol.* **2013**, *33*, 589–592. [CrossRef]
24. Clempson, A.M.; Pollott, G.E.; Brickell, J.S.; Bourne, N.E.; Munce, N.; Wathes, D.C. Evidence that leptin genotype is associated with fertility, growth, and milk production in Holstein cows. *J. Dairy Sci.* **2011**, *94*, 3618–3628. [CrossRef] [PubMed]
25. Su, Y.Q.; Wu, X.; O'Brien, M.J.; Pendola, F.L.; Denegre, J.N.; Matzuk, M.M.; Eppig, J.J. Synergistic roles of BMP15 and GDF9 in the development and function of the oocyte-cumulus cell complex in mice: Genetic evidence for an oocyte-granulosa cell regulatory loop. *Dev. Biol.* **2004**, *276*, 64–73. [CrossRef]
26. Petro, E.M.; Leroy, J.L.; Van Cruchten, S.J.; Covaci, A.; Jorssen, E.P.; Bols, P.E. Endocrine disruptors and female fertility: Focus on (bovine) ovarian follicular physiology. *Theriogenology* **2012**, *78*, 1887–1900. [CrossRef]
27. Bettaieb, A.; Averill-Bates, D.A. Thermotolerance induced at a fever temperature of 40 degrees C protects cells against hyperthermia-induced apoptosis mediated by death receptor signalling. *Biochem. Cell Biol.* **2008**, *86*, 521–538. [CrossRef]
28. Eppig, J.J. Oocyte control of ovarian follicular development and function in mammals. *Reproduction* **2001**, *122*, 829–838. [CrossRef] [PubMed]
29. He, C.; Holme, J.; Anthony, J. *SNP Genotyping: The KASP Assay*; Springer: New York, NY, USA, 2014; pp. 75–86.
30. Barrett, J.C.; Fry, B.; Maller, J.; Daly, M.J. Haploview: Analysis and visualization of LD and haplotype maps. *Bioinformatics* **2005**, *21*, 263–265. [CrossRef]
31. Gabriel, S.B.; Schaffner, S.F.; Nguyen, H.; Moore, J.M.; Roy, J.; Blumenstiel, B.; Higgins, J.; DeFelice, M.; Lochner, A.; Faggart, M.; et al. The structure of haplotype blocks in the human genome. *Science* **2002**, *296*, 2225–2229. [CrossRef] [PubMed]
32. Ettema, J.F.; Santos, J.E.P. Impact of age at calving on lactation, reproduction, health, and income in first-parity Holsteins on commercial farms. *J. Dairy Sci.* **2004**, *87*, 2730–2742. [CrossRef] [PubMed]
33. Nayeri, S.; Sargolzaei, M.; Abo-Ismail, M.K.; Miller, S.; Schenkel, F.; Moore, S.S.; Stothard, P. Genome-wide association study for lactation persistency, female fertility, longevity, and lifetime profit index traits in Holstein dairy cattle. *J. Dairy Sci.* **2017**, *100*, 1246–1258. [CrossRef]
34. Liu, A.; Lund, M.S.; Wang, Y.; Guo, G.; Dong, G.; Madsen, P.; Su, G. Variance components and correlations of female fertility traits in Chinese Holstein population. *J. Anim. Sci. Biotechnol.* **2017**, *8*, 56. [CrossRef] [PubMed]
35. Hill, W.G.; Mackay, T.F. Falconer and Introduction to quantitative genetics. *Genetics* **2004**, *167*, 1529–1536. [CrossRef] [PubMed]
36. Huang, D.W.; Sherman, B.T.; Lempicki, R.A. Systematic and integrative analysis of large gene lists using DAVID bioinformatics resources. *Nat. Protoc.* **2009**, *4*, 44–57. [CrossRef] [PubMed]
37. Vailati-Riboni, M.; Farina, G.; Batistel, F.; Heiser, A.; Mitchell, M.D.; Crookenden, M.A.; Walker, C.G.; Kay, J.K.; Meier, S.; Roche, J.R.; et al. Far-off and close-up dry matter intake modulate indicators of immunometabolic adaptations to lactation in subcutaneous adipose tissue of pasture-based transition dairy cows. *J. Dairy Sci.* **2017**, *100*, 2334–2350. [CrossRef]
38. Mandour, M.A.B.; Al-Shami, S.A.; Al-Eknah, M.M. Body condition scores at calving and their association with dairy cow performance and health in semiarid environment under two cooling systems. *Ital. J. Anim. Sci.* **2015**, *14*, 2015–3690. [CrossRef]
39. Trakovická, A.; Moravčíková, N.; Kasarda, R. Genetic polymorphisms of leptin and leptin receptor genes in relation with production and reproduction traits in cattle. *Acta Biochim. Pol.* **2013**, *60*, 783. [CrossRef]
40. Asadollahpour, N.H.; Ansari, M.S.; Edriss, M.A. Effect of LEPR, ABCG2 and SCD1 gene polymorphisms on reproductive traits in the Iranian Holstein cattle. *Reprod. Domest. Anim.* **2014**, *49*, 769–774. [CrossRef]
41. Baghenoey, S.G.; Mahyari, S.A.; Nanaei, H.A.; Rostami, M.; Edriss, M.A. Association of LEPR gene polymorphism with milk yield and age at first calving in the Iranian Holstein dairy cows. *J. Livest. Sci. Technol.* **2014**, *2*, 39–42.
42. Hill, R.; Canal, A.; Bondioli, K.; Morell, R.; Garcia, M.D. Molecular markers located on the DGAT1, CAST, and LEPR genes and their associations with milk production and fertility traits in Holstein cattle. *Genet. Mol. Res.* **2016**, *15*, gmr.15017794. [CrossRef]
43. Al-Janabi, H.R.A.; Al-Rawi, A.J.; Al-Anbari, N.N. Association of leptin receptor gene polymorphism with some productive and reproductive traits in Holstein primiparous cows. *J. Entomol. Zool. Stud.* **2018**, *6*, 1353–1358.
44. Pandey, V. Association analyses of single nucleotide polymorphism in the leptin receptor gene with reproduction and production traits in high yielding Indian cow breed. *J. Anim. Res.* **2021**, *11*, 241–247. [CrossRef]
45. Banos, G.; Woolliams, J.A.; Woodward, B.W.; Forbes, A.B.; Coffey, M.P. Impact of single nucleotide polymorphisms in leptin, leptin receptor, growth hormone receptor, and diacylglycerol acyltransferase (DGAT1) gene loci on milk Production, feed, and body energy traits of UK dairy cows. *J. Dairy Sci.* **2008**, *91*, 3190–3200. [CrossRef]
46. Komisarek, J. Impact of LEP and LEPR gene polymorphisms on functional traits in Polish Holstein-Friesian cattle. *Anim. Sci. Pap. Rep.* **2010**, *10*, 133–141.
47. Giblin, L.; Butler, S.T.; Kearney, B.M.; Waters, S.M.; Callanan, M.J.; Berry, D.P. Association of bovine leptin polymorphisms with energy output and energy storage traits in progeny tested Holstein-Friesian dairy cattle sires. *BMC Genet.* **2010**, *11*, 73. [CrossRef]

48. Guo, Y.; Chen, H.; Lan, X.; Zhang, B.; Pan, C.; Zhang, L.; Zhang, C.; Zhao, M. Novel SNPs of the bovine LEPR gene and their association with growth traits. *Biochem. Genet.* **2008**, *46*, 828–834. [CrossRef] [PubMed]
49. Raza, S.; Liu, G.Y.; Zhou, L.; Gui, L.S.; Khan, R.; Jinmeng, Y.; Chugang, M.; Schreurs, N.M.; Ji, R.; Zan, L. Detection of polymorphisms in the bovine leptin receptor gene affects fat deposition in two Chinese beef cattle breeds. *Gene* **2020**, *758*, 144957. [CrossRef] [PubMed]
50. Samir, M.; Glister, C.; Mattar, D.; Laird, M.; Knight, P.G. Follicular expression of pro-inflammatory cytokines tumour necrosis factor-α (TNFα), interleukin 6 (IL6) and their receptors in cattle: TNFα, IL6 and macrophages suppress thecal androgen production in vitro. *Reproduction* **2017**, *154*, 35–49. [CrossRef]
51. Baskind, N.E.; Orsi, N.M.; Sharma, V. Follicular-phase ovarian follicular fluid and plasma cytokine profiling of natural cycle in vitro fertilization patients. *Fertil. Steril.* **2014**, *102*, 410–418. [CrossRef]
52. Salmassi, A.; Lu, S.; Hedderich, J.; Oettinghaus, C.; Jonat, W.; Mettler, L. Interaction of interleukin-6 on human granulosa cell steroid secretion. *J. Endocrinol.* **2001**, *170*, 471–478. [CrossRef] [PubMed]
53. Zhang, J.; Liu, Y.; Yao, W.; Li, Q.; Liu, H.; Pan, Z. Initiation of follicular atresia: Gene networks during early atresia in pig ovaries. *Reproduction* **2018**, *156*, 23–33. [CrossRef] [PubMed]
54. Sarkar, M.; Schilffarth, S.; Schams, D.; Meyer, H.H.; Berisha, B. The expression of leptin and its receptor during different physiological stages in the bovine ovary. *Mol. Reprod. Dev.* **2010**, *77*, 174–181. [CrossRef]
55. Boelhauve, M.; Sinowatz, F.; Wolf, E.; Paula-Lopes, F.F. Maturation of bovine oocytes in the presence of leptin improves development and reduces apoptosis of in vitro-produced blastocysts. *Biol. Reprod.* **2005**, *73*, 737–744. [CrossRef] [PubMed]

agriculture

MDPI

*Communication*

# Regulation of bta-miRNA29d-3p on Lipid Accumulation via *GPAM* in Bovine Mammary Epithelial Cells

**Xin Zhao [1,2,†], Jun Li [3,†], Shuying Zhao [4], Lili Chen [1], Man Zhang [1], Yi Ma [1] and Dawei Yao [1,*]**

1 Tianjin Key Laboratory of Animal Molecular Breeding and Biotechnology, Tianjin Engineering Research Center of Animal Healthy Farming, Institute of Animal Science and Veterinary, Tianjin Academy of Agricultural Sciences, Tianjin 300381, China
2 College of Life Science, Nankai University, Tianjin 300071, China
3 College of Animal Science and Technology, Henan University of Animal Husbandry and Economy, Zhengzhou 450046, China
4 College of Animal Science and Animal Medicine, Tianjin Agricultural University, Tianjin 300384, China
* Correspondence: yaodawei2008@126.com
† These authors contributed equally to this work.

**Abstract:** MicroRNAs (miRNAs) are small RNA molecules consisting of approximately 22 nucleotides that are engaged in the regulation of various bio-processes. There is growing evidence that miR-29 is a key regulator of hepatic lipid metabolism. Mimics and inhibitors of bta-miRNA29d-3p were transiently transfected in bovine mammary epithelial cells (BMECs) to reveal the regulation of bta-miRNA29d-3p on lipid accumulation in BMECs. Results showed that overexpression of bta-miRNA29d-3p significantly inhibited the expression of genes related to triglyceride (TAG) synthesis, namely *DGAT1* and mitochondrial glycerol-3-phosphate acyltransferase (*GPAM*, $p < 0.01$) and down-regulated TAG levels in cells ($p < 0.05$). The expression of fatty acid synthesis and desaturation-related genes *FASN*, *SCD1*, and *ACACA*, and transcription factor *SREBF1* also decreased. Interference of bta-miRNA29d-3p significantly increased the expression of *GPAM*, *DGAT1*, *FASN*, *SCD1*, *ACACA*, and *SREBF1* ($p < 0.01$), and significantly upregulated the concentration of TAG in cells. Furthermore, a luciferase reporter assay confirmed that *GPAM* is a direct target of bta-miRNA29d-3p. In summary, bta-miRNA29d-3p modulates fatty acid metabolism and TAG synthesis by regulating genes related to lipid metabolism in BMECs and targeting *GPAM*. Thus, bta-miRNA29d-3p plays an important role in controlling mammary lipid synthesis in cows.

**Keywords:** miR-29; lipid metabolism; *GPAM*; milk fat; cow mammary cells

**Citation:** Zhao, X.; Li, J.; Zhao, S.; Chen, L.; Zhang, M.; Ma, Y.; Yao, D. Regulation of bta-miRNA29d-3p on Lipid Accumulation via *GPAM* in Bovine Mammary Epithelial Cells. *Agriculture* **2023**, *13*, 501. https://doi.org/10.3390/agriculture13020501

Academic Editor: Michael Schutz

Received: 9 December 2022
Revised: 13 February 2023
Accepted: 16 February 2023
Published: 20 February 2023



## 1. Introduction

Milk has gradually become an indispensable part of the daily diet of humans. Milk is high in fat, and it contains unsaturated fatty acids that are effective in preventing high blood cholesterol and cardiovascular disease in humans [1]. The composition and content of fatty acids are also strong correlates of the nutrition of milk [2]. Increasing lipid content and ameliorating lipid composition could help improve milk quality. One of the essential organs of lactating animals is the breast. Its development and lactation are regulated by cytokines, hormones, and some natural biological functional substances [3]. MicroRNA (miRNA) research in terms of breast development and secretion control in mammals (cows, mice, etc.) has also progressed to a certain extent.

MiRNAs are small, siRNA-like molecules that bind to targeted mRNAs, inhibiting their translation and accelerating their degradation at the post-transcriptional level [4]. MiRNA regulates the biometabolism of most lactating animal tissues, such as the liver, fat, and skeletal muscle in humans, dairy cows, and dairy goats, as well as in mice [5]. The contributions of miRNAs to milk gland evolution are mainly achieved by regulating the proliferation and differentiation of mammary epithelial cells, the differentiation of mammary stem cells, and the development of glandular vesicles and milk ducts [6]. The miR-29 family mainly includes miR-29a, miR-29b, miR-29c, miR-29d-3p, and miR-29e. Reports on the structure, function, and regulation of the miR-29 family are mainly focused on research related to humans and mice at present.

In ruminants, miR-29 was shown to target *PTX3* in goat granulosa cells through activation of PI3K/AKT/mTOR and Erk1/2 signaling pathways to stimulate granulosa cell multiplication and suppress steroidogenesis and apoptosis [7]. However, studies in non-ruminants have shown that miR-29 not only participates in fatty acid oxidation but also leads to the synthesis of TAG and, thus, performs an instrumental role in controlling fatty acid metabolism. He et al. used microarray technology to study rat skeletal muscle and discovered that miR-29 expression was significantly higher in a type II diabetic group than in healthy rats [8]. In addition, reduced fatty acid content and significantly decreased plasma cholesterol and triglyceride (TAG) levels were observed in the livers of miR-29 knockdown mice [9]. These findings offer the possibility that bta-miRNA29d-3p regulates fatty acid composition and TAG content in dairy cows.

MiRNAs have become significant regulators of glycolipid metabolism among many tissues [5]. However, to date, studies on bta-miRNA29d-3p in the dairy mammary gland have not been reported. This study explored the effect of bta-miRNA29d-3p on the expression of genes related to lipid production and TAG content in mammary epithelial cells of dairy cows and revealed its role in regulating milk lipids.

## 2. Materials and Methods

### 2.1. Cell Culture and Transfection

The BMECs used in this experiment were provided by Mr. Chen Zhi from the College of Animal Science and Technology, Yangzhou University. The mammary gland tissue samples used for BMEC isolation were collected from three healthy Holstein cows (4 years old, second parity, non-pregnant) at peak lactation. The psiCHECK-2 vector (psiCHECK-2 plasmid) was presented by Professor Jun Luo from Northwest A&S University.

On the basis of the available literature [10], BMECs were cultured by 60 mm cell dishes in a 5% $CO_2$ incubator at 37 °C. The medium contained 10% fetal bovine serum, 5 μg/mL insulin, 100 U/mL penicillin and streptomycin, 10 ng/mL epidermal growth factor, 1 μg/mL hydrocortisone, and 90% DMEM/F12. The medium was replaced once in 24 h.

When the cell confluence reached 80%, the cells were transfected with Lipofectamine® RNAiMAX Transfection Reagent, and the transfection complex was prepared by referring to the product's instructions. The total volume of transfection complexes was 100 μL per well. miR-29d mimic (miR-29d-3p) and its negative control (Con miR) and miR-29d inhibitor (anti-miR-29d-3p) and its negative control (Con Inh) were transiently transfected into mammary epithelial cells at a final concentration of 50 nM, and the remaining volume was replenished with DMEM/F-12. The transfection complexes were mixed well and incubated at room temperature for 20 min. The cultured BMECs were digested into separate suspensions and added to a 24-well plate. Finally, the incubated transfection complexes were added drop-wise and incubated at 37 °C in a 5% $CO_2$ incubator. After 48 h, intracellular miR-29d-3p mRNA levels were examined using real-time fluorescence quantification (RT-qPCR). Each treatment was performed in three biological replicates.

### 2.2. RNA Extraction and Real-Time Quantitative PCR

Cells were lysed to carry out RNA extraction after 48 h of transfection, and the RNA extraction method was carried out using the cultured cell/bacterial total RNA extraction kit from Tiangen Biochemical Technology (Beijing, China) Co., Ltd. (DP430). The extracted RNA was reverse transcribed with the PrimeScriptTM RT kit (Perfect Real time, Takara Bio Inc., Kusatsu, Japan), where the bta-miRNA29d-3p reverse transcription primers were designed using the stem–loop method. The bta-miRNA29d-3p reverse-transcribed cDNA was used as a template, and the 18s gene was used as an internal reference gene for RT-qPCR amplification of the gene by using SYBR Green (SYBR Premix Ex Taq II, Perfect Real Time, Takara Bio Inc., Kusatsu, Japan). Then, a 20 μL system was designed as follows: SYBR GreenMix, 10 μL; PCR Forward Primer, 0.8 μL; PCR Reverse Primer, 0.8 μL; DNA template, 2 μL; and sterilized water, 6.0 μL. The PCR reaction program was 40 cycles of 95 °C pre-denaturation for 30 s, 95 °C for 5 s, and 60 °C for 30 s.

The genes detected included fatty acid synthase (*FASN*), glycerol-3-phosphate acyltransferase (*GPAM*), peroxisome proliferator-activated receptor γ (*PPARG1*), diacylglycerol acyltransferase 1 (*DGAT1*), sterol regulatory element-binding factor 1 (*SREBF1*), liver X receptor alpha (*LXRA*), and acetyl-coenzyme A carboxylase alpha (*ACACA*). The internal controls were ubiquitously expressed transcript (*UXT*) and mitochondrial ribosomal protein L39 (*MRPL39*). The sequences of primers with qPCR are shown in Table 1.

**Table 1.** Characteristics of primers used of the RT-qPCR reaction.

| Gene1 | Accession No. | Primer Sequences (5′-3′) | Source |
|---|---|---|---|
| *UXT* | XM_005700842.1 | F: TGTGGCCCTTGGATATGGTT<br>R: GGTTGTCGCTGAGCTCTGTG | [11] |
| *MRPL39* | XM_005674737.1 | F: AGGTTCTCTTTTGTTGGCATCC<br>R: TTGGTCAGAGCCCCAGAAGT | [11] |
| *GPAM* | AY515690 | F: GCAGGTTTATCCAGTATGGCATT<br>R: GGACTGATATCTTCCTGATCATCTTG | [12] |
| *PPARG1* | NM_001290893.1 | F: CGTTTCCTTAAACAAGTG<br>R: TCCCTCAAAATAATAGTGC | [13] |
| *DGAT1* | NM_174693 | F: CCACTGGGACCTGAGGTGTC<br>R: GCATCACCACACACCAATTCA | [14] |
| *LXRA* | NM_001014861.1 | F: CATGCCTACGTCTCCATCCA<br>R: TCACCAGTTTCATCAGCATCCT | [15] |
| *SREBF1* | TC263657 | F: CCAGCTGACAGCTCCATTGA<br>R: TGCGCGCCACAAGGA | [12] |
| *FASN* | CR552737 | F: ACCTCGTGAAGGCTGTGACTCA<br>R: TGAGTCGAGGCCAAGGTCTGAA | [14] |
| *ACACA* | AJ132890 | F: CATCTTGTCCGAAACGTCGAT<br>R: CCCTTCGAACATACACCTCCA | [14] |
| *SCD1* | GU947654 | F: CCATCGCCTGTGGAGTCAC<br>R: GTCGGATAAATCTAGCGTAGCA | [16] |

### 2.3. Detection of Triglycerides in BMECs

The transfection operation was the same as above. Analyses were performed in triplicate. After the cells were transfected for 48 h and replaced with new medium, the cells were washed twice with PBS, and 1 mL of lysis buffer was added to each well. An appropriate amount of lysate was reserved for BCA protein quantitative determination. TAG content was detected and calculated in accordance with the instructions of the TAG test kit from Beijing Applygen. A TAG standard curve was constructed to specify the triglyceride content of the samples, after which the TAG content was corrected for total cellular protein concentration per mg.

### 2.4. Construction of psiCHECK-2-GPAM 3′UTR Recombinant Plasmid

To generate a reporter gene for luciferase assays, we amplified bovine *GPAM* 3′-UTR fragments by PCR and cloned them into the XhoI/NotI digest site of the dual-luciferase reporter gene vector psiCHECK-2 to construct a dual-luciferase reporter gene vector containing the wild-type bovine *GPAM* 3′UTR sequence. Meanwhile, a dual-luciferase expression vector containing a mutated *GPAM* 3′-UTR sequence was constructed by overlap PCR. The primer sequences are given in Table 2. All constructs were confirmed by sequencing.

**Table 2.** Primers used for site-directed mutation of bovine *GPAM* 3′UTR constructs.

| Primer Name | Primer Sequences (5′-3′) |
|---|---|
| *GPAM*-F | TTGCCCTTTCAGTTGGTTT |
| *GPAM*-R | TTGCCCTTTCAGTTGGTTT |
| MiR-29d-3p-binding site | TGGTGCTTTG |
| Mutation site | CACGTACCCT |
| Overlap-F | GTAATTAACCACGTACCCTAA |
| Overlap-R | GGTTTTAGGGTACGTGGTTAATTAC |

### *2.5. Dual Luciferase Assays*

After mammary epithelial cells were transfected for 48 h, they were lysed and a dual-luciferase assay was performed by referring to the Promega dual reporter gene assay kit. The cells were washed twice using PBS. In each well of the 24-well opaque assay plate, 65 µL of cell lysate was added. The firefly luciferase activity value (F value) was first obtained, and then the firefly luciferase activity was terminated, the Renilla luciferase activity was activated, and the Renilla luciferase activity (R value) was read. Relative luciferase activity = R value/F value, which is the relative transcriptional activity.

### *2.6. Statistical Analysis*

All data were analyzed in SAS 9.4 (Cary, NC, USA) using reusable two-factor analysis of variance and expressed as mean ± standard error of the mean (SEM). Each experiment was repeated at least three times. Relative quantitative analysis was performed using the $2^{-\Delta\Delta Ct}$ method, and the data were analyzed for significance. $p < 0.05$ indicated a significant difference (*, $p < 0.05$; **, $p < 0.01$).

## 3. Results

### *3.1. Detection of Overexpression and Interference Efficiency of bta-miRNA29d-3p in BMECs*

According to Figure 1, the expression of bta-miRNA29d-3p increased by approximately 300-fold compared with that of Con miR ($p < 0.01$, Figure 1A), and the expression of anti-miR-29d-3p significantly decreased by approximately 80% compared with that of Con Inh ($p < 0.01$, Figure 1B). This result indicated that miR-29d-3p and anti-miR-29d-3p were transfected with good efficiency.

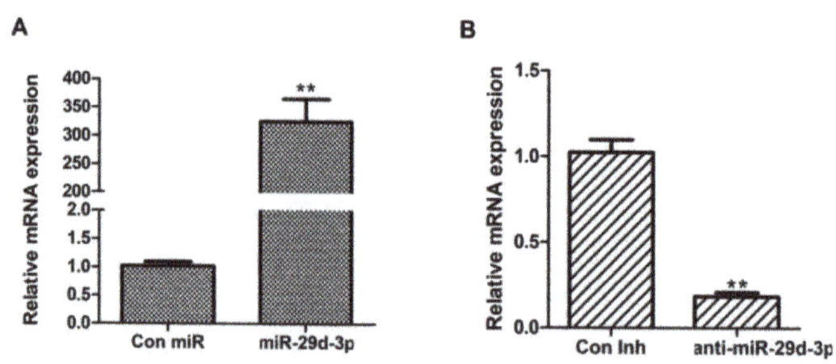


**Figure 1.** Detection of overexpression and interference efficiency of bta-miRNA29d-3p. Results are the mean ± SEM of three separate experiments. ** $p < 0.01$ versus control.

### *3.2. Overexpression of bta-miRNA29d-3p Affects the Expression of Genes Related to Lipid Metabolism in BMECs*

On the basis of Figure 2, after bta-miRNA29d-3p was overexpressed and the expression of SREBF1 was clearly decreased, whereas that of LXRA was markedly upregulated ($p < 0.01$, Figure 2A). The synthesis and desaturation of fatty acid-related gene (ACACA, FASN, and SCD1) expression were remarkably reduced ($p < 0.01$, Figure 2B), and that of TAG synthesis-related genes DGAT1 and GPAM was significantly down-regulated ($p < 0.01$, Figure 2C).

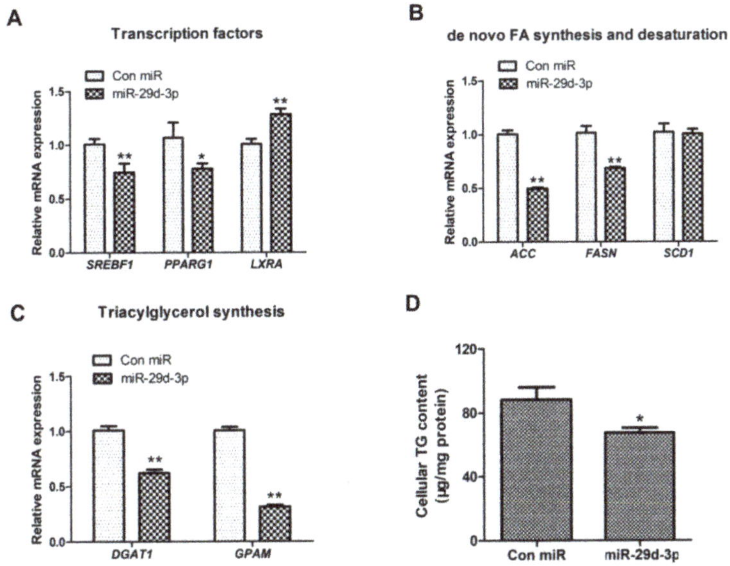


**Figure 2.** Effect of bta-miRNA29d-3p overexpression on genes related to lipid metabolism in BMECs. Panel (**A**): Effect of bta-miRNA29d-3p overexpression on the expression of transcription factors SREBF1, PPARG1, and LXRA. Panel (**B**): Effect of bta-miRNA29d-3p overexpression on the expression of fatty acid de novo synthesis and desaturation-related genes (ACACA, FASN, and SCD1). Panel (**C**): Effect of bta-miRNA29d-3p overexpression on TAG synthesis and expression of lipid droplet formation-related genes (DGAT1 and GPAM). Panel (**D**): Effect of bta-miRNA29d-3p overexpression on intracellular TAG content. Results are the mean ± SEM of three separate experiments. ** $p < 0.01$ versus control, * $p < 0.05$ versus control (Con miR).

### *3.3. Interference of bta-miRNA29d-3p Promotes the Expression of Genes Related to Lipid Metabolism in BMECs*

As shown in Figure 3, interference with bta-miRNA29d-3p increased GPAM expression, and the expression of FASN, SCD1, ACACA, DGAT1, and PPARG1 was also significantly elevated ($p < 0.01$, Figure 3).

### *3.4. Effect of bta-miRNA29d-3p on TAG Content in BMECs*

As illustrated in Figures 2 and 3, when bta-miRNA29d-3p was overexpressed, the TAG content in BMECs was significantly down-regulated ($p < 0.05$, Figure 2D). However, when bta-miRNA29d-3p was disrupted, the TAG content increased significantly ($p < 0.01$, Figure 3D).

### *3.5. Prediction of bta-miRNA29d-3p Target Genes*

As shown in Figure 4, the target gene associated with fatty acid metabolism was GPAM, and bta-miRNA29d-3p was more conserved in human and chimpanzee sequences (Figure 4A). When the GPAM 3′-UTR was wild-type, bta-miRNA29d-3p overexpression significantly decreased the activity of the dual-luciferase reporter gene ($p < 0.01$, Figure 4B), whereas interference with the bta-miRNA29d-3p gene significantly increased the activity of the dual-luciferase reporter ($p < 0.01$, Figure 4C). When the GPAM 3′-UTR was mutant, bta-miRNA29d-3p showed an absence of effect towards the activity of dual-luciferase reporter gene expression. Therefore, bta-miRNA29d-3p acts directly on the cow GPAM 3′-UTR target site.

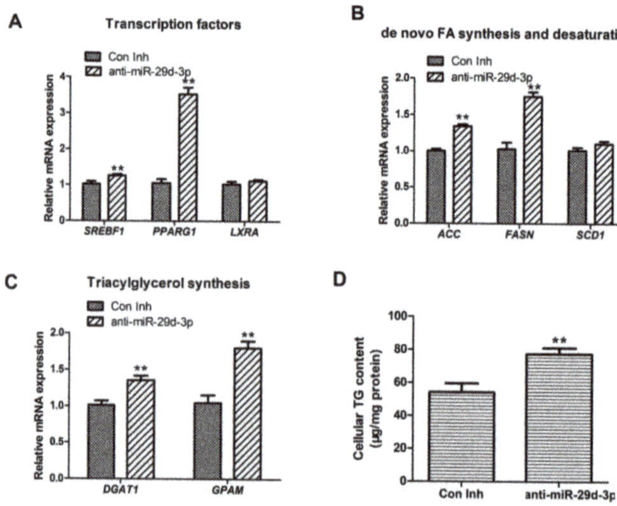


**Figure 3.** Effect of interference with bta-miRNA29d-3p with respect to lipid metabolism-related genes in BMECs. Panel (**A**): Effect of interfering with bta-miRNA29d-3p on the expression of transcription factors SREBF1, PPARG1, and LXRA. Panel (**B**): Effect of interfering with bta-miRNA29d-3p with reference to the expression of fatty acid ab initio synthesis and desaturation-related genes (ACACA, FASN, and SCD1). Panel (**C**): Effect of interfering with bta-miRNA29d-3p on the analysis of TAG synthesis and lipid droplet formation-related genes (DGAT1 and GPAM) expression. Panel (**D**): Effect of interfering with bta-miRNA29d-3p on intracellular TAG content. Results are the mean ± SEM of three separate experiments. ** $p < 0.01$ versus control (Con Inh).

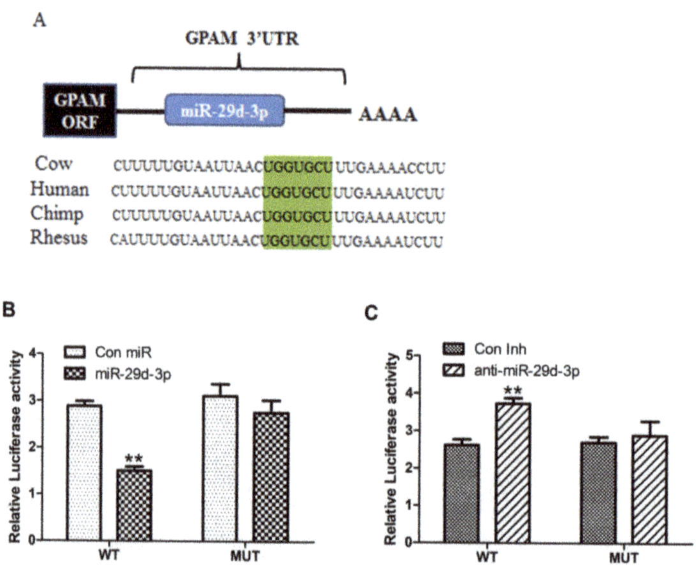


**Figure 4.** Bta-miRNA29d-3p target gene prediction and dual luciferase assay. Panel (**A**): bta-miRNA29d-3p target gene prediction and conservativeness analysis. Panel (**B**): Effect of bta-miRNA29d-3p overexpression on dual luciferase activity. Panel (**C**): Effect of interfering with bta-miRNA29d-3p on dual-luciferase activity. Results are the mean ± SEM of three separate experiments. ** $p < 0.01$ versus control (Con miR, Con Inh).

## 4. Discussion

The miR-29 family plays a core role in the etiology and pathogenesis of osteoarthritis (OA), osteoporosis, cardiac and renal diseases, and immune disorders [17]. Furthermore, miR-29 participates in the modulation of multiple cellular functions and in lipid metabolism [7,18,19]. In this study, combined with previous research of miR-29's effects on lactation in ruminants, the hypothesis was that bta-miRNA29d-3p has an effect on lipid metabolism in BMECs. In vitro experiments confirmed that pharmacological inhibition of miR-29 significantly down-regulated hepatic cholesterol and adipose synthesis de novo [9]. Moreover, miR-29 was found to be an important regulator of lipid oxidation by overexpression or deletion in human or primary human skeletal muscle cells [5]. Consequently, we investigated the regulatory mechanism of bta-miRNA29d-3p on milk lipids to provide a basis for further improvement of lactation performance of cows and their progeny.

Transcription factors (TF) are an important class of proteins involved in the regulation of gene transcription. They are key regulators of various signaling pathways. In this study, overexpression of bta-miRNA29d-3p dramatically down-regulated the expression of *SREBF1*, a member of a family of TFs that binds to the ER and acts centrally involved in fatty acid, phospholipid, and cholesterol synthesis [20]. Overexpression of nSREBP-1a or -1c was suggested to significantly upregulate the expression of its target genes *ACC*, *FASN*, and stearoyl coenzyme A desaturase 1 (*SCD1*) and remarkably enhanced the pri-miR-29 and mature miR-29 expression in glioblastoma (GBM) cells [21]. Overall, miR-29 suppressed SREBP-1 and the lipid synthesis pathway in GBM cells as a negatively regulated factor. The results of this research are in agreement with the findings reported in the literature above. Accordingly, bta-miRNA29d-3p may control lipid metabolism-related gene expression through *SREBF1*, which consequently impacts fatty acid synthesis.

Interestingly, we found extremely significant upregulation of *PPARG1* expression by interfering with bta-miRNA29d-3p. However, it has been demonstrated that inhibition of miR-29s can cause hypermethylation of overall DNA and augments the levels of methylation in the promoters of lactation-associated genes, including PPARG [22]. The peroxisome proliferator-activated receptor (PPAR) family is an important member of the nuclear hormone receptor superfamily, and their physiological functions are mainly associated with fatty acid metabolism, glucose metabolism, and cell proliferation and differentiation [16]. There is evidence that *PPARA* controls not only fatty acid oxidation-related genes, but also other genes concerned with its synthesis, including *ACACA*, *FASN*, and *SCD1* [16]. While *PPAR*$\gamma$ is expressed at a high level in adipose tissue, it is a key coordinator of the transcriptional cascade response for adipocyte differentiation [23]. There are two homotypes of *PPARG*, *PPARG1* and *PPARG2*, which are splice variants of the *PPARG* gene via alternative promoters [24]. The results of the in vitro culture of mammary epithelial cell lines showed that PPARG-specific agonist ROSI treatment of mammary epithelial cells was able to upregulate fatty acid metabolism-related genes, such as *ACACA*, *FASN*, *DGAT1*, and *SREBF1*, and clarified that *PPARG* directly regulates the expression of some genes in mammary tissue and affects the fatty acid metabolic network [25–27]. To summarize the above reports, we presume that bta-miRNA29d-3p may have a vital effect in the control of lipid metabolism in BMECs by affecting the expression of *PPARG*.

Changes in *DGAT1* expression were also found in the current study. *DGAT1* is found to be expressed at a high level in adipose tissue, the liver, and the small intestine, all of which are more active in TAG synthesis [28]. For dairy cows, the non-conservative substitution of alanine for lysine in the *DGAT1* gene could have a significant effect on milk composition and yield. Evidence showed that mice with *DGAT1* gene deletion have reduced absorption of triacylglycerols [29]. Through interference with bta-miRNA29d-3p expression, *DGAT1* mRNA expression was significantly increased, along with an increase in intracellular TAG content. A combination of the above series of studies on lipid metabolism by bta-miRNA29d-3p showed that the synergistic effect of bta-miRNA29d-3p and *DGAT1* affects lipid synthesis in BMECs.

Screening and identifying miRNAs and their target genes are the key steps in studying the functional mechanism of miRNAs. The bioinformatics approach could provide theoretical guidance for the identification of miRNA target genes, and it has an essential role in studying the functional mechanisms of miRNAs. In this study, *GPAM* was identified as a target gene related to lipid metabolism by online study of the bta-miRNA29d-3p gene. *GPAM* is the limiting enzyme in the first step of TAG synthesis, and intracellular TAG levels are regulated by *GPAM* enzyme content. Knockdown of *GPAM* significantly reduced the expression of genes associated with TAG synthesis and lipid metabolism in bovine embryonic fibroblasts [30]. Furthermore, *GPAM* plays a dynamic role in TAG synthesis. For example, a 10-fold increase in *GPAM* activity in 3T3-L1 adipocyte differentiation and a 5-fold increase in *GPAM* activity during TAG synthesis were observed in neonatal liver hepatocytes [31]. The present findings on TAG synthesis have mainly focused on both *PPAR* and *SREBP* regarding the transcriptional regulation of the gene, but there are also reports confirming that *GPAM* may have a greater regulatory role on TAG synthesis [31,32]. To summarize, GPAM, as a target gene of bta-miRNA29d-3p, performs a key role in TAG regulation in cells, and its mutation can be used as an effective marker for selective breeding of dairy cows.

Taken together, the bta-miRNA29d-3p gene has a significant function in the control of lipid accumulation in dairy cows, and its regulatory mechanism is linked to the expression of lipid synthesis-associated genes in BMECs. However, the specific regulatory mechanism of the bta-miRNA29d-3p gene on lipid synthesis in mammary epithelial cells of dairy cows needs to be studied in depth. Additional studies on relevant non-coding RNAs in the mammary lipid metabolism pathway in dairy cows could help to better utilize functional miRNAs to regulate the fatty acid composition of milk in the future, which has important application value for dairy development.

## 5. Conclusions

In this study, overexpression of bta-miRNA29d-3p down-regulated the expression of *DGAT1* and *GPAM*, which are related to TAG synthesis ($p < 0.01$), and that of *FASN*, *ELOVL4*, *ACACA*, and *SREBF1*, which are related to FA synthesis and prolongation, significantly down-regulated intracellular TAG levels ($p < 0.05$). Interfering with the expression of the bta-miRNA29d-3p gene showed the opposite result. The downstream target gene of bta-miRNA29d-3p was predicted to be *GPAM*. The results provide a theoretical basis for the regulatory mechanism of the bta-miRNA29d-3p gene in the lipid synthesis of BMECs.

**Author Contributions:** Conceptualization, D.Y.; funding acquisition, D.Y.; supervision, Y.M.; formal analysis, M.Z. and L.C.; validation, S.Z.; writing—original draft, X.Z.; writing—review and editing, J.L. All authors have read and agreed to the published version of the manuscript.

**Funding:** This research was jointly supported by the Innovative Research and Experimental Projects for Young Researchers of Tianjin Academy of Agricultural Sciences (No. 2021010 to D.Y.); the National Natural Science Foundation of China (No.31702095 to D.Y.) and Natural Science Foundation of Tianjin (No.18JCYBJC30200 to D.Y.); the Breeding innovation research project (2022ZYCX010 to Y.M.); the Tianjin Science and Technology project (22ZXZYSN00020 to Y.M.); and the Tianjin "131" Innovative Talents Team (No.20180338 to Y.M.).

**Institutional Review Board Statement:** All animal collection and protocols were authorized by the Animal Care and Use Committee of the Tianjin Institute of Animal Husbandry and Veterinary Medicine (2022015). All procedures were performed with minimal pain.

**Data Availability Statement:** The data are available online at https://doi.org/10.6084/m9.figshare.21393204.v1, accessed on 25 October 2022.

**Acknowledgments:** We thank Yi Ma for supporting the experiments, Dawei Yao for technical guidance, Zhi Chen for providing bovine mammary epithelial cells, and Jun Luo for providing the plasmids.

**Conflicts of Interest:** The authors declare no conflict of interest. The funders had no role in the design of the study; in the collection, analyses, or interpretation of data; in the writing of the manuscript; or in the decision to publish the results.

## References

1. Curadi, M.C.; Leotta, R.; Contarini, G.; Orlandi, M. Milk fatty acids from different horse breeds compared with cow, goat and human milk. *Maced. J. Anim. Sci.* **2012**, *2*, 79–82. [CrossRef]
2. Hanuš, O.; Samková, E.; Křížová, L.; Hasoňová, L.; Kala, R. Role of Fatty Acids in Milk Fat and the Influence of Selected Factors on Their Variability—A Review. *Molecules* **2018**, *23*, 1636. [CrossRef] [PubMed]
3. Hennighausen, L.; Robinson, G.W. Information networks in the mammary gland. *Nat. Rev. Mol. Cell Biol.* **2005**, *6*, 715–725. [CrossRef] [PubMed]
4. Sassi, Y.; Avramopoulos, P.; Ramanujam, D.P.; Grüter, L.; Werfel, S.; Giosele, S.; Brunner, A.-D.; Esfandyari, D.; Papadopoulou, A.-S.; De Strooper, B.; et al. Cardiac myocyte miR-29 promotes pathological remodeling of the heart by activating Wnt signaling. *Nat. Commun.* **2017**, *8*, 1614. [CrossRef] [PubMed]
5. Massart, J.; Sjögren, R.J.O.; Lundell, L.S.; Mudry, J.M.; Franck, N.; O'Gorman, D.J.; Egan, B.; Zierath, J.R.; Krook, A. Altered miR-29 Expression in Type 2 Diabetes Influences Glucose and Lipid Metabolism in Skeletal Muscle. *Diabetes* **2017**, *66*, 1807–1818. [CrossRef]
6. Zhang, S.; Zhang, Q.; Zhang, M.; Yang, J.C. Research Progress on Regulation Effect of microRNA on Mammary Gland Development and Lactation. *China Anim. Husb. Vet. Med.* **2015**, *42*, 663–667. [CrossRef]
7. Wang, P.; Liu, S.; Zhu, C.; Duan, Q.; Jiang, Y.; Gao, K.; Bu, Q.; Cao, B.; An, X. MiR-29 regulates the function of goat granulosa cell by targeting PTX3 via the PI3K/AKT/mTOR and Erk1/2 signaling pathways. *J. Steroid Biochem. Mol. Biol.* **2020**, *202*, 105722. [CrossRef]
8. He, A.; Zhu, L.; Gupta, N.; Chang, Y.; Fang, F. Overexpression of Micro Ribonucleic Acid 29. highly Up-Regulated in Diabetic Rats, Leads to Insulin Resistance in 3T3-L1 Adipocytes. *Mol. Endocrinol.* **2007**, *21*, 2785–2794. [CrossRef]
9. Kurtz, C.L.; Fannin, E.E.; Toth, C.L.; Pearson, D.S.; Vickers, K.C.; Sethupathy, P. Inhibition of miR-29 has a significant lipid-lowering benefit through suppression of lipogenic programs in liver. *Sci. Rep.* **2015**, *5*, 12911. [CrossRef]
10. Chen, Z.; Cao, X.; Lu, Q.; Zhou, J.; Wang, Y.; Wu, Y.; Mao, Y.; Xu, H.; Yang, Z. circ01592 regulates unsaturated fatty acid metabolism through adsorbing miR-218 in bovine mammary epithelial cells. *Food Funct.* **2021**, *12*, 12047–12058. [CrossRef]
11. Bionaz, M.; Loor, J.J. Identification of reference genes for quantitative real-time PCR in the bovine mammary gland during the lactation cycle. *Physiol. Genom.* **2007**, *29*, 312–319. [CrossRef]
12. Loor, J.J.; Dann, H.M.; Everts, R.E.; Oliveira, R.; Green, C.A.; Guretzky, N.A.J.; Rodriguez-Zas, S.L.; Lewin, H.A.; Drackley, J.K. Temporal gene expression profiling of liver from periparturient dairy cows reveals complex adaptive mechanisms in hepatic function. *Physiol. Genom.* **2005**, *23*, 217–226. [CrossRef]
13. Zhou, F.; Teng, X.; Wang, P.; Zhang, Y.; Miao, Y. Isolation, identification, expression and subcellular localization of PPARG gene in buffalo mammary gland. *Gene* **2020**, *759*, 144981. [CrossRef]
14. Bionaz, M.; Loor, J.J. Gene networks driving bovine milk fat synthesis during the lactation cycle. *BMC Genom.* **2008**, *9*, 366. [CrossRef]
15. Harvatine, K.J.; Boisclair, Y.R.; Bauman, D.E. Liver x receptors stimulate lipogenesis in bovine mammary epithelial cell culture but do not appear to be involved in diet-induced milk fat depression in cows. *Physiol. Rep.* **2014**, *2*, e00266. [CrossRef]
16. Christofides, A.; Konstantinidou, E.; Jani, C.; Boussiotis, V.A. The role of peroxisome proliferator-activated receptors (PPAR) in immune responses. *Metabolism* **2021**, *114*, 154338. [CrossRef]
17. Horita, M.; Farquharson, C.; Stephen, L.A. The role of miR-29 family in disease. *J. Cell. Biochem.* **2021**, *122*, 696–715. [CrossRef]
18. Kurtz, C.L.; Peck, B.C.; Fannin, E.E.; Beysen, C.; Miao, J.; Landstreet, S.R.; Ding, S.; Turaga, V.; Lund, P.K.; Turner, S.; et al. MicroRNA-29 Fine-tunes the Expression of Key FOXA2-Activated Lipid Metabolism Genes and Is Dysregulated in Animal Models of Insulin Resistance and Diabetes. *Diabetes* **2014**, *63*, 3141–3148. [CrossRef]
19. Xu, J.; Xu, P.; Wu, M.; Chen, H.; Xu, J.; Wu, M.; Li, M.; Qian, F. Bioinformatics analysis of hepatitis C virus genotype 2a-induced human hepatocellular carcinoma in Huh7 cells. *OncoTargets Ther.* **2016**, *9*, 191–202. [CrossRef]
20. Jeon, T.-I.; Osborne, T.F. SREBPs: Metabolic integrators in physiology and metabolism. *Trends Endocrinol. Metab.* **2012**, *23*, 65–72. [CrossRef]
21. Ru, P.; Hu, P.; Geng, F.; Mo, X.; Cheng, C.; Yoo, J.Y.; Cheng, X.; Wu, X.; Guo, J.Y.; Nakano, I.; et al. Feedback Loop Regulation of SCAP/SREBP-1 by miR-29 Modulates EGFR Signaling-Driven Glioblastoma Growth. *Cell Rep.* **2016**, *16*, 1527–1535. [CrossRef] [PubMed]
22. Bian, Y.; Lei, Y.; Wang, C.; Wang, J.; Wang, L.; Liu, L.; Liu, L.; Gao, X.; Li, Q. Epigenetic Regulation of miR-29s Affects the Lactation Activity of Dairy Cow Mammary Epithelial Cells. *J. Cell. Physiol.* **2015**, *230*, 2152–2163. [CrossRef] [PubMed]
23. Semple, R.K.; Chatterjee, V.K.K.; O'Rahilly, S. PPAR and human metabolic disease. *J. Clin. Investig.* **2006**, *116*, 581–589. [CrossRef] [PubMed]
24. Kidani, Y.; Bensinger, S.J. Liver X receptor and peroxisome proliferator-activated receptor as integrators of lipid homeostasis and immunity. *Immunol. Rev.* **2012**, *249*, 72–83. [CrossRef]

25. Kadegowda, A.K.; Bionaz, M.; Piperova, L.S.; Erdman, R.; Loor, J. Peroxisome proliferator-activated receptor-γ activation and long-chain fatty acids alter lipogenic gene networks in bovine mammary epithelial cells to various extents. *J. Dairy Sci.* **2009**, *92*, 4276–4289. [CrossRef]
26. Moyes, K.M.; Drackley, J.K.; Morin, D.; Bionaz, M.; Rodriguez-Zas, S.L.; Everts, R.E.; Lewin, H.A.; Loor, J.J. Gene network and pathway analysis of bovine mammary tissue challenged with Streptococcus uberis reveals induction of cell proliferation and inhibition of PPARγ signaling as potential mechanism for the negative relationships between immune response and lipid metabolism. *BMC Genom.* **2009**, *10*, 542. [CrossRef]
27. Thering, B.J.; Graugnard, D.E.; Piantoni, P.; Loor, J. Adipose tissue lipogenic gene networks due to lipid feeding and milk fat depression in lactating cows. *J. Dairy Sci.* **2009**, *92*, 4290–4300. [CrossRef]
28. Swanton, E.M.; Saggerson, E.D. Glycerolipid metabolizing enzymes in rat ventricle and in cardiac myocytes. *Biochim. Biophys. Acta (BBA) Lipids Lipid Metab.* **1997**, *1346*, 93–102. [CrossRef]
29. Buhman, K.K.; Smith, S.J.; Stone, S.J.; Repa, J.J.; Wong, J.S.; Knapp, F.; Burri, B.J.; Hamilton, R.L.; Abumrad, N.A.; Farese, R.V., Jr. DGAT1 Is Not Essential for Intestinal Triacylglycerol Absorption or Chylomicron Synthesis. *J. Biol. Chem.* **2002**, *277*, 25474–25479. [CrossRef]
30. Yu, H.; Zhao, Z.; Yu, X.; Li, J.; Lu, C.; Yang, R. Bovine lipid metabolism related gene GPAM: Molecular characterization, function identification, and association analysis with fat deposition traits. *Gene* **2017**, *609*, 9–18. [CrossRef]
31. Yu, H.; Zhao, Y.; Iqbal, A.; Xia, L.; Bai, Z.; Sun, H.; Fang, X.; Yang, R.; Zhao, Z. Effects of polymorphism of the GPAM gene on milk quality traits and its relation to triglyceride metabolism in bovine mammary epithelial cells of dairy cattle. *Arch. Anim. Breed.* **2021**, *64*, 35–44. [CrossRef]
32. Yu, H.; Iqbal, A.; Fang, X.; Jiang, P.; Zhao, Z. Transcriptome analysis of CRISPR/Cas9-mediated GPAM−/− in bovine mammary epithelial cell-line unravelled the effects of GPAM gene on lipid metabolism. *Gene* **2022**, *834*, 146574. [CrossRef]

*agriculture*

Review

# Diagnostic Use of Serum Amyloid A in Dairy Cattle

**Michał Trela, Dominika Domańska and Olga Witkowska-Piłaszewicz ***

Department of Large Animals Diseases and Clinic, Institute of Veterinary Medicine, Warsaw University of Life Sciences, Nowoursynowska 166, 02-787 Warsaw, Poland; michal_trela@sggw.edu.pl (M.T.); dominika_domanska@sggw.edu.pl (D.D.)
* Correspondence: olga_witkowska_pilaszewicz@sggw.edu.pl

**Abstract:** Checking the health status of the individual animal and/or herd in a farm is one of the most important factors in diary production. Because of its high economical value, the early detection of ongoing disease is of high interest in breeders and veterinary clinical practitioners. The acute phase response (APR) is a non-specific systemic reaction for any type of tissue injury leading to disturbances in homeostasis. During this reaction, the production of acute-phase proteins (APPs) is changed. APPs may act as biomarkers of inflammation, allowing researchers to study the progression of the inflammatory response. One of the major APPs in cows is serum amyloid A (SAA). Due to its short half-life and the fast dynamic of changes in blood concentration, SAA seems to be a reliable indicator of several pathologies and treatment effectiveness. Because the blood-based and milk protein biomarkers of the herd's health status are of great interest, this article reviews the current information about changes in SAA concentrations in the blood and milk of cattle in health and disease. It summarizes its clinical usefulness as a health status indicator in diary production.

**Keywords:** cow; milk production; heard health status; APPs; SAA

**Citation:** Trela, M.; Domańska, D.; Witkowska-Piłaszewicz, O. Diagnostic Use of Serum Amyloid A in Dairy Cattle. *Agriculture* **2022**, *12*, 459. https://doi.org/10.3390/agriculture12040459

Academic Editors: Cong Li and Zhi Chen

Received: 16 February 2022
Accepted: 22 March 2022
Published: 25 March 2022



## 1. Introduction

The unspecific response to a variety of stimuli including trauma, infection, surgery, neoplasia, presence of a chronic disease, or ongoing inflammatory processes is called the acute phase response (APR) [1]. The main task of this early defense system is to eliminate the agent(s) which disturb the animal's homeostasis. The integral part of the APR is acute-phase protein (APP) synthesis. Depending on the fold and the dynamic of the changes, the APPs are classified as major (10- to 100-fold increase within the first 48 h with a rapid decline), moderate (2- to 10-fold increase with prolonged in duration), or minor (slight increase with prolonged duration) [1]. The pattern of changes in the concentrations of APPs in blood is species specific.

In cattle, the one of the major APPs is serum amyloid A (SAA), the concentration of which starts to increase within the first 4 h with a pick after 24–48 h [1]. In the dairy cow, seven isoforms have been recognized in the blood, but SAA1 and SAA2, which are mainly expressed in the liver, are the major ones that are overproduced during the APR [1]. SAA3 is expressed extrahepatically by adipose tissue, the mammary gland, intestinal epithelial cells, the lungs, ovarian granulosa, skeletal muscles, synovial membrane, the thymus, the thyroid gland, and the uterus but also by macrophages; thus, it is present in the blood but at a very low level in healthy animals [2]. SAA belongs to a family of apolipoproteins which have been very well conserved throughout evolution and have a wide range of functions (Figure 1).

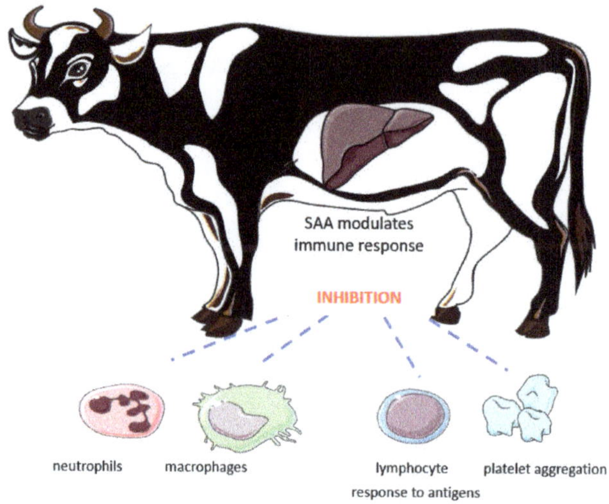


**Figure 1.** Systemic inflammation stimulates SAA production in cattle.

In healthy dairy cattle, the reference range of SAA concentration is variable (0–70 mg/L) [3]. It was suggested to use as a cut point the value of >600 mg/L for detecting the presence of the inflammatory process [3]. However, this cut-off value may be set up too high taking into account other reports (ex. 162.19 or 447.8 mg/L during mastitis) [4]. It is known that SAA is over expressed during several disorders; thus, changes in its concentration are considered as blood-based protein biomarkers in adult cows (Table 1) and calves. It is commonly used in clinical practice [1]. However, there is no APP which is specific for a particular disease. The APP concentration is elevated in animals with many different diseases, having very poor diagnostic specificity in detecting the cause. Thus, they need to be interpreted in the full clinical context. Despite the challenges in the determination of APPs, their usefulness has been proven, especially in the early detection of sub-clinical disorders or alterations of the health status/individuals.

**Table 1.** Examples of diseases/environmental conditions upregulating SAA blood concentration in adult cattle.

| Diseases/Environmental Conditions | SAA Reported Values/Fold Increase [mg/L] (Reference Values: 0–70 Blood, Milk 0.6–50) [3,5] |
|---|---|
| high grain concentration dry matter (30–45%) | 19–46 [6] |
| heat stress | 4 fold [7] |
| long-term transportation (4–6 h) | 1–2 fold [7] |
| ketosis | 89.2 [8] |
| subacute ruminal acidosis | 313.9 (at least one fold) [9] |
| fatty liver | 80 [10] |
| abomasal displacement | >126 [3] |
| traumatic reticuloperitonitis | 159 [11] |
| paratuberculosis | one fold [12] |
| brucellosis | 123.75 [13] |
| babesiosis | mild 169.8; severe 183.4 [14] |
| *Anaplasma marginale* | 167–180 [15] |
| one week postpartum | 94 [16] |
| endometritis | low 20.25; mild 28.17; severe 34.62 [17] |
| pyometra | 62.5 [18] |
| acute laminitis | 281.26–590 [19] |
| mastitis | 447.9 (milk 1315.9) [4] |

## 2. SAA Concentrations in Healthy Cattle—Impact of Nutrition and Stress

Intensive dairy cow management supporting high milk yields requires the inclusion of large amounts of concentrates in the diet. Thus, significant economic losses in the dairy industry may be caused by unsuitable feeding. The maintenance of proper levels of fiber, dry matter intake, and the energy density of the diet are the most important factors in formulating diets to prevent problems connected with feeding. However, slight incompatibility in cows' production levels and improper feeding levels may go unnoticed. Cows with concentrations of nonesterified fatty acids (NEFA) above 0.35 mmol/L have higher mean serum concentrations of SAA (90.3 mg/L) than cows with NEFA concentrations less than (0.35 mmol/L to 27.0 mg/L) [20], but it is still within reference value. SAA concentration was found to be enhanced in cows fed with high grain concentration [6,21]. The higher dry matter (30–45%) of barley grain influences higher SAA concentrations (19–46 mg/L) in comparison to lower dry matter contents (0–15%; SAA 2–12 mg/L) in cows blood [6]. In addition, the concentrate level and neutral detergent fiber (NDF) content in the diet are connected with the development of inflammatory conditions characterized by increased SAA blood levels [22]. Probably, it is caused by an increased endotoxin level in the rumen. The results of Zebeli et al.'s study indicate that the feeding of more than 44% concentrate or less than 39.2% NDF in the diet linearly increases the SAA blood concentration. The one percent increase in NDF content is connected with decreased concentration of plasma SAA of 9.2 mg/L. Worth noting is that the level of the systemic inflammation resulting from improper diets vary in different cattle breeds (ex. Holstein vs. Jersey) [21].

Fortunately, some feeding supplements may downregulate the systemic inflammatory reaction. Saccharomyces cerevisiae fermentation product (SCFP) decreases the SAA blood levels by 33% after 5 days after beginning the supplementation in cows with heat stress [23]. Heat stress, which has a negative impact on animal welfare and productivity, stimulates the APR characterized by a four-fold increase in SAA blood concentration.

Other stressors in addition to heat stress are weaning and transport. It was documented that the concentration of SAA and other APPs, such as haptoglobin (Hp), increase significantly in response to long-term transportation (4–6 h) [7]. In addition, in this study, SAA response occurred in every cow, whereas only five of the eight animals showed Hp responses. Thus, improper management may strongly influence SAA concentration.

## 3. SAA as a Diagnostic Tool in Metabolic Disorders in Adult Dairy Cows

In dairy cows, one of the most common metabolic diseases is ketosis, which often occurs during the parturition transition period. Thus, the early identification of this disease at a subclinical level is important in dairy cow health prevention. It was documented that concentration of SAA is increased in blood (41.2 vs. 89.2 mg/L) but reduced in urine in cattle during ketosis [8]. It may be connected with changed high-density lipoproteins metabolism. Another common metabolic disease (in up to 19% of early lactation dairy cows and 26% of mid-lactation) is subacute ruminal acidosis (SARA). It is caused by inappropriate feeding, leading to rumenitis caused by low rumen pH (below 5.6 lasting at least 3 h/day). It is documented that SAA blood concentration is increased at least one-fold in cows with SARA (168 vs. 313.9 mg/L) [9]. The concentration of SAA in the bloodstream of cows affected by fatty liver reached a peak value of 80 mg/L and is higher than average concentrations in healthy cows [10].

## 4. SAA as a Diagnostic Tool in Gastrointestinal Disorders in Adult Dairy Cows

Common gastrointestinal disorders in high-producing dairy cows is abomasal displacement (AD). It is connected with extensive dietary, metabolic, endocrine, and immunological changes in dairy cows during the period occurring two weeks prepartum through two to four weeks. In cows below 2, lactation SAA is 126, and above 2, lactation is at 132 mg/L during left AD (LAD). In comparison, in healthy animals it was 59 mg/L and 60 mg/L, respectively. In another study, the values of the SAA varied: In LAD it was 55 mg/L and right AD-90 mg/L in comparison to healthy ones (12 mg/L) [3]. It was postulated that

APR during AD may be connected with a negative energy balance in postparturient dairy cows with LAD, which may lead to the release of proinflammatory cytokines.

Traumatic reticuloperitonitis is a relatively common disease caused by a foreign body ingested by the cow and migrated to the reticulum. Also during this disorder, the SAA concentration increases to 159 mg/L and has 100% sensitivity and 86.1% specificity. It was suggested that the optimal cut-off point is set at 68 mg/L for SAA and at 0.74 mg/L for Hp in this condition [11].

In dairy cattle production, a popular gastrointestinal infectious disease is paratuberculosis (PTB), which is a chronic disorder caused by *Mycobacterium avium* subspecies *paratuberculosis* (Map). It also effects the inflammatory response. In a large scale study in which 190 cows were examined, it was documented that in animals with different types of lesions, SAA increases regardless of the type (focal 13.10; multifocal 13.56; diffuse paucibacillary 19.82; diffuse multibacillary 6.44 mg/L) in comparison to healthy controls (6.91 mg/L) [12].

## 5. SAA as a Diagnostic Tool in Respiratory Tract Infections

The relative usefulness of changes in the leukocyte number and some APPs such as fibrinogen and Hp in predicting infectious diseases during the incubation period is not always satisfactory, especially in cattle. In this case, SAA seems to be more sensitive. In viral pneumonias, SAA responded more rapidly to infection than Hp [1]. Also during brucellosis, the SAA blood concentration is increased in comparison to healthy controls (32.92 vs. 123.75 mg/L) [13], whereas Hp concentration is not changed during this disorder. However, most studies connected with SAA changes in respiratory tract infections are performed in calves; this part is discussed in paragraph 10.

## 6. SAA as a Diagnostic Tool in Parasitic Infections

In addition, SAA may be a good indicator of the inflammatory process in cattle naturally infected with several species of parasites. It was documented that in bovine babesiosis (*Babesia bigemina*), SAA testing had 100% specificity and 100% sensitivity, with values for mild (169.8) and severe (183.4 mg/L) parasitemia (healthy animals 4.79 mg/L) [14]. *Anaplasma marginale* infection also causes the APR connected with a high concentration of several APPs (Hp, SAA, ceruloplasmin, and fibrinogen); however, SAA is suggested to be the most sensitive diagnostic factor in this case (167–180 vs. healthy 5.17 mg/L) [15]. In another study, the infected cows showed a significant, but as high as the previous one, increase in SAA (134 mg/L) [24]. During cystic echinocccosis in the liver and/or lungs caused by the larval stage of *Echinococcus granulosus* and connected with infection with microbial agents such as *Staphylococcus aureus, Escherichia coli, Corynebacterium* spp. and *Campylobacter* spp., *Candida* spp., *Streptococcus* spp., *Mannheimia haemolytica, Corynebacterium* spp., and *Micrococcus* spp., the concentration of SAA was increased twofold (7.51 $\pm$ 0.41 mg/L) in the infected cattle blood [25]. However, this is within reference value.

## 7. SAA in Reproduction

SAA serum concentration in healthy cows is not related with the physiological estrus phase. High values of SAA have been found in cows with ovarian cysts such as luteal (>35 mg/L) and follicular cysts (>50 mg/L) (healthy cows < 10 mg/L) [1,26]. Alsemgeest et al. (1993) reported that, in cows, 48 h before delivery, the mean SAA concentration was low. According to results obtained at 24, 48, 72, and 96 h after delivery, the mean SAA concentrations were significantly increased [27]; the highest concentration of SAA was reached 48 h after calving. The results of Ametaj et al. (2005) are in contrast with these findings, as immediately after parturition, increases in two main bovine APPs, hepcidin and SAA, were confirmed in cows [10]. Thus, higher concentrations of SAA in the blood and cervico-vaginal mucus of healthy animals in late postpartum may be important to modulate physiological inflammation and prevent tissue damage caused by severe inflammation [28].

In healthy fresh lactating cows (within one week postpartum) from 10 farms, the serum SAA concentration was 94 mg/L [16]. Differences among farms have been observed. These may be caused by different sensitivity in detecting the farm-specific inflammatory reactions induced be environmental factors such as regrouping and farm-specific diseases, such as mastitis and metritis [29]. Cows with metritis between the first and third week postpartum had greater (11.44 mg/L) concentrations of SAA compared to healthy cows in this time (10.40 mg/L). SAA levels in metritic cows at 8 (17.95 mg/L) and 4 (60.03 mg/L) weeks prior to parturition were also greater than in the control groups during the same time (84.47 and 34.61 mg/L, respectively). The study of Zhang et al. showed that both the protein and mRNA of SAA were increased in endometritis or in LPS-stimulated cells, and the increases were positively correlated with the severity of endometritis in vivo or LPS stimulation strength in vitro [30]. In another study, the correlation between the level of endometritis and SAA plasma concentration was observed. The average levels of SAA were noted: in healthy cows—14.24 mg/L; cows with low level endometritis—20.25 mg/L; mild endometritis—28.17 mg/L; and severe endometritis—34.62 mg/L [17]. It was also confirmed that blood and uterine concentrations of SAAs might be used as a diagnostic tool for subclinical endometritis during postpartum [18]. Thus, SAA might be used as useful parameter to evaluate the condition of the reproductive system and its problems. In addition, data suggest that serum SAA can be used to screen cows for the potential occurrence of uterine infections prior to calving [31].

The level of SAA was significantly higher in the serum of cows with pyometra ($62.5 \pm 9.01$ mg/L) compared to healthy animals ($35.4 \pm 6.90$ mg/L). In addition, in healthy cows, the level of SAA was significantly higher in the serum than in uterine washings ($27.5 \pm 6.62$ mg/L) [18]. Thus, the SAA uterine/plasma ratio may be a valuable tool for pyometra detection.

## 8. SAA in Limb Diseases

Limb diseases in dairy herds are a serious clinical problem that directly affects the milk yield of cows. A lame cow loses on average 350 kg of milk (from 160 to 550 kg) during its lactation. Lameness in dairy cows means that these animals have a problem with moving, take longer to return from the milking parlor, and experience pain and discomfort. The most common causes for lameness in dairy cattle are inflammatory claw diseases such as sole ulcer, papilomatous digital dermatitis, interdigital necrobacilosis, septic pododermatitis, and white line disease [32]. In general, the clinical form of limb diseases can be easily diagnosed through a clinical examination, while the diagnosis of subclinical limb diseases remains a challenge. However, very often, the only symptom is lower milk production, which can persist for as much as four months before lameness is diagnosed and treatment applied, and may even last up to five months after treatment. That is why it is so important to look for a more sensitive diagnostic marker of lameness.

In previous studies, SAA seems to be a sensitive marker of lameness. Acute laminitis, sole ulcer, and digital dermatitis are responsible for the increase in SAA concentration of 442.76 mg/L, 590.00 mg/L, and 281.26 mg/L, respectively [19]. In contrast, diseases with an absence of lameness, such as heel erosions and white line disease in the primary stage (without secondary infections), did not induce a significant increase in APPs. However, in another study, SAA performed in cows 4 and 8 weeks before parturition the level of SAA was 2–3 times higher in lame animals [33]. Similar findings were also confirmed in another study in which SAA concentration increased approximately 3 times in lame cows in comparison with healthy controls (22.19 vs. 8.89 mg/L); however, Hp was 20 times higher (217.1 vs. 1.17 mg/L) [34]. Thus, SAA may be a marker for lameness connected mostly with inflammatory process.

In addition, SAA is useful for treatment monitoring [35]. In this study there were three groups: the first one in which systematic decreases in APPs levels in subsequent blood collections were recorded; the second in which an increase in the level of one or more APPs in the second or third blood collection was recorded; and the control group. In both study groups, the SAA concentration decreased during treatment (220.8 to 130.7 and 130.1 to 99.2 mg/L, respectively). Thus, SAA seems to be a promising marker in lameness detection and recovery in dairy herds.

## 9. SAA in Mastitis

Bovine mastitis is a major cause of economic loss in the dairy industry. The consequence of mastitis is a decrease in milk yield, quality reduction, and an influence on the composition of the milk and the processing properties of milk [36]. Methods that will enable fast and precise diagnosis, especially of subclinical mastitis, are still being sought. Among such methods, the determination of APPs deserves attention. SAA is produced in the liver in response to acute phase stimulus but also in other extra-hepatic tissues, including the mammary gland. Among the main three isoforms, $SAA_3$ is produced predominantly in milk and has been called mammary-associated amyloid A (MAA-S3) [37].

MAA-S3 has at the N-end of its sequence four amino-acid motif amino acids (TFLK), regardless of species (horse, cow, sheep), that are absent in the isoform of the liver synthesis protein [37]. MAA-S3 contains 83% of the same amino acid sequence as SAA. The highest concentration of MAA-S3 (267 mg/L) occurs in the colostrum of healthy cows on the first day after parturition. MAA-S3 stimulates the expression of mucin in enterocytes, indicating a possible protective action on the digestive tract of newborns, especially in the case of necrotizing enterocolitis [38]. During the days following birth, the concentration gradually drops to a value of 2.63 mg/L. In a more recent study, the MAA-S3 concentration in milk in healthy cows with a normal somatic cell count (SCC) was <0.6–50 mg/L [5].

The secretion of MAA-S3 is stimulated by prolactin, LPS contained in the wall of Gram-negative bacteria, and also by lipoteichoic acid (TLA) contained in the wall of Gram-positive bacterial [37] and may influence the development of local immunity in the mammary gland. Thus, thanks to ability to stimulate integral bacterial elements to MAA-S3 synthesis, MAA-S3 seems to be a good indicator of udder inflammation. It was documented that MAA-S3 is a marker of early inflammation of the mammary gland, especially subclinical inflammations [2]. The MAA-S3 increases faster than SAA in blood because it occurs after 12 h, whereas, in plasma, it is after 24 h. Additionally, the concentration of MAA-S3 during mastitis in the cows inoculated with the *E. coli* is much higher compared to blood (1315.9 vs. 447.9 mg/L), unlike Hp [4]. This was also confirmed in very recent study [39]. In addition, the highest MAA-S3 levels were found in cows with chronic mastitis and the lowest in cows with subclinical mastitis, but still, the obtained values were several times higher than in healthy cows [40]. It is hypothesized that faster and higher increase in MAA-S3 is connected with the early phase of inflammation when the systemic defense response of the organism has not yet been initiated. There is also a possibility that SAA from blood is transported to the mammary gland, where the appearance of pathogens triggered an APR, thus causing the significant increase in MAA-S3 and resulting in a high level of MAA-S3 in milk. In addition, MAA-S3 may be mostly produced locally, which is the most likely cause, as there is a different amino acid sequence from SAA [37]. Eckersall (2010) proved that in cows with mastitis, the sensitivity of the determination of SAA in serum is 83%, and in milk, 93%, while the specificity is 90% in serum and 100% in milk [2].

It was documented that the increase in the MAA-S3 concentration in milk depends on the type of pathogenic factor. In infections caused by *E. coli*, the increase in APPs concentrations was the greatest compared to infections caused by streptococci (*Strep. agalactiae, Strep. uberis, Strep. dysgalacyiae*) and *Staphylococcus aureus*. In contrast, significant differences in the level of SAA concentration in the serum depending on the pathogens in more recent study was not detected (*Strep. agalactiae* 210 mg/L, *Strep. dysgalactiae* 210 mg/L *and Strep. uberis* 229 mg/L) [41]. However, measuring the SAA or MAA-S3 should not be consider as method for evaluating the need of antibiotic use. Susceptibility testing and the targeting of pathogens should always be performed to counteract increased global problem with antibiotic resistance.

## 10. SAA in Calves Diseases

In available publications, there are many studies describing SAA and its concentration in calves. It was documented that SAA concentration is higher in calves than adult cows, which may be caused by various physiological needs and challenges faced by the calves during maturation. The average concentration of SAA in one-month-old healthy calves is 59 mg/L, and at the age of six months, it is 19 mg/L [42]. Dudek et al. (2014) tested the influence of vaccination in pregnant cows with inactivated vaccine composed of BRSV-PI3V-*M. haemolytica* antigens on the SAA serum concentration in their calves [43]. The SAA concentration was significantly higher in the experimental calves (>150 mg/L) compared to the control calves (from nonvaccinated dams until the ninth week). Then, the SAA concentration increased in the control group (>300 mg/L) while the experimental group declined, giving statistically significant differences until week 12 [43].

Moreover, during respiratory tract infections, there is an increase in SAA. In calves with bronchopneumonia, Coskun et al. (2012) found higher concentrations of SAA in bronchoalveolar lavage fluid (BALF) than in healthy calves [24,44]. In viral infection such as syncytial virus and secondary bacterial infections, the increase in SAA in the first week of infection was observed with the maximal level at the third week and a decrease up to week six (>20 mg/L) [45]. In the other study, the SAA level was used for treatment monitoring. In calves with respiratory diseases (SAA = 725.11 $\pm$ 7.19 mg/L vs. healthy ones 304.00 $\pm$ 25.69 mg/L), the SAA level started to drop during antibiotic therapy from 824.55 $\pm$ 14.73 mg/L on day 0 to 526.05 $\pm$ 27.39 mg/L on day 10 [46]. However, it should be highlighted that because of the increased global problem with antibiotic resistance, antibiogram and the pathogen(s) testing should be performed as often as possible.

One of the more costless problems in calf health management are diarrheas. SAA blood concentration as response to stress, inflammation, infection, or tissue injury during diarrhea in calves has been developed. In post-weaned calves with diarrhea caused by bovine coronavirus (BCoV), SAA blood concentration was higher (46.2 $\pm$ 7.6 mg/L) than in recovered animals (28.5 $\pm$ 7.4 mg/L), but this increase was not statistically significant [47]. This study showed that changes in SAA were less pronounced in post-weaned calves (aged 117–155 days) with diarrhea.

The blood concentration of SAA seems to be a more sensitive marker of stress than disease in calves, but several factors can also affect SAA results [1]. In one study, the SAA concentration was checked in calves in physical stress conditions. Alsemgeest et al. observed that the plasma SAA concentrations were significantly raised in animals housed on the floor-type associated with the highest level of physical stress (17.9 $\pm$ 1.8 mg/L vs. 2.8 $\pm$ 0.9 mg/L), although the concentrations were within the normal range for healthy cattle [27]. Changes in aldolase and cortisol concentrations have not been observed, which suggest the presence only physical stress, not neuro-hormonal. Other authors reported significant elevations in SAA concentrations (from 3.2 $\pm$ 2.9 mg/L) after dehorning procedures in calves during 48 h (to 36.9 $\pm$ 30.3 mg/L) [48].

The measurement of SAA concentrations may also be used in the diagnosis of bacterial infections in stillborn calves (53.3 mg/L) in comparison to unexplained stillbirth (10.7 mg/L) [49]. In perinatal dead calves' plasma, authors observed lower SAA concentration when uterine infection was not present in the dams (6.3 mg/L) in comparison to dead calves where infection was present in the dam's uterus (13.8 mg/L) [50]. Summarizing, the changes in SAA levels may be an unspecific indicator of health status in calves; however, the different values than in adult cows should be taken into consideration. It should be highlighted that there is a shift toward more productive cows and larger herds [51]. Thus, it is associated with more health problems.

## 11. Conclusions

As non-specific markers of inflammation, changes in SAA blood concentration may be a beneficial tool to help recognize the inflammatory process in dairy cattle. One of the most important conclusions that changes in SAA concentration are very dynamic in connection to very short half-life time. Thus, it may be useful in some cases in the quantification of inflammatory activity. However, in most cases, it does not allow one to recognize the etiology of disease. However, monitoring diseases and their treatments by means of SAA may allow cattle farmers to determine the efficiency and efficacy of a specific treatment.

SAA testing is a useful tool for the assessment of health in general, to monitor the health state, and prevent the spread of infection in the whole herd. Very often, the research is performed on large cow populations. Thus, SAA seems to be a promising unspecific biomarker of the inflammatory process. In addition, it may also help in antemortem inspections of large herds in slaughter houses. Thus, the early diagnosis of several diseases is extremely important because of their significant economical and zoonotic impact.

However, the clinical application of SAA has some practical limitations associated with measurement methods. Mostly, they are time-consuming and relatively expensive, such as ELISA tests; thus, rapid field tests that allow the determination of SAA are still needed. Moreover, in veterinary practice, it is not always obvious that one isolated condition needs to be assessed. Thus, clinical examination should be performed, and SAA cannot be evaluated alone. Future research should be focused on creation rather the APPs diagnostic profiles.

**Author Contributions:** Conceptualization, M.T., D.D. and O.W.-P.; writing—original draft preparation, M.T., D.D. and O.W.-P.; writing—review and editing, M.T., D.D. and O.W.-P.; supervision, O.W.-P. All authors have read and agreed to the published version of the manuscript.

**Funding:** This research received no external funding.

**Institutional Review Board Statement:** Not applicable.

**Informed Consent Statement:** Not applicable.

**Data Availability Statement:** Not applicable.

**Conflicts of Interest:** The authors declare no conflict of interest.

## References


1. Cray, C.; Zaias, J.; Altman, N.H. Acute phase response in animals: A review. *Comp. Med.* **2009**, *6*, 517–526.
2. Eckersall, P.D.; Bell, R. Acute phase proteins: Biomarkers of infection and inflammation in veterinary medicine. *Vet. J.* **2010**, *185*, 23–27. [CrossRef] [PubMed]
3. Guzelbektes, H.; Sen, I.; Ok, M. Serum amyloid A and haptoglobin concentrations and liver fat percentage in lactating dairy cows with abomasal displacement. *J. Vet. Intern. Med.* **2010**, *24*, 213–219. [CrossRef] [PubMed]
4. Suojala, L.; Orro, T.; Järvinen, H. Acute phase response in two consecutive experimentally induced *E. coli* intramammary infections in dairy cows. *Acta Vet. Scand.* **2008**, *50*, 18. [CrossRef]
5. Thomas, F.C.; Waterston, M.; Hastie, P. The major acute phase proteins of bovine milk in a commercial dairy herd. *BMC Vet. Res.* **2015**, *11*, 207. [CrossRef]
6. Emmanuel, D.G.; Dunn, S.M.; Ametaj, B.N. Feeding high proportions of barley grain stimulates an inflammatory response in dairy cows. *J. Dairy Sci.* **2008**, *91*, 606–614. [CrossRef]
7. Lomborg, S.R.; Nielsen, L.R.; Heegaard, P.M. Acute phase proteins in cattle after exposure to complex stress. *Vet. Res. Commun.* **2008**, *7*, 575–582. [CrossRef]
8. Du, X.; Chen, L.; Huang, D. Elevated Apoptosis in the Liver of Dairy Cows with Ketosis. *Cell. Physiol. Biochem.* **2017**, *2*, 568–578. [CrossRef]
9. Zhao, C.; Liu, G.; Li, X. Inflammatory mechanism of Rumenitis in dairy cows with subacute ruminal acidosis. *BMC Vet. Res.* **2018**, *14*, 135. [CrossRef]
10. Ametaj, B.N.; Bradford, B.J.; Bobe, G. Strong Relationships between Mediators of the Acute Phase Response and Fatty Liver in Dairy Cows. *Can. J. Anim. Sci.* **2005**, *85*, 165–175. [CrossRef]
11. Nazifi, S.; Rezakhani, A.; Moaddeli, A. Study on diagnostic values of haptoglobin and serum amyloid A concentration in bovine heart diseases. *Comp. Clin. Pathol.* **2009**, *18*, 47–51. [CrossRef]
12. Espinosa, J.; de la Morena, R.; Benavides, J. Assessment of Acute-Phase Protein Response Associated with the Different Pathological Forms of Bovine Paratuberculosis. *Animals* **2020**, *10*, 1925. [CrossRef] [PubMed]
13. Sharifiyazdia, H.; Nazifi, S.; Nikseresht, K. Evaluation of Serum Amyloid A and Haptoglobin in Dairy Cows Naturally Infected with Brucellosis. *J. Bacteriol. Parasitol.* **2012**, *3*, 157–161. [CrossRef]
14. Mohammadi, S.; Mohammadi, V.; Esmaeilnejad, B. Evaluation of some acute phase proteins in cattle naturally infected with Babesia bigemina. *Comp. Immunol. Microbiol. Infect. Dis.* **2021**, *76*, 101642. [CrossRef] [PubMed]

15. Nazifi, S.; Razavi, S.M.; Kaviani, F.; Rakhshandehroo, E. Acute phase response in cattle infected with Anaplasma marginale. *Vet. Microbiol.* **2012**, *155*, 267–271. [CrossRef] [PubMed]
16. Schmitt, R.; Pieper, L.; Gonzalez-Grajales, L.A. Evaluation of different acute-phase proteins for herd health diagnostics in early postpartum Holstein Friesian dairy cows. *J. Dairy Res.* **2021**, *88*, 33–37. [CrossRef] [PubMed]
17. Kaya, S.; Merhan, O.; Kacar, C. Determination of ceruloplasmin, some other acute phase proteins, and biochemical parameters in cows with endometritis. *Vet. World* **2016**, *10*, 1056–1062. [CrossRef]
18. Brodzki, P.; Kostro, K.; Brodzki, A. The concentrations of inflammatory cytokines and acute-phase proteins in the peripheral blood and uterine washings in cows with pyometra. *Reprod. Domest. Anim.* **2015**, *3*, 417–422. [CrossRef]
19. Ilievska, K.; Atanasov, B.; Dovenski, T. Acute phase proteins—As indicators of claw diseases in dairy cattle. *Mac. Vet. Rev.* **2019**, *42*, 95–100. [CrossRef]
20. Tóthová, C.; Nagy, O.; Kovác, G. Changes in the concentrations of selected acute phase proteins and variables of energetic profile in dairy cows after parturition. *J. Appl. Anim. Res.* **2014**, *3*, 278–283. [CrossRef]
21. Xu, T.; Cardoso, F.C.; Pineda, A. Grain challenge affects systemic and hepatic molecular biomarkers of inflammation, stress, and metabolic responses to a greater extent in Holstein than Jersey cows. *J. Dairy Sci.* **2017**, *11*, 9153–9162. [CrossRef]
22. Zebeli, Q.; Metzler-Zebeli, B.U.; Ametaj, B.N. Meta-analysis reveals threshold level of rapidly fermentable dietary concentrate that triggers systemic inflammation in cattle. *J. Dairy Sci.* **2012**, *5*, 2662–2672. [CrossRef]
23. Al-Qaisi, M.; Horst, E.A.; Mayorga, E.J. Effects of a Saccharomyces cerevisiae fermentation product on heat-stressed dairy cows. *J. Dairy Sci.* **2020**, *10*, 9634–9645. [CrossRef]
24. Coskun, A.; Guzelbektes, H.; Simsek, A. Haptoglobin and SAA concentrations and enzyme activities in bronchoalveolar lavage fluids from calves with bronchopneumonia. *Rev. Méd. Vét.* **2012**, *12*, 615–620.
25. Sevimli, A.; Sevimli, F.K.; Şeker, E. Acute-phase responses in cattle infected with hydatid cysts and microbial agents. *J. Helminthol.* **2015**, *4*, 471–479. [CrossRef]
26. Brodzki, P.; Kostro, K.; Brodzki, A. Inflammatory cytokines and acute-phase proteins concentrations in the peripheral blood and uterus of cows that developed endometritis during early postpartum. *Theriogenology* **2015**, *84*, 11–18. [CrossRef]
27. Alsemgeest, S.P.; Lambooy, I.E.; Wierenga, H.K. Influence of physical stress on the plasma concentration of serum amyloid-A (SAA) and haptoglobin (Hp) in calves. *Vet. Q.* **1995**, *17*, 9–12. [CrossRef]
28. Adnane, M.; Chapwanya, A.; Kaidi, R. Profiling inflammatory biomarkers in cervico-vaginal mucus (CVM) postpartum: Potential early indicators of bovine clinical endometritis? *Theriogenology* **2017**, *103*, 117–122. [CrossRef]
29. Giannetto, C.; Fazio, F.; Casella, S. Acute phase protein response during road transportation and lairage at a slaughterhouse in feedlot beef cattle. *J. Vet. Med. Sci.* **2011**, *11*, 1531–1534. [CrossRef]
30. Zhang, S.; Yang, F.; Oguejiofor, C.F. Endometrial expression of the acute phase molecule SAA is more significant than HP in reflecting the severity of endometritis. *Res. Vet. Sci.* **2018**, *121*, 130–133. [CrossRef]
31. Dervishi, E.; Zhang, G.; Hailemariam, D. Alterations in innate immunity reactants and carbohydrate and lipid metabolism precede occurrence of metritis in transition dairy cows. *Res. Vet. Sci.* **2016**, *104*, 30–39. [CrossRef] [PubMed]
32. Weaver, D.A.; Jean, S.G.; Steiner, A. Lameness. In *Bovine Surgery and Lameness*, 2nd ed.; Blackwell Publishing: Hoboken, NJ, USA, 2005; pp. 198–258.
33. Zhang, G.; Hailemariam, D.; Dervishi, E. Alterations of Innate Immunity Reactants in Transition Dairy Cows before Clinical Signs of Lameness. *Animals* **2015**, *3*, 717–747. [CrossRef] [PubMed]
34. Bagga, A.; Randhawa, S.S.; Sharma, S.; Bansal, B.K. Acute phase response in lame crossbred dairy cattle. *Vet. World* **2016**, *11*, 1204–1208. [CrossRef] [PubMed]
35. Jawor, S.; Steiner, T.; Stefaniak, W. Determination of selected acute phase proteins during the treatment of limb diseases in dairy cows. *Vet. Med.* **2008**, *4*, 173–183. [CrossRef]
36. Akerstedt, M.; Waller, K.P.; Larsen, L.B. Relationship between haptoglobin and serum amyloid A in milk and milk quality. *Int. Dairy J.* **2008**, *6*, 669–674. [CrossRef]
37. McDonald, T.L.; Larson, M.A.; Mack, D.R. Elevated extrahepatic expression and secretion of mammary-associated serum amyloid A 3 (M-SAA3) into colostrum. *Vet. Immunol. Immunopathol.* **2001**, *83*, 203–211. [CrossRef]
38. Mack, D.R.; McDonald, T.L.; Larson, M.A. The conserved TFLK motif of mammary-associated serum amyloid A3 is responsible for up-regulation of intestinal MUC3 mucin expression in vitro. *Pediatr. Res.* **2003**, *53*, 137–142. [CrossRef]
39. Otsuka, M.; Sugiyama, M.; Ito, T. Diagnostic utility of measuring serum amyloid A with a latex agglutination turbidimetric immunoassay in bovine mastitis: Comparison with haptoglobin and alpha 1 acid glycoprotein. *J. Vet. Med. Sci.* **2021**, *83*, 329–332. [CrossRef]
40. Szczubiał, M.; Dabrowski, R.; Kankofer, M.; Bochniarz, M.; Komar, M. Concentration of serum amyloid A and ceruloplasmin activity in milk from cows with subclinical mastitis caused by different pathogens. *Pol. J. Vet. Sci.* **2012**, *2*, 291–296. [CrossRef]
41. Bochniarz, M.; Szczubiał, M.; Brodzki, P.; Krakowski, L.; Dąbrowski, R. Serum amyloid A as an marker of cow` s mastitis caused by *Streptococcus* sp. *Comp. Immunol. Microbiol. Infect. Dis.* **2020**, *72*, 101498. [CrossRef]
42. Tothova, C.; Nagy, Y.; Naygova, V. Changes in the concentrations of acute phase proteins in calves during the first months of life. *Acta Vet.-Beograd.* **2015**, *65*, 260–270. [CrossRef]
43. Dudek, K.; Bednarek, D.; Ayling, R.D. Stimulation and analysis of the immune response in calves from vaccinated pregnant cows. *Res. Vet. Sci.* **2014**, *97*, 32–37. [CrossRef] [PubMed]

44. Prohl, A.; Schroedl, W.; Rhode, H. Acute phase proteins as local biomarkers of respiratory infection in calves. *BMC Vet. Res.* **2015**, *11*, 167. [CrossRef] [PubMed]
45. Orro, T.; Pohjanvirta, T.; Rikula, U. Acute phase protein changes in calves during an outbreak of respiratory disease caused by bovine respiratory syncytial virus. *Comp. Immunol. Microbiol. Infect. Dis.* **2011**, *34*, 23–29. [CrossRef]
46. Joshi, V.; Gupta, V.K.; Bhanuprakash, A.G. Haptoglobin and serum amyloid A as putative biomarker candidates of naturally occurring bovine respiratory disease in dairy calves. *Microb. Pathog.* **2018**, *116*, 33–37. [CrossRef]
47. Chae, J.B.; Park, J.; Jung, S.H. Acute phase response in bovine coronavirus positive post-weaned calves with diarrhea. *Acta Vet. Scand.* **2019**, *61*, 36. [CrossRef]
48. Tsukano, K.; Shimamori, T.; Fukuda, T. Serum iron concentration as a marker of inflammation in young cows that underwent dehorning operation. *J. Vet. Med. Sci.* **2019**, *4*, 626–628. [CrossRef]
49. Jawor, P.; Mee, J.F.; Stefaniak, T. Perinatal immuno/inflammatory responses in the presence or absence of bovine fetal infection. *BMC Vet. Res.* **2018**, *14*, 322. [CrossRef]
50. Jawor, P.; Stefaniak, T.; Mee, J.F. Immune and inflammatory biomarkers in cases of bovine perinatal mortality with and without infection in utero. *J. Dairy Sci.* **2017**, *100*, 1408–1416. [CrossRef]
51. Trela, M.; Witkowska-Piłaszewicz, O.; Domańska, D.; Kaczmarek, M.M.; Pawliński, B.; Gajewski, Z.; Domino, M. The Influence of Intravaginal Gestagens Treatment on the Morphological Features and Endometrial Steroid Hormone Receptors Content during Anestrus Type II in Dairy Cattle. *Int. J. Mol. Sci.* **2022**, *23*, 1235. [CrossRef]

agriculture

Review

# The Overlooked Transformation Mechanisms of VLCFAs: Peroxisomal β-Oxidation

Qinyue Lu [1,2,†], Weicheng Zong [3,†], Mingyixing Zhang [1,2], Zhi Chen [1,2,*] and Zhangping Yang [1,2]

1 College of Animal Science and Technology, Yangzhou University, Yangzhou 225009, China; mz120201330@yzu.edu.cn (Q.L.); zhangmingyixing@126.com (M.Z.); yzp@yzu.edu.cn (Z.Y.)
2 Joint International Research Laboratory of Agriculture & Agri-Product Safety, Ministry of Education, Yangzhou University, Yangzhou 225009, China
3 College of Veterinary Medcine, Yangzhou University, Yangzhou 225009, China; wczong@foxmail.com
* Correspondence: zhichen@yzu.edu.cn; Tel.: +86-182-5271-3205
† These authors contributed equally to this work.

**Abstract:** Beta-oxidation(β-oxidation) is an important metabolic process involving multiple steps by which fatty acid molecules are broken down to produce energy. The very long-chain fatty acids (VLCFAs), a type of fatty acid (FA), are usually highly toxic when free in vivo, and their oxidative metabolism depends on the peroxisomal β-oxidation. For a long time, although β-oxidation takes place in both mitochondria and peroxisomes, most studies have been keen to explore the mechanism of β-oxidation in mitochondria while ignoring the importance of peroxisomal β-oxidation. However, current studies indicate that it is hard to provide effective treatment for diseases caused by the disorder of peroxisomal β-oxidation, such as X-ALD, SCOX deficiency, and D-BP deficiency; thus, actions should be taken to solve this problem. Based on existing research results, this review will summarize the importance of peroxisomal β-oxidation and help further learning.

**Keywords:** peroxisome; very long-chain fatty acids; beta-oxidation

**Citation:** Lu, Q.; Zong, W.; Zhang, M.; Chen, Z.; Yang, Z. The Overlooked Transformation Mechanisms of VLCFAs: Peroxisomal β-Oxidation. *Agriculture* **2022**, *12*, 947. https://doi.org/10.3390/agriculture12070947

Academic Editor: Maria Grazia Cappai

Received: 21 May 2022
Accepted: 29 June 2022
Published: 30 June 2022



## 1. Introduction

Fatty acids (FAs) can be divided into four types according to the length of the carbon chain, namely short-chain fatty acids (containing 2–4 carbon atoms, SCFAs), medium-chain fatty acids (containing 6–12 carbon atoms, MCFAs), long-chain fatty acids (containing 13–18 carbon atoms, LCFAs) and very long-chain fatty acids (containing 20 or more carbon atoms, VLCFAs) [1]. In this review, we will refer to VLCFAs, including C20:0, C20:1, C22:0, and others. Any of these VLCFAs have important functions that cannot be substituted by LCFAs, such as skin barrier formation, retinal functions, resolution of inflammation, maintenance of myelin, sperm development and maturation, and liver homeostasis.

Most carbon saturated fatty acids (SFAs) are metabolized by β-oxidation in mitochondria. However, some specific FAs, such as unsaturated fatty acids (UFAs), branched-chain fatty acids (BCFAs), and VLCFAs, require different oxidation processes, including isomerization, alpha-oxidation (α-oxidation), omega-oxidation (ω-oxidation), and the oxidation process in peroxisomes. Similar to β-oxidation in mitochondria, four sequential reactions also occur in peroxisomal β-oxidation [2]. Despite similarities in the reactions, mitochondria and peroxisomes still have different catalytic proteins, electron transport chains, and orientations of metabolites, all of these suggest that research on mitochondria cannot be applied to peroxisomes (Figure 1).

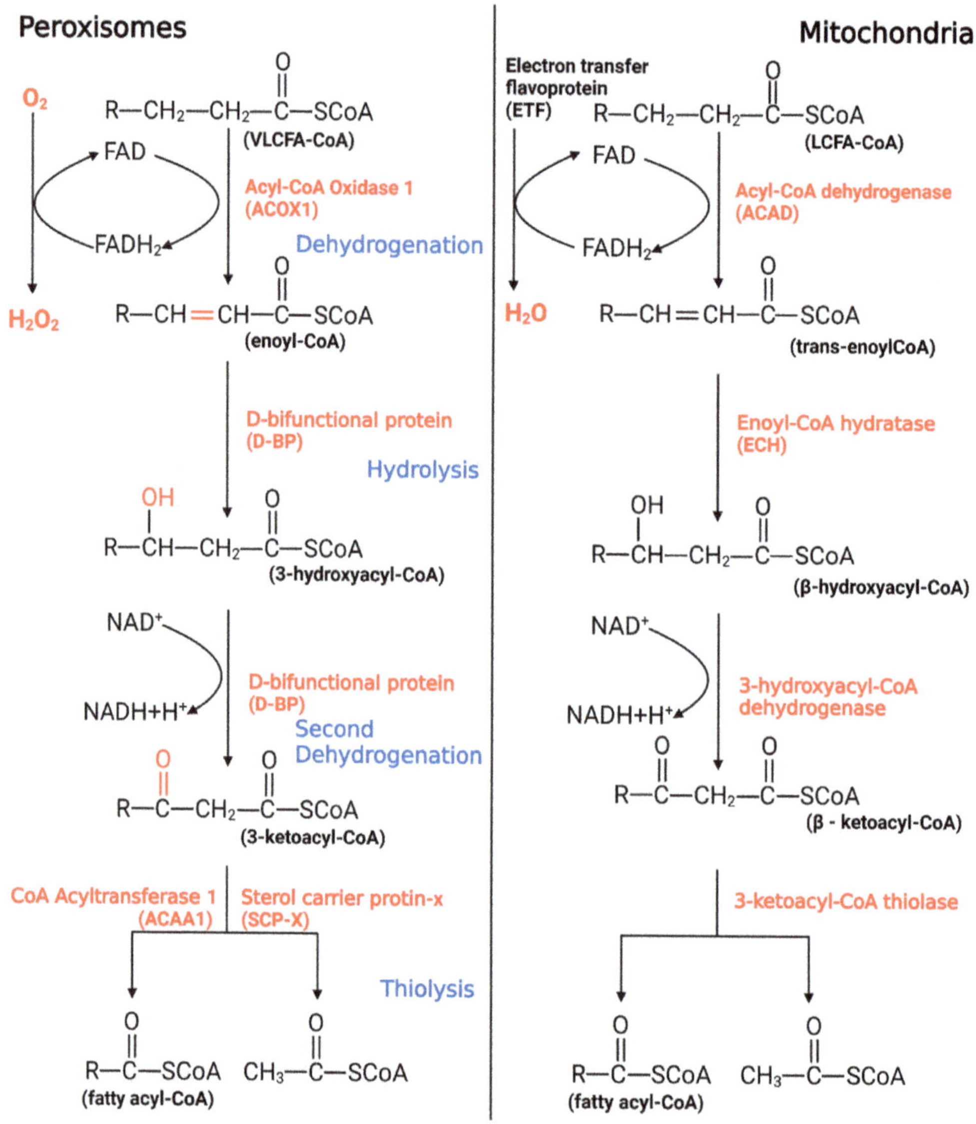


**Figure 1.** The process of peroxisomal β-oxidation.

VLCFAs enter the peroxisome in the form of -CoA, start the β-oxidation from the first oxidation, undergo hydrolysis, begin the second oxidation, undergo thiolysis, and remove two C atoms. If VLCFAs are still present, they will enter the cycle again until the number of carbon chains is less than 18, after which the products will enter the mitochondria to continue metabolism.

While peroxisomal β-oxidation plays a role in fat metabolism, researchers habitually look for solutions in mitochondria when encountering problems associated with fat metabolism, especially cancer. Taking prostate cancer (PCa) as an example, multiple studies have demonstrated that the occurrence of PCa is related to free FAs and oxidative stress in the body [3], whereas most studies focus on regulating mitochondria to avoid or treat

PCa; thus, peroxisomes have not been taken seriously yet. Interestingly, since 2015, certain studies have proposed to locate some relevant biomarkers in PCa cells peroxisome, such as monocarboxylate transporter 2 [4]. Some studies have also indicated that the expression of peroxisomal β-oxidation changes with PCa proliferation, and the rate of β-oxidation might affect the homeostasis of PCa cells [5]. By contrast, the specific mechanism of action between peroxisome and PCa and how to treat PCa through peroxisome remained unclear. It can be seen that peroxisome β-oxidation may also be the target scheme in many problems, but it is always overlooked. These days most studies focus on the interaction between peroxisome and mitochondria, whereas fewer studies have been done on its independent role. So, does it have research significance?

## 2. The Significance of Peroxisomal β-Oxidation

As alluded to above, it is not paradoxical that what has been overlooked tends to be of great importance. Peroxisomes, the widely distributed organelles in the body, play irreplaceable roles in cellular metabolism, especially in fatty acid oxidation (FAO) and the generation and elimination of reactive oxygen species (ROS).

In FAO, the oxidation of VLCFAs is an aspect of peroxisome that differs from mitochondria. The mitochondrial β-oxidation pathway has long been considered to play a central role in lipid degradation [6,7], and any blockage of the oxidative pathway leads to increased lipid levels in tissues, yet the role of peroxisomes has been considered. Dysregulation of VLCFAs, essential components in the body, can lead to the occurrence of many diseases [8]. In humans, studies have shown that the accumulation of VLCFAs is the main cause of many neurological diseases, such as Alzheimer's disease, multiple sclerosis, and dementia. In addition, studies of sexually transmitted diseases have found that VLCFAs are associated with ichthyosis, myopathy, and demyelination [9]. To be specific, VLCFAs accumulate in the plasma and tissue of patients, resulting in a fatal neurodegenerative phenotype, including childhood-onset cerebral adrenoleukodystrophy (CCALD) and adrenomyeloneuropathy (AMN) [10]. AMN is the milder phenotype characterized by a slowly progressive axonopathy. Thus, VLCFAs are not only an indispensable part of the body but also a substance that, if dysregulated in vivo, may result in strong toxicity.

In living organisms, free FAs, generally with low concentration, are mainly bound to fatty acid-binding proteins [11,12]. In this case, FAs in vivo are usually produced by the degradation of deposited fat and basically do not contain VLFCAs, and mitochondrial β-oxidation is the primary way of FAO. However, when the toxic VLCFAs enter the body or are in a free state, the peroxisome immediately activates the transport capacity through the transporter; in turn, acyl-CoA oxidase 1 (ACOX1) functions to ensure that peroxisomes can preferentially process VLCFAs that are not suitable for the internal environment. It is this timely processing mechanism that makes the intoxication caused by VLCFAs rare in the body. Of course, this may also be part of the reason why peroxisomal β-oxidation is easily overlooked.

Peroxisomal β-oxidation also plays an irreplaceable role in coping with oxidative stress. When the body is subjected to various harmful stimuli, highly active molecules such as ROS and reactive nitrogen species (RNS) generate excessive free radicals, and the oxidation degree exceeds the antioxidant capacity of cells to remove oxides. The oxidative and antioxidant systems are unbalanced, leading to tissue damage. This is related to the ratio of $FADH_2$/NADH (F/N) (an electron transport chain involved in the transfer of free hydrogen ions and electrons) entering the electron transport chain. In short, the length of the FA carbon chain will affect the saturation and the F/N ratio, which in turn affects ROS production. If VLCFAs are metabolized in mitochondria, the ROS formed can cause severe oxidative stress in mitochondria. Special oxidation products generated during peroxisomal β-oxidation effectively reduced the occurrence of oxidative stress (Figure 2). In peroxisomes, the high-energy electrons stored in $FADH_2$ are directly transferred from $O_2$ to $H_2O_2$, and subsequently decomposed into $H_2O$ and $O_2$. Therefore, the oxidation of VLCFAs in peroxisomes can reduce the amount of β-oxidation in mitochondria, thereby reducing the

F/N ratio and ROS formation. Of course, peroxisomal β-oxidation inevitably causes ATP loss, and the energy carried by $FADH_2$ is not used to synthesize ATP but is dissipated in the form of thermal energy. However, this heat loss is not a physiologically ineffective behavior, since studies have shown that peroxisomes accelerate the decomposition of FAs in BAT under cold stress conditions to help the body quickly adapt [13]. Therefore, peroxisomal β-oxidation is an important metabolic activity during nonshivering thermogenesis. Correspondingly, there is noise stimulation. In an extremely noisy environment, ROS production increases and causes oxidative damage, and peroxisomes can also regulate disorders caused by these stimuli through β-oxidation. Current research shows that this oxidative feedback regulation mechanism is activated by PEX5 [14].

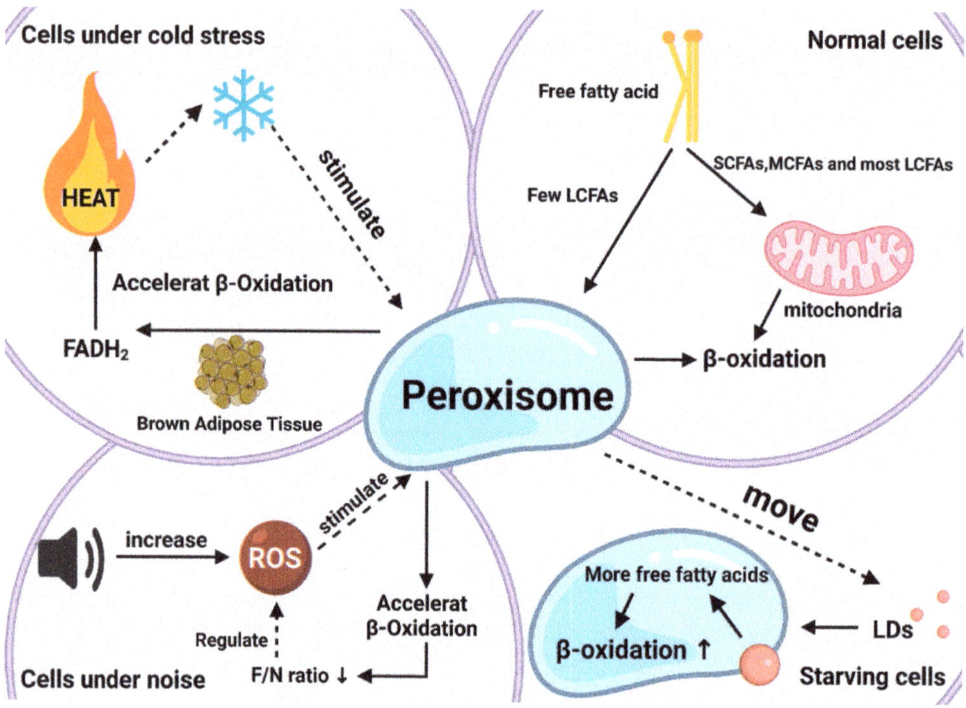


**Figure 2.** The roles of peroxisomal β-oxidation under various conditions. Responses and effects of peroxisomal β-oxidation in cells under normal (**top right**), starvation (**bottom right**), cold stress (**top left**), and noise (**bottom left**) conditions.

Another neglected effect is related to lipid droplets (LDs). LDs are lipid-storing organelles present in nearly all organisms, from bacteria to mammals, and their degradation provides metabolic energy for different cellular processes, such as membrane synthesis and molecular signaling [15]. Studies have shown that LDs and peroxisomes are generated at the same place in the endoplasmic reticulum with close subcellular localization after maturation, implying the possibility of interaction between the two organelles [16]. Indeed, researchers have found that when the body is starving, peroxisomes can move to and contact LDs with the help of kinesin KLFC3, and then transfer lipids from lipid droplets into the β-oxidation process more quickly to promote their degradation and maintain energy balance [17]. The latest research also authenticates this conclusion and gives a more specific explanation for "starvation" this condition arises to protect the body from ROS, since starvation increases fatty acid peroxidation as does the production of ROS [18].

Among mammals, peroxisomes also play an important role in ruminants, especially dairy cows. In other non-ruminant mammals, where FAO occurs mainly in mitochondria

(76%), in ruminants, FAO occurs in mitochondria and peroxisomes (approximately 50% in each organelle) [19]. As dairy livestock are important, many studies have focused on the different lactation stages and butterfat percentage of dairy cows. Consumer choices today are based not only on the nutritional aspects of the food but also on products known to promote better health or prevent disease [20,21]. In this regard, the proportion of VLCFAs in milk is a concern. Dairy cows have a complete pathway for the synthesis and utilization of VLCFAs, through ELOVL protein synthesis and peroxisome utilization [22]. In actual production, high-producing dairy cows are subject to constant oxidative stress due to a high metabolic rate and physiological adaptation to intensive farming [23]. During the perinatal period of dairy cows, the body will undergo complex physiological changes, among which ketosis often occurs [24]. The occurrence of ketosis is often accompanied by fat deposition in the liver [25]. The essence is that after excess NEFAs (nonesterified fatty acid, fatty acids above C10, mainly VLCFAs) enter the liver, part of NEFAs enter the ketone body synthesis pathway to generate ketone bodies [26,27]. Study showed a greater level of ROS in mammary epithelial cells of ketotic cows, and greater oxidant indices, indicating increased oxidative stress status [28,29]. Although there is no direct research to prove that the occurrence of ketosis is related to peroxisomes, many studies have proved that factors related to peroxisomal β-oxidation are involved in the occurrence and control of the disease, such as PPARα, and AMPK, combined with the presence of VLCFAs, we think this is very possibly related to peroxisomes [30,31]. Also, many diseases in dairy cows can be attributed to oxidative stress, such as mastitis and breast edema [32–35]. Although some studies use extrinsic drugs to treat diseases from oxidative stress [36–38], authors believe that the harm caused by oxidative stress can be alleviated by intrinsically regulating the rate of peroxisomal β-oxidation. Unfortunately, at present, few studies have linked peroxisomal β-oxidation to these diseases.

Overall, current research on peroxisomal β-oxidation has demonstrated its importance as a major factor in regulating lipid metabolism disorders in the internal environment and maintaining the balance of lipids and ROS. Unfortunately, most research on these functions has focused on understanding how they operate, and the current understanding of molecular-level mechanisms of functions remains limited.

## 3. Factors Involved in the Regulation of the Peroxisomal β-Oxidation

Although it attracts less research than mitochondria, the importance of the peroxisomal β-oxidation molecular mechanism can still be spotted from some mechanisms involved in upstream regulation.

### 3.1. PPAR

Peroxisome proliferation activating nuclear receptors (PPARs), members of the steroid hormone nuclear receptor ligand-dependent transcription factor superfamily, can regulate peroxisome proliferation by regulating the peroxisome proliferation β-oxidation process [39]. PPARs have three subunits, namely PPARα, PPARβ/δ, and PPARγ (Figure 3), which differ in tissue distribution, ligand affinity, and target genes. Surprisingly, they are all related to peroxisomal β-oxidative metabolism [40]. Activation of PPARα promotes fatty acid entry into peroxisomes, and PPARα participates in the regulation of energy homeostasis. By contrast, activation of PPARγ causes insulin sensitization and enhances glucose metabolism, whereas activation of PPARβ/δ enhances FAO.

#### 3.1.1. PPARα

The effect of peroxisome proliferators on the synergistic induction of the peroxisomal β-oxidation system enzymes has been well established, which is regulated by PPARα. As a lipid sensor, PPARα can coordinate and promote the expression of a large number of target genes in the process of FAO, especially in starvation or high-fat diet conditions [41]. Studies have shown that PPARα induces the translation of downstream genes under the stimulation of VLCFAs: it promotes the proliferation of peroxisomes, and it specifically upregulates the expression of peroxisomal β-oxidative proteins and coenzymes [42]. At present, research on PPARα mainly focuses on its targeted ligands. For example, synthetic lipid-lowering drugs have been shown to activate PPARα and then regulate the peroxisomal β-oxidation [43]. And in the ruminants, some studies have found that PPARα has different expression levels in the mammary between the dry and lactation period [19,44]. Further, the study also showed that PPARα plays an important role in regulating milk fat synthesis in ruminants [45]. In addition to this, PPARα is also involved in the regulation of fat in the liver [46,47]. Interestingly, PPARα also plays an important role in ruminant reproduction [48,49], but it is not known whether this role is related to peroxisomal β-oxidation, which is a direction worth exploring. Simply put, regardless of whether the drugs are endogenous or exogenous ligands, their activation mechanism to bind the PPARα further increases transcriptional activity.

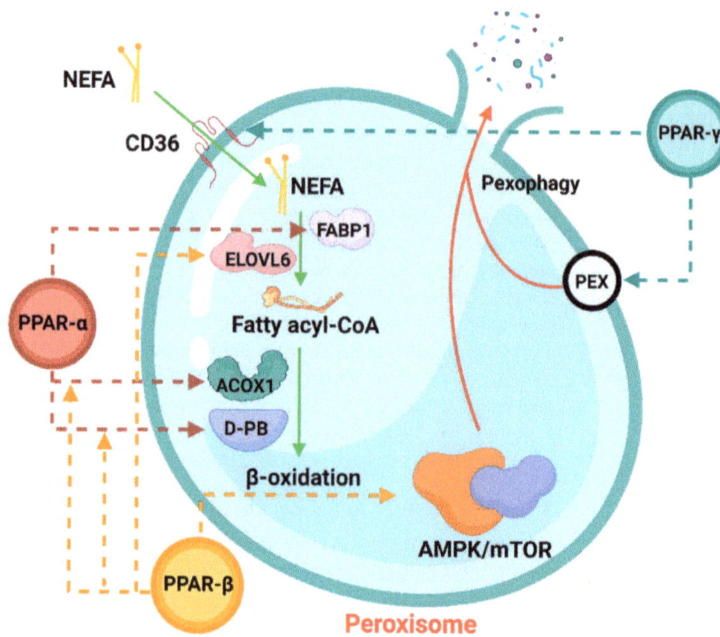


**Figure 3.** The role of PPARs in peroxisomal β-oxidation.

In the process of peroxisomal β-oxidation, PPARs are actively involved in various processes. The dotted arrows in the figure indicate that the corresponding PPAR members participate in the regulation of the process. Among them, NEFA: non-esterified fatty acid, FABP1: fatty acid binding protein 1, ELOVL6: ELOVL Fatty Acid Elongase 6, ACOX1: Acyl-CoA Oxidase 1. The process involving pexophagy was complicated, and "PEX" was only a general reference. For details, please refer to Section 3.3.

### 3.1.2. PPARγ

As one of the key transcriptional regulators of adipocyte differentiation, PPARγ also plays an important role in mediating peroxisomal β-oxidation and lipid metabolisms [50]. Studies have shown that a variety of genes involved in FA transport and metabolism are regulated by PPARγ at the transcriptional level, such as FA translocases, implying that PPARγ can stimulate peroxidase by increasing the expression of FA transporters and FA transportases [51,52]. The initiation of β-oxidation in vivo has also been proven. By knocking out PPARγ in mice, it was found that they had obvious lipid metabolism disorders [53].

In dairy cows, PPARγ plays an important role which is the critical mediator of lipogenesis [54]. One of the significant roles it plays in dairy cows from the point of view of economic interest is controlling the synthesis of milk fat in dairy cows, not only in mitochondria but also with ELVOL participating in the regulation of peroxisomal β-oxidation. PPARγ expression increases during changes in the lactation period [55,56], thereby regulating peroxisomal β-oxidation to prevent metabolic stress. Research further confirmed the importance of PPARγ in regulating milk fat production [57,58]. Similarly, as milk-producing ruminants, goats and sheep are also regulated by PPARγ. Many studies indicate that PPARγ is involved in adipocyte differentiation and adipogenesis in sheep/goats [59–63], though, in addition to fat regulation, PPARγ was previously thought to be involved in the regulation of hormones in goats and sheep [64,65]. However, there has been some relevant research in recent years. Some studies have pointed out that the role of PPARγ and hormones is only a servo-assist mechanism [66]. In the author's opinion, there is no obvious evidence to prove either statement; thus, this is also a worthy question.

#### 3.1.3. PPARβ/δ

Compared with PPARα and PPARγ, PPARβ/δ is more widely distributed in various tissues in vivo. Aline et al. showed that PPARβ/δ could participate in the activation of FAO in BATs, but genes involved in processes such as lipogenesis were not significantly correlated [67]. This demonstrates that PPARβ/δ is a thermogenic transcription factor in vivo, which bears great resemblance to the purpose of peroxisomal β-oxidation. Furthermore, Tong et al. studied PPARβ/δ-induced autophagy, although the study assessed expression changes in mitochondria and revealed the PPARβ/δ-AMPK/mTOR pathway by searching for a signaling pathway (mTOR, one of the important factors of peroxisomal β-oxidation), which further demonstrated that PPARβ/δ has a regulatory effect on peroxisomal β-oxidation [68]. Recent studies have shown that PPARβ/δ is indispensable for the upregulation of autophagic behavior [43,69–72], making the relationship between PPARβ/δ and peroxisomal β-oxidation more explicit. When the body is under certain conditions, PPARβ/δ frees more VLCFAs by upregulating cellular autophagy, which in turn regulates the initiation and enhancement of peroxisomal β-oxidation. However, a lot of research is still needed to confirm what certain conditions are and whether this speculation is correct.

Regardless of which subunit of PPARs is involved in the regulation of peroxisomal β-oxidation, as PPAR is often reported to be associated with the occurrence of diseases like type 2 diabetes, PPAR-associated peroxisomal β-oxidation changes may also be involved. A study pointed out that inhibition of peroxisome biosynthesis can interrupt β-oxidation through the action of PPAR, thereby effectively preventing the occurrence of type 2 diabetes [13]. Nevertheless, the molecular mechanism underlying this phenomenon has not been elucidated. In addition, the prevalence of some diseases is different by gender [73], and many experiments also show that PPARs are different in function by gender. PPARα expression is more abundant in native and activated male T cells than in female cells [74,75], suggesting that PPARα has a more substantial role in male T cells than in female T cells. Likewise, multiple studies have shown that the male hormone androgen has been suggested to influence the expression of PPARα in male T cells [76–78]. A study of sex-specific differences in the role of PPARγ in T cell survival has shown that male PPARγ-deficient T cells have increased apoptosis and contain a greater proportion of apoptotic cells than female PPARγ-deficient T cells [79]. Some studies have also validated similar conclusions, suggesting that PPARγ plays an important role in T cell survival [80–83]. Although more convincing data are needed to resolve this discrepancy, PPARγ may act as a survival factor in female T cells. Unlike others, the regulatory role of PPARβ/δ in T cells has not been well studied, but sex-specific differences in PPARβ/δ regulation should be considered in future studies.

### 3.2. *PGC-1α*

Peroxisome proliferator-activated receptor-gamma coactivator-1α (PGC-1α) is a powerful transcriptional coactivator that regulates a broad range of physiological and energy homeostasis responses at the transcriptional level in diverse mammalian tissues. In the past, scholars focused more on the regulatory mechanism of PGC-1α in controlling mitochondrial function [84,85]. In 2010, Bagattin et al. found that PGC-1α coordinates not only mitochondrial remodeling but also peroxisome specialization and biogenesis [86], which stimulated an upsurge in the function study of peroxisome PGC-1α. Moreover, Huang et al. studied peroxisomes in human skeletal muscle cells. They found that overexpression of PGC-1α induced the expression of AOCX1, and that the levels of some proteins associated with peroxisome activity also substantially increased [87]. In addition, some studies have also demonstrated the importance of PGC-1α in peroxisomes [88,89].

In conclusion, the regulatory relationship between PGC-1α and peroxisomal β-oxidation deserves further research because of the critical functionality of PGC-1α, and we wonder whether PGC-1α could be used as a novel chemical modulator for the treatment of Zellweger syndrome symptoms and other diseases. PGC-1α has not been studied much in ruminants and has mostly focused on regulating fatty acids [90,91]. But there is one direction worth mentioning. Zhou et al. found that PGC-1α seems to play a role in the skeletal muscles of goats, and it can also maintain metabolic rhythm through the phosphorylation of upstream regulators [92]. We speculate that this is related to peroxisomal β-oxidation.

### 3.3. *PEX*

In 1996, the term peroxin was coined for proteins in peroxisome biogenesis, including peroxisomal matrix protein import, membrane biogenesis, peroxisome proliferation, and peroxisome inheritance [93]. Peroxins are encoded by PEX genes, also known as PEX proteins. To date, 37 PEX proteins have been discovered and studied. Some are highly conserved, while others only occur in a limited number of species, such as PEX17, which only distributes in Fungi, and PEX35, which only distributes in *Saccharomycetaceae* [94]. Most PEX proteins have been shown to play a significant role in peroxisomal β-oxidation (Figure 4).

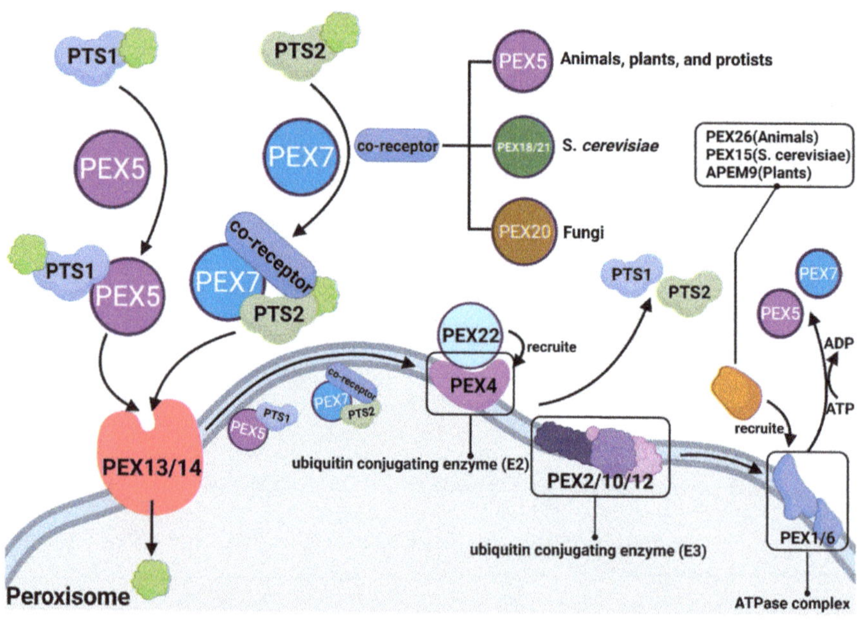


**Figure 4.** The role of different PEX proteins in peroxisomal β-oxidation.

PEX5 is the most commonly studied object in eukaryotes. As mentioned above, peroxisomes are very sensitive to ROS, whereas multiple organelles in the internal environment can produce ROS. Consequently, peroxisomes themselves are susceptible to the influences of other metabolic processes and become dysfunctional. A study found that PEX5 can respond to the expression of ROS and induce cell autophagy after ubiquitination to prevent damage from excessive or defective peroxisomes from cellular β-oxidation [95].

More research on PEX has focused on diseases caused by mutation or dysfunction of the PEX gene, including peroxisome biogenesis disorders in the Zellweger spectrum (PBD-ZSD) and rhizomelic chondrodysplasia punctata (RCDP). Studies have pointed out that PBD-ZSD is caused by PEX gene mutation that results in insufficient peroxisomal β-oxidation, which increases levels of VLCFAs in plasma and cells. However, so far, there is no treatment for PBD-ZSD, and researchers have tried to treat it from the perspective of autophagy, but the effect is not ideal [96]. Does this imply that there are unknown mechanisms other than autophagy regulating the relationship between PEX and β-oxidation? It reminds us that long-term neglect makes it impossible for us to have precise treatment for diseases caused by peroxisomal β-oxidation disorders, that we can't find the therapeutic target, and thus further exploration is urgently needed.

Substances entering peroxisomes for β-oxidation need to form receptor cargo complexes with PTS (PTS: peroxisome targeting sequence, whose role is to locate peroxisomes to ensure accurate transport of carried substances) and PEX, in which PTS1 is transported by PEX5, and PTS2 is transported by PEX7 and coreceptors (coreceptors are PEX5, PEX18/21 and PEX20, depending on the species). After the substance enters the peroxisome, PTS and PEX must be ubiquitinated and recovered.

### 3.4. ATP-Binding Cassette (ABC)

In peroxisomes, VLCFAs are mainly introduced as CoA through ABCD1–3 (Figure 5). Studies have shown that peroxisomal β-oxidation is dysfunctional after the deletion of ABCD1/ALDP in vivo, resulting in the expression levels of VLCFAs in both plasma and tissues being increased [97,98]. ABCD1 and ABCD2 share a high degree of sequence homology, except that ABCD2 plays a central role in the metabolism of monounsaturated and polyunsaturated VLCFAs, rather than saturated VLCFAs, and may be involved in the regulation of oxidative stress and DHA synthesis [99]. ABCD3/PMP70 plays a major role in transporting 2-methylacyl-CoA esters [100]. In addition, despite the distinct functions of peroxisomal ABC transporters, in vitro and in vivo studies have clearly identified that there is at least partial functional redundancy between these transporters [101,102]. In ruminants, ABC is equally important. Ahmad et al. found that the expression of ABC varies significantly among different milk yields by RNA-seq [103]. Since the expression of ABC is also different between different breeds of cattle, the difference in immunity may also be related to ABC. And Lopez et al. further verified this conclusion, ABC is indeed related to the immune system of cattle [104]. Although this conclusion was obtained by RNA-seq, we speculate that this is caused by the different content of VLCFAs in vivo and the autophagy function of peroxisome; the latest research also suggests that ABC is involved in intracellular cholesterol-mediated autophagy [105]. In addition, the role of ABC was also found in the reproductive system of sheep [106], revealing that ABC may be related to the degeneration of germ cells. Although researchers have paid more attention to the transport function of ABCD, we would like to emphasize that the X-linked adrenoleukodystrophy (X-ALD) remains a matter of concern because it lacks peroxisomal β-oxidation caused by ABCD deletion. At present, the only effective treatment is hematopoietic stem cell transplantation (HSCT), but the risk of death remains high. Some studies have pointed out that the β-oxidation defects can be restored by overexpressing ABCD1/2 in cells, whereas the understanding of its mechanism is still incomplete [28,30,35]. As mentioned above, a comprehensive understanding of peroxisomal β-oxidation is urgently needed.

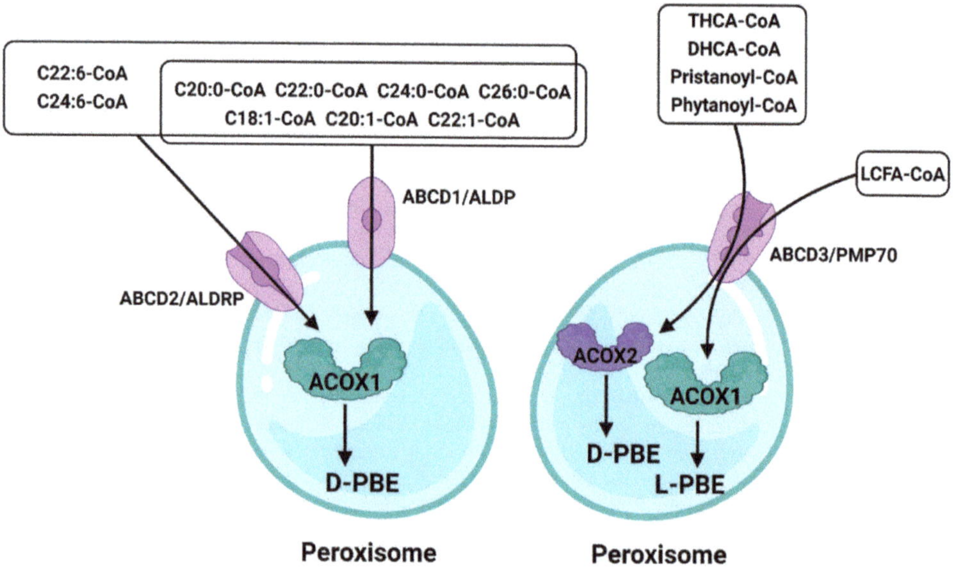


**Figure 5.** The role of the ATP-binding cassette in peroxisomal β-oxidation.

Different fatty acids enter peroxisome through different ATP-binding cassettes, and ABCD2 has a broader role. Fatty acids that enter peroxisomes through ABCD3 are processed by different enzymes to achieve oxidation results.

### 3.5. Others

New regulators involved in the regulation of peroxisomal β-oxidation have constantly been identified. AMPK, for example, promotes FAO by activating the expression of PPARα and ACOX1 [107,108], whereas mTOR promotes the accumulation of FAs by shutting down β-oxidation [109,110]. Hence, these are considered the AMPK-mTOR pathway for regulating β-oxidation [111]. The NAD-dependent protein lysine deacylases of the sirtuin family regulate various physiological functions, from energy metabolism to stress responses [112]. Numerous studies have found that Sirtuin has a variety of catalytic activities [113]. These pleiotropic enzymatic activities give sirtuins their far-reaching functions in maintaining genome integrity, regulating metabolism homeostasis, and promoting organismal longevity. And a recent study found that sirtuin 5 (SIRT5) functions similarly to mTOR, which shuts down peroxisomal β-oxidation by inhibiting the activity of ACOX1 [114]. Other reports have studied CoA in the process of β-oxidation and found that Nudt7 and Nudt19 in the Nudt superfamily can exert positive effects on CoA substrates in certain metabolic processes to promote the metabolic process [115,116]. With the upsurge of research on non-coding RNAs (ncRNAs), some play positive or inhibitory roles in peroxisomal β-oxidation, such as miR-222, miR-25-3p, circ_0005379, and others [117–122], which mainly target the rate-limiting enzyme ACOX1. Recently, Li et al. analyzed the ACOX1 transcript and found that miR-532-3P could regulate the expression of ACOX1 by targeting the complementary sequence in the 3′-UTR, thereby participating in lipid metabolism [117].

Moreover, other enzymes involved in peroxisomal β-oxidation were also regulated. Several studies have demonstrated that phosphatidylserine (PS) can bind to D-bifunctional protein (D-BP) and localize to peroxisomes, implying that PS can also affect the β-oxidation process [123]. It is worth noting that some studies have pointed out that acetylation is important for the normal functioning of D-BP [124,125], but related studies are not common; thus, we do not know which factors affect its acetylation process. Another key regulatory enzyme of peroxisomal β-oxidation, Acetyl-CoA acyltransferase 1 (ACAA1), has been extensively studied. In the breeding of livestock and poultry, researchers often explore its effects on adipocytes; for example, in goats and sheep, ACAA1 is involved in regulating adipogenesis. Studies have shown that ACAA1 deficiency increased lipid accumulation and the triglyceride content and promoted sheep preadipocyte differentiation [126]. At the same time, the study also proved a regulatory relationship between ACAA1 and PPARγ. Although the study does not point out the relationship between such a phenomenon and peroxisomal β-oxidation, it is not difficult to speculate that peroxisomal β-oxidation plays an important role. And in humans or model animals, researchers have focused on the diseases caused by it [127]. At present, it is generally believed that the expression of PPAR affects the expression of ACCA1, which in turn affects the process of β-oxidation.

## 4. Discussion and Future Perspectives

In 1954, peroxisomes were first identified as important organelles capable of β-oxidation in living organisms. Compared to mitochondria, there have been few studies on peroxisomes, and the reason why they are always ignored may be that they have fewer tasks in the body. As the entire process runs correctly, it makes people ignore their presence. However, disease can quickly emerge when peroxisomal β-oxidation is abnormal, particularly in human-related studies, such as X-ALD, SCOX deficiency, and D-BP deficiency [128–130] (Table 1). It is undeniable that fewer therapeutic options once again demonstrate that many functional and molecular mechanisms in metabolic processes remain undetermined. Of course, due to its unique subcellular localization, its extraction may not be carried out smoothly. Fortunately, peroxisomes are attracting the attention of scholars, and an increasing number of related studies have begun to appear. In a new study, McGill University studied the origin of the peroxisome and found two origins, suggesting that in addition to the endoplasmic reticulum, mitochondria may also be involved in the formation of new peroxisomes [131]. We do not yet know whether this finding points to a compensatory mechanism on the other side when peroxisomal or mitochondrial β-oxidation is defective. Still, it is most likely that this research will have a major impact on the field of peroxisome biology and ultimately on understanding human disease progression. Ding et al. showed peroxisomal β-oxidation functions as a novel sensor of FAs that regulates lipolysis through a complex pathway and modulates the interaction between peroxisomes and LDs. They described a previously unimaginable interpretation of the relationship [132]. This also means that the relationship between peroxisomes and lipid droplets has not been thoroughly studied, and their interdependent functions remain to be determined (Figures 6 and 7).

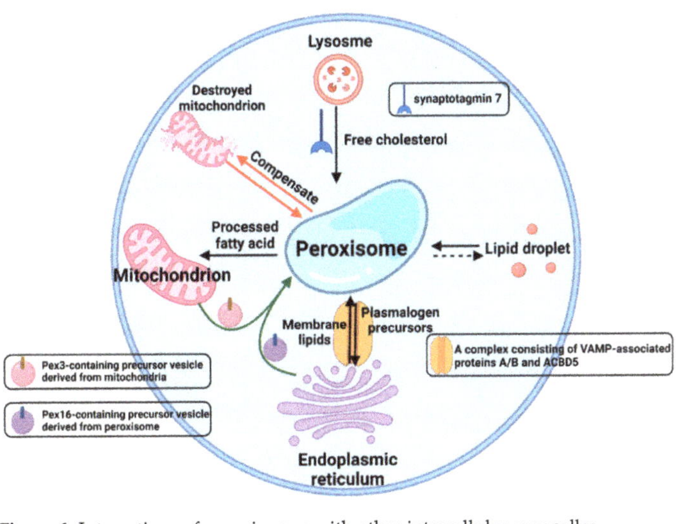


**Figure 6.** Interactions of peroxisomes with other intracellular organelles.

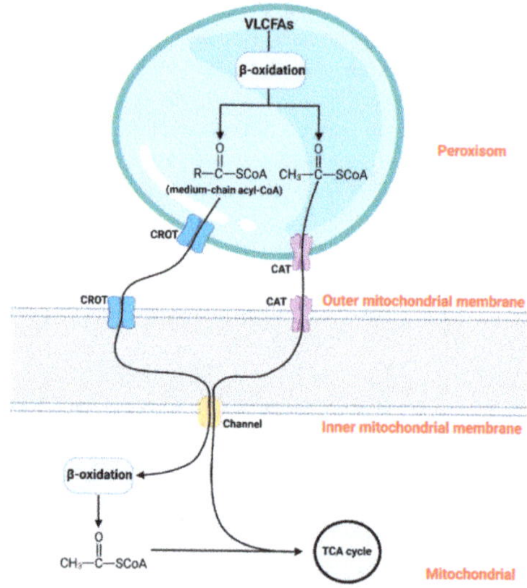


**Figure 7.** The transfer of fatty acids from peroxisomes to mitochondria.

With the continuous emergence of new technologies (such as employing omics) and the expansion of research fields [30,35,133], this is indeed an exciting time to comprehensively explore the peroxisome itself and its β-oxidation mechanism, a research hotspot with plenty of opportunities and challenges ahead.

The converging green arrow lines in the figure indicate the biogenesis of the peroxisome, and the red arrow lines indicate that when one side of the peroxisome/mitochondrion undergoes oxidative damage/other damage, the other side performs functional compensation. The black arrow lines represent normal material transport.

When fatty acids are processed in peroxisomes into fatty acyl-CoA with 18 carbons or less, they enter the mitochondria through CROT for further processing until they finally become acetyl-CoA, and enter the TCA cycle to provide energy for the body. Among them are CROT: Carnitine O-Octanoyltransferase, CAT: Catalase from Micrococcus lysodeikticus, TCA cycle: tricarboxylic acid cycle.

**Table 1.** Major diseases in peroxisome.

| Peroxisome Biogenesis Disorders (PBDs) | Incentive | References |
|---|---|---|
| Zellweger spectrum disorders (ZSDs) | Genetic disorders caused by mutations in PEX genes | [134–139] |
| Zellweger syndrome (ZS) | Mutations in peroxisome biogenesis or mutations in PEX gene | [140–147] |
| Neonatal adrenoleukodystrophy (NALD) | Mutations in PEX gene | [148,149] |
| Infantile Refsum disease (IRD) | A medical condition within the ZSDs | [150–153] |
| Heimler syndrome (HS) | Biallelic mutations in PEX1 or PEX6 | [154–156] |
| Rhizomelic chondrodysplasia punctata (RCDP) | A peroxisome biogenesis disorder, may be related to PEX gene | [157–161] |
| X-linked adrenoleukodystrophy (X-ALD) | Mutations in ABCD1 gene | [130,162–166] |
| Acyl-CoA oxidase deficiency | Deletion or mutation of ACOX1 gene | [167–169] |
| D-Bifunctional protein deficiency | Deletion or mutation of D-BP protein | [170,171] |
| 3-Ketoacyl-CoA thiolase deficiency | Deletion or mutation of THIO enzyme | [172,173] |
| α-Methylacyl-CoA racemase deficiency | Biallelic mutations in AMACR gene | [174–177] |
| Mevalonate kinase deficiency | Mutations in peroxisome biogenesis or deletion or mutation of MK | [178–182] |
| Glutaric aciduria type 3 (glutaryl-CoA oxidase deficiency) | Deficiency of succinyl-CoA | [183–185] |
| Acatalasemia | Homozygous mutations in the catalase gene | [186–188] |

**Author Contributions:** Q.L. and W.Z. contributed equally to this work. Writing—original draft preparation, Q.L.; writing—review and editing, W.Z., M.Z., Z.C. and Z.Y. All authors have read and agreed to the published version of the manuscript.

**Funding:** This review was supported by the Independent Innovation in Jiangsu Province of China (CX (21) 3119), Yangzhou University "Blue Project".

**Institutional Review Board Statement:** Not applicable.

**Informed Consent Statement:** Not applicable.

**Data Availability Statement:** Not applicable.

**Conflicts of Interest:** The authors declare no conflict of interest.

## References

1. Chen, Z.; Lu, Q.; Liang, Y.; Cui, X.; Wang, X.; Mao, Y.; Yang, Z. Circ11103 Interacts with miR-128/PPARGC1A to Regulate Milk Fat Metabolism in Dairy Cows. *J. Agric. Food Chem.* **2021**, *69*, 4490–4500. [CrossRef]
2. Hiltunen, J.K.; Filppula, S.A.; Häyrinen, H.M.; Koivuranta, K.T.; Hakkola, E.H. Peroxisomal beta-oxidation of polyunsaturated fatty acids. *Biochimie* **1993**, *75*, 175–182. [CrossRef]
3. Zhou, X.; Mei, H.; Agee, J.; Brown, T.; Mao, J. Racial differences in distribution of fatty acids in prostate cancer and benign prostatic tissues. *Lipids Health Dis.* **2019**, *18*, 189. [CrossRef]
4. Valença, I.; Pértega-Gomes, N.; Vizcaino, J.R.; Henrique, R.M.; Lopes, C.; Baltazar, F.; Ribeiro, D. Localization of MCT2 at peroxisomes is associated with malignant transformation in prostate cancer. *J. Cell. Mol. Med.* **2015**, *19*, 723–733. [CrossRef]
5. Ha, X.; Wang, J.; Chen, K.; Deng, Y.; Zhang, X.; Feng, J.; Li, X.; Zhu, J.; Ma, Y.; Qiu, T.; et al. Free Fatty Acids Promote the Development of Prostate Cancer by Upregulating Peroxisome Proliferator-Activated Receptor Gamma. *Cancer Manag. Res.* **2020**, *12*, 1355–1369. [CrossRef]
6. Cadenas, S. Mitochondrial uncoupling, ROS generation and cardioprotection. *Biochim. Biophys. Acta. Bioenerg.* **2018**, *1859*, 940–950. [CrossRef]
7. Angelova, P.R.; Abramov, A.Y. Role of mitochondrial ROS in the brain: From physiology to neurodegeneration. *FEBS Lett.* **2018**, *592*, 692–702. [CrossRef]
8. Erdbrügger, P.; Fröhlich, F. The role of very long chain fatty acids in yeast physiology and human diseases. *Biol. Chem.* **2020**, *402*, 25–38. [CrossRef]
9. Mueller, N.; Sassa, T.; Morales-Gonzalez, S.; Schneider, J.; Salchow, D.J.; Seelow, D.; Knierim, E.; Stenzel, W.; Kihara, A.; Schuelke, M. De novo mutation in ELOVL1 causes ichthyosis, acanthosis nigricans, hypomyelination, spastic paraplegia, high frequency deafness and optic atrophy. *J. Med. Genet.* **2019**, *56*, 164–175. [CrossRef]
10. Hama, K.; Fujiwara, Y.; Takashima, S.; Hayashi, Y.; Yamashita, A.; Shimozawa, N.; Yokoyama, K. Hexacosenoyl-CoA is the most abundant very long-chain acyl-CoA in ATP binding cassette transporter D1-deficient cells. *J. Lipid Res.* **2020**, *61*, 523–536. [CrossRef]
11. Chen, Z.; Cao, X.; Lu, Q.; Zhou, J.; Wang, Y.; Wu, Y.; Mao, Y.; Xu, H.; Yang, Z. circ01592 regulates unsaturated fatty acid metabolism through adsorbing miR-218 in bovine mammary epithelial cells. *Food Funct.* **2021**, *12*, 12047–12058. [CrossRef] [PubMed]
12. Chen, Z.; Zhou, J.; Wang, M.; Liu, J.; Zhang, L.; Loor, J.J.; Liang, Y.; Wu, H.; Yang, Z. Circ09863 Regulates Unsaturated Fatty Acid Metabolism by Adsorbing miR-27a-3p in Bovine Mammary Epithelial Cells. *J. Agric. Food Chem.* **2020**, *68*, 8589–8601. [CrossRef] [PubMed]
13. Park, H.; He, A.; Tan, M.; Johnson, J.M.; Dean, J.M.; Pietka, T.A.; Chen, Y.; Zhang, X.; Hsu, F.F.; Razani, B.; et al. Peroxisome-derived lipids regulate adipose thermogenesis by mediating cold-induced mitochondrial fission. *J. Clin. Investig.* **2019**, *129*, 694–711. [CrossRef] [PubMed]
14. Defourny, J.; Aghaie, A.; Perfettini, I.; Avan, P.; Delmaghani, S.; Petit, C. Pejvakin-mediated pexophagy protects auditory hair cells against noise-induced damage. *Proc. Natl. Acad. Sci. USA* **2019**, *116*, 8010–8017. [CrossRef] [PubMed]
15. Zechner, R.; Madeo, F.; Kratky, D. Cytosolic lipolysis and lipophagy: Two sides of the same coin. *Nat. Rev. Mol. Cell Biol.* **2017**, *18*, 671–684. [CrossRef]
16. Joshi, A.S.; Nebenfuehr, B.; Choudhary, V.; Satpute-Krishnan, P.; Levine, T.P.; Golden, A.; Prinz, W.A. Lipid droplet and peroxisome biogenesis occur at the same ER subdomains. *Nat. Commun.* **2018**, *9*, 2940. [CrossRef]
17. Kong, J.; Ji, Y.; Jeon, Y.G.; Han, J.S.; Han, K.H.; Lee, J.H.; Lee, G.; Jang, H.; Choe, S.S.; Baes, M.; et al. Spatiotemporal contact between peroxisomes and lipid droplets regulates fasting-induced lipolysis via PEX5. *Nat. Commun.* **2020**, *11*, 578. [CrossRef]
18. Karabiyik, C.; Vicinanza, M.; Son, S.M.; Rubinsztein, D.C. Glucose starvation induces autophagy via ULK1-mediated activation of PIKfyve in an AMPK-dependent manner. *Dev. Cell* **2021**, *56*, 1961–1975.e1965. [CrossRef]
19. Angeli, E.; Trionfini, V.; Gareis, N.C.; Matiller, V.; Huber, E.; Rey, F.; Salvetti, N.R.; Ortega, H.H.; Hein, G.J. Protein and gene expression of relevant enzymes and nuclear receptor of hepatic lipid metabolism in grazing dairy cattle during the transition period. *Res. Vet. Sci.* **2019**, *123*, 223–231. [CrossRef]

20. Harwood, W.S.; Drake, M.A. Validation of fluid milk consumer segments using qualitative multivariate analysis. *J. Dairy Sci.* **2020**, *103*, 10036–10047. [CrossRef]
21. Flowers, S.; McFadden, B.R.; Carr, C.C.; Mateescu, R.G. Consumer preferences for beef with improved nutrient profile1. *J. Anim. Sci.* **2019**, *97*, 4699–4709. [CrossRef] [PubMed]
22. Faria, O.A.C.; Kawamoto, T.S.; Dias, L.R.O.; Fidelis, A.A.G.; Leme, L.O.; Caixeta, F.M.C.; Gomes, A.; Sprícigo, J.F.W.; Dode, M.A.N. Maturation system affects lipid accumulation in bovine oocytes. *Reprod. Fertil. Dev.* **2021**, *33*, 372–380. [CrossRef] [PubMed]
23. Mavangira, V.; Sordillo, L.M. Role of lipid mediators in the regulation of oxidative stress and inflammatory responses in dairy cattle. *Res. Vet. Sci.* **2018**, *116*, 4–14. [CrossRef] [PubMed]
24. Song, Y.; Loor, J.J.; Li, C.; Liang, Y.; Li, N.; Shu, X.; Yang, Y.; Feng, X.; Du, X.; Wang, Z.; et al. Enhanced mitochondrial dysfunction and oxidative stress in the mammary gland of cows with clinical ketosis. *J. Dairy Sci.* **2021**, *104*, 6909–6918. [CrossRef] [PubMed]
25. Cainzos, J.M.; Andreu-Vazquez, C.; Guadagnini, M.; Rijpert-Duvivier, A.; Duffield, T. A systematic review of the cost of ketosis in dairy cattle. *J. Dairy Sci.* **2022**, *105*, 6175–6195. [CrossRef] [PubMed]
26. Huang, Y.; Zhao, C.; Kong, Y.; Tan, P.; Liu, S.; Liu, Y.; Zeng, F.; Yuan, Y.; Zhao, B.; Wang, J. Elucidation of the mechanism of NEFA-induced PERK-eIF2α signaling pathway regulation of lipid metabolism in bovine hepatocytes. *J. Steroid Biochem. Mol. Biol.* **2021**, *211*, 105893. [CrossRef]
27. Mohsin, M.A.; Yu, H.; He, R.; Wang, P.; Gan, L.; Du, Y.; Huang, Y.; Abro, M.B.; Sohaib, S.; Pierzchala, M.; et al. Differentiation of Subclinical Ketosis and Liver Function Test Indices in Adipose Tissues Associated with Hyperketonemia in Postpartum Dairy Cattle. *Front. Vet. Sci.* **2021**, *8*, 796494. [CrossRef] [PubMed]
28. Xu, Q.; Fan, Y.; Loor, J.J.; Liang, Y.; Sun, X.; Jia, H.; Zhao, C.; Xu, C. Adenosine 5′-monophosphate-activated protein kinase ameliorates bovine adipocyte oxidative stress by inducing antioxidant responses and autophagy. *J. Dairy Sci.* **2021**, *104*, 4516–4528. [CrossRef] [PubMed]
29. Li, Y.; Ding, H.; Liu, L.; Song, Y.; Du, X.; Feng, S.; Wang, X.; Li, X.; Wang, Z.; Li, X.; et al. Non-esterified Fatty Acid Induce Dairy Cow Hepatocytes Apoptosis via the Mitochondria-Mediated ROS-JNK/ERK Signaling Pathway. *Front. Cell Dev. Biol.* **2020**, *8*, 245. [CrossRef] [PubMed]
30. Xu, T.; Lu, X.; Arbab, A.A.I.; Wu, X.; Mao, Y.; Loor, J.J.; Yang, Z. Metformin acts to suppress β-hydroxybutyric acid-mediated inflammatory responses through activation of AMPK signaling in bovine hepatocytes. *J. Anim. Sci.* **2021**, *99*, skab153. [CrossRef]
31. Wu, Z.L.; Chen, S.Y.; Qin, C.; Jia, X.; Deng, F.; Wang, J.; Lai, S.J. Clinical Ketosis-Associated Alteration of Gene Expression in Holstein Cows. *Genes* **2020**, *11*, 219. [CrossRef] [PubMed]
32. Sammad, A.; Umer, S.; Shi, R.; Zhu, H.; Zhao, X.; Wang, Y. Dairy cow reproduction under the influence of heat stress. *J. Anim. Physiol. Anim. Nutr.* **2020**, *104*, 978–986. [CrossRef] [PubMed]
33. Wyck, S.; Herrera, C.; Requena, C.E.; Bittner, L.; Hajkova, P.; Bollwein, H.; Santoro, R. Oxidative stress in sperm affects the epigenetic reprogramming in early embryonic development. *Epigenet. Chromatin* **2018**, *11*, 60. [CrossRef] [PubMed]
34. Tabatabaee, N.; Heidarpour, M.; Khoramian, B. Milk metabolites, proteins and oxidative stress markers in dairy cows suffering from Staphylococcus aureus subclinical mastitis with or without spontaneous cure. *J. Dairy Res.* **2021**, *88*, 326–329. [CrossRef] [PubMed]
35. Zhang, Y.; Xu, Y.; Chen, B.; Zhao, B.; Gao, X.J. Selenium Deficiency Promotes Oxidative Stress-Induced Mastitis via Activating the NF-κB and MAPK Pathways in Dairy Cow. *Biol. Trace Elem. Res.* **2022**, *200*, 2716–2726. [CrossRef] [PubMed]
36. Li, C.; Wang, Y.; Li, L.; Han, Z.; Mao, S.; Wang, G. Betaine protects against heat exposure-induced oxidative stress and apoptosis in bovine mammary epithelial cells via regulation of ROS production. *Cell Stress Chaperones* **2019**, *24*, 453–460. [CrossRef]
37. Ciampi, F.; Sordillo, L.M.; Gandy, J.C.; Caroprese, M.; Sevi, A.; Albenzio, M.; Santillo, A. Evaluation of natural plant extracts as antioxidants in a bovine in vitro model of oxidative stress. *J. Dairy Sci.* **2020**, *103*, 8938–8947. [CrossRef]
38. Kumar, S.; Adjei, I.M.; Brown, S.B.; Liseth, O.; Sharma, B. Manganese dioxide nanoparticles protect cartilage from inflammation-induced oxidative stress. *Biomaterials* **2019**, *224*, 119467. [CrossRef]
39. Wagner, N.; Wagner, K.D. The Role of PPARs in Disease. *Cells* **2020**, *9*, 2367. [CrossRef]
40. Christofides, A.; Konstantinidou, E.; Jani, C.; Boussiotis, V.A. The role of peroxisome proliferator-activated receptors (PPAR) in immune responses. *Metab. Clin. Exp.* **2021**, *114*, 154338. [CrossRef]
41. Cai, D.; Li, Y.; Zhang, K.; Zhou, B.; Guo, F.; Holm, L.; Liu, H.Y. Co-option of PPARα in the regulation of lipogenesis and fatty acid oxidation in CLA-induced hepatic steatosis. *J. Cell. Physiol.* **2021**, *236*, 4387–4402. [CrossRef] [PubMed]
42. Jia, Y.; Liu, N.; Viswakarma, N.; Sun, R.; Schipma, M.J.; Shang, M.; Thorp, E.B.; Kanwar, Y.S.; Thimmapaya, B.; Reddy, J.K. PIMT/NCOA6IP Deletion in the Mouse Heart Causes Delayed Cardiomyopathy Attributable to Perturbation in Energy Metabolism. *Int. J. Mol. Sci.* **2018**, *19*, 1485. [CrossRef] [PubMed]
43. Liu, H.Y.; Gu, H.; Li, Y.; Hu, P.; Yang, Y.; Li, K.; Li, H.; Zhang, K.; Zhou, B.; Wu, H.; et al. Dietary Conjugated Linoleic Acid Modulates the Hepatic Circadian Clock Program via PPARα/REV-ERBα-Mediated Chromatin Modification in Mice. *Front. Nutr.* **2021**, *8*, 711398. [CrossRef] [PubMed]
44. Zhang, X.; Zhang, S.; Ma, L.; Jiang, E.; Xu, H.; Chen, R.; Yang, Q.; Chen, H.; Li, Z.; Lan, X. Reduced representation bisulfite sequencing (RRBS) of dairy goat mammary glands reveals DNA methylation profiles of integrated genome-wide and critical milk-related genes. *Oncotarget* **2017**, *8*, 115326–115344. [CrossRef]
45. Chen, Z.; Luo, J.; Sun, S.; Cao, D.; Shi, H.; Loor, J.J. miR-148a and miR-17-5p synergistically regulate milk TAG synthesis via PPARGC1A and PPARA in goat mammary epithelial cells. *RNA Biol.* **2017**, *14*, 326–338. [CrossRef]

46. Li, X.; Li, Y.; Ding, H.; Dong, J.; Zhang, R.; Huang, D.; Lei, L.; Wang, Z.; Liu, G.; Li, X. Insulin suppresses the AMPK signaling pathway to regulate lipid metabolism in primary cultured hepatocytes of dairy cows. *J. Dairy Res.* **2018**, *85*, 157–162. [CrossRef]
47. Zou, D.; Liu, R.; Shi, S.; Du, J.; Tian, M.; Wang, X.; Hou, M.; Duan, Z.; Ma, Y. BHBA regulates the expressions of lipid synthesis and oxidation genes in sheep hepatocytes through the AMPK pathway. *Res. Vet. Sci.* **2021**, *140*, 153–163. [CrossRef]
48. Socha, B.M.; Łada, P.; Szczepańska, A.A.; Łupicka, M.; Korzekwa, A.J. The influence of experimentally induced endometritis on the PPAR expression profile in the bovine endometrium. *Theriogenology* **2018**, *122*, 74–83. [CrossRef]
49. Cañón-Beltrán, K.; Cajas, Y.N.; Peréz-Cerezales, S.; Leal, C.L.V.; Agirregoitia, E.; Gutierrez-Adán, A.; González, E.M.; Rizos, D. Nobiletin enhances the development and quality of bovine embryos in vitro during two key periods of embryonic genome activation. *Sci. Rep.* **2021**, *11*, 11796. [CrossRef]
50. Cai, D.; Li, H.; Zhou, B.; Han, L.; Zhang, X.; Yang, G.; Yang, G. Conjugated linoleic acid supplementation caused reduction of perilipin1 and aberrant lipolysis in epididymal adipose tissue. *Biochem. Biophys. Res. Commun.* **2012**, *422*, 621–626. [CrossRef]
51. El Ouarrat, D.; Isaac, R.; Lee, Y.S.; Oh, D.Y.; Wollam, J.; Lackey, D.; Riopel, M.; Bandyopadhyay, G.; Seo, J.B.; Sampath-Kumar, R.; et al. TAZ Is a Negative Regulator of PPARγ Activity in Adipocytes and TAZ Deletion Improves Insulin Sensitivity and Glucose Tolerance. *Cell Metab.* **2020**, *31*, 162–173.e165. [CrossRef] [PubMed]
52. Cai, D.; Liu, H.; Zhao, R. Nuclear Receptors in Hepatic Glucose and Lipid Metabolism During Neonatal and Adult Life. *Curr. Protein Pept. Sci.* **2017**, *18*, 548–561. [CrossRef] [PubMed]
53. Dean, J.M.; He, A.; Tan, M.; Wang, J.; Lu, D.; Razani, B.; Lodhi, I.J. MED19 Regulates Adipogenesis and Maintenance of White Adipose Tissue Mass by Mediating PPARγ-Dependent Gene Expression. *Cell Rep.* **2020**, *33*, 108228. [CrossRef] [PubMed]
54. Hassan, F.U.; Nadeem, A.; Li, Z.; Javed, M.; Liu, Q.; Azhar, J.; Rehman, M.S.; Cui, K.; Rehman, S.U. Role of Peroxisome Proliferator-Activated Receptors (PPARs) in Energy Homeostasis of Dairy Animals: Exploiting Their Modulation through Nutrigenomic Interventions. *Int. J. Mol. Sci.* **2021**, *22*, 2463. [CrossRef] [PubMed]
55. Gao, S.T.; Girma, D.D.; Bionaz, M.; Ma, L.; Bu, D.P. Hepatic transcriptomic adaptation from prepartum to postpartum in dairy cows. *J. Dairy Sci.* **2021**, *104*, 1053–1072. [CrossRef]
56. Schmitt, E.; Ballou, M.A.; Correa, M.N.; DePeters, E.J.; Drackley, J.K.; Loor, J.J. Dietary lipid during the transition period to manipulate subcutaneous adipose tissue peroxisome proliferator-activated receptor-γ co-regulator and target gene expression. *J. Dairy Sci.* **2011**, *94*, 5913–5925. [CrossRef]
57. Fan, Y.; Han, Z.; Lu, X.; Zhang, H.; Arbab, A.A.I.; Loor, J.J.; Yang, Y.; Yang, Z. Identification of Milk Fat Metabolism-Related Pathways of the Bovine Mammary Gland during Mid and Late Lactation and Functional Verification of the ACSL4 Gene. *Genes* **2020**, *11*, 1357. [CrossRef]
58. Sandri, E.C.; Camêra, M.; Sandri, E.M.; Harvatine, K.J.; De Oliveira, D.E. Peroxisome proliferator-activated receptor gamma (PPARγ) agonist fails to overcome trans-10, cis-12 conjugated linoleic acid (CLA) inhibition of milk fat in dairy sheep. *Anim. Int. J. Anim. Biosci.* **2018**, *12*, 1405–1412. [CrossRef]
59. Zhang, Y.; Wang, Y.; Wang, X.; Ji, Y.; Cheng, S.; Wang, M.; Zhang, C.; Yu, X.; Zhao, R.; Zhang, W.; et al. Acetyl-coenzyme A acyltransferase 2 promote the differentiation of sheep precursor adipocytes into adipocytes. *J. Cell. Biochem.* **2019**, *20*, 8021–8031. [CrossRef]
60. Xu, X.; Zhao, R.; Ma, W.; Zhao, Q.; Zhang, G. Comparison of lipid deposition of intramuscular preadipocytes in Tan sheep co-cultured with satellite cells or alone. *J. Anim. Physiol. Anim. Nutr.* **2021**, *10*, 1111. [CrossRef]
61. Wallace, J.M.; Milne, J.S.; Aitken, B.W.; Aitken, R.P.; Adam, C.L. Ovine prenatal growth-restriction and sex influence fetal adipose tissue phenotype and impact postnatal lipid metabolism and adiposity in vivo from birth until adulthood. *PLoS ONE* **2020**, *15*, e0228732. [CrossRef] [PubMed]
62. Fan, Y.; Ren, C.; Meng, F.; Deng, K.; Zhang, G.; Wang, F. Effects of algae supplementation in high-energy dietary on fatty acid composition and the expression of genes involved in lipid metabolism in Hu sheep managed under intensive finishing system. *Meat Sci.* **2019**, *157*, 107872. [CrossRef] [PubMed]
63. Vargas-Bello-Pérez, E.; Zhao, W.; Bionaz, M.; Luo, J.; Loor, J.J. Nutrigenomic Effect of Saturated and Unsaturated Long Chain Fatty Acids on Lipid-Related Genes in Goat Mammary Epithelial Cells: What Is the Role of PPARγ? *Vet. Sci.* **2019**, *6*, 54. [CrossRef] [PubMed]
64. Fensterseifer, S.R.; Austin, K.J.; Ford, S.P.; Alexander, B.M. Effects of maternal obesity on maternal and fetal plasma concentrations of adiponectin and expression of adiponectin and its receptor genes in cotyledonary and adipose tissues at mid- and late-gestation in sheep. *Anim. Reprod. Sci.* **2018**, *197*, 231–239. [CrossRef]
65. Yao, Y.C.; Song, X.T.; Zhai, Y.F.; Liu, S.; Lu, J.; Xu, X.; Qi, M.Y.; Zhang, J.N.; Huang, H.; Liu, Y.F.; et al. Transcriptome analysis of sheep follicular development during prerecruitment, dominant, and mature stages after FSH superstimulation. *Domest. Anim. Endocrinol.* **2021**, *74*, 106563. [CrossRef]
66. Hassanpour, H.; Khalaji-Pirbalouty, V.; Adibi, M.; Nazari, H. Involvement of peroxisome proliferator-activated receptors in the estradiol production of ovine Sertoli cells. *Vet. Res. Forum Int. Q. J.* **2017**, *8*, 251–257.
67. Penna-de-Carvalho, A.; Graus-Nunes, F.; Rabelo-Andrade, J.; Mandarim-de-Lacerda, C.A.; Souza-Mello, V. Enhanced pan-peroxisome proliferator-activated receptor gene and protein expression in adipose tissue of diet-induced obese mice treated with telmisartan. *Exp. Physiol.* **2014**, *99*, 1663–1678. [CrossRef]
68. Tong, L.; Wang, L.; Yao, S.; Jin, L.; Yang, J.; Zhang, Y.; Ning, G.; Zhang, Z. PPARδ attenuates hepatic steatosis through autophagy-mediated fatty acid oxidation. *Cell Death Dis.* **2019**, *10*, 197. [CrossRef]

69. Kim, N.H.; Kim, S.G. Fibrates Revisited: Potential Role in Cardiovascular Risk Reduction. *Diabetes Metab. J.* **2020**, *44*, 213–221. [CrossRef]
70. Yamashita, S.; Masuda, D.; Matsuzawa, Y. Pemafibrate, a New Selective PPARα Modulator: Drug Concept and Its Clinical Applications for Dyslipidemia and Metabolic Diseases. *Curr. Atheroscler. Rep.* **2020**, *22*, 5. [CrossRef]
71. De Vries, E.; Bolier, R.; Goet, J.; Parés, A.; Verbeek, J.; de Vree, M.; Drenth, J.; van Erpecum, K.; van Nieuwkerk, K.; van der Heide, F.; et al. Fibrates for Itch (FITCH) in Fibrosing Cholangiopathies: A Double-Blind, Randomized, Placebo-Controlled Trial. *Gastroenterology* **2021**, *160*, 734–743.e736. [CrossRef] [PubMed]
72. Rubio, B.; Mora, C.; Pintado, C.; Mazuecos, L.; Fernández, A.; López, V.; Andrés, A.; Gallardo, N. The nutrient sensing pathways FoxO1/3 and mTOR in the heart are coordinately regulated by central leptin through PPARβ/δ. Implications in cardiac remodeling. *Metab. Clin. Exp.* **2021**, *115*, 154453. [CrossRef] [PubMed]
73. Moreira-Pais, A.; Ferreira, R.; Neves, J.S.; Vitorino, R.; Moreira-Gonçalves, D.; Nogueira-Ferreira, R. Sex differences on adipose tissue remodeling: From molecular mechanisms to therapeutic interventions. *J. Mol. Med.* **2020**, *98*, 483–493. [CrossRef]
74. Park, H.J.; Choi, J.M. Sex-specific regulation of immune responses by PPARs. *Exp. Mol. Med.* **2017**, *49*, e364. [CrossRef]
75. Hara, S.; Furukawa, F.; Mukai, K.; Yazawa, T.; Kitano, T. Peroxisome proliferator-activated receptor alpha is involved in the temperature-induced sex differentiation of a vertebrate. *Sci. Rep.* **2020**, *10*, 11672. [CrossRef]
76. Pierrot, N.; Ris, L.; Stancu, I.C.; Doshina, A.; Ribeiro, F.; Tyteca, D.; Baugé, E.; Lalloyer, F.; Malong, L.; Schakman, O.; et al. Sex-regulated gene dosage effect of PPARα on synaptic plasticity. *Life Sci. Alliance* **2019**, *2*, e201800262. [CrossRef]
77. De Felice, M.; Melis, M.; Aroni, S.; Muntoni, A.L.; Fanni, S.; Frau, R.; Devoto, P.; Pistis, M. The PPARα agonist fenofibrate attenuates disruption of dopamine function in a maternal immune activation rat model of schizophrenia. *CNS Neurosci. Ther.* **2019**, *25*, 549–561. [CrossRef]
78. Luft, C.; Levices, I.P.; Pedrazza, L.; de Oliveira, J.R.; Donadio, M.V.F. Sex-dependent metabolic effects of pregestational exercise on prenatally stressed mice. *J. Dev. Orig. Health Dis.* **2021**, *12*, 271–279. [CrossRef]
79. Bapat, S.P.; Whitty, C.; Mowery, C.T.; Liang, Y.; Yoo, A.; Jiang, Z.; Peters, M.C.; Zhang, L.J.; Vogel, I.; Zhou, C.; et al. Obesity alters pathology and treatment response in inflammatory disease. *Nature* **2022**, *604*, 337–342. [CrossRef]
80. Xin, G.; Chen, Y.; Topchyan, P.; Kasmani, M.Y.; Burns, R.; Volberding, P.J.; Wu, X.; Cohn, A.; Chen, Y.; Lin, C.W.; et al. Targeting PIM1-Mediated Metabolism in Myeloid Suppressor Cells to Treat Cancer. *Cancer Immunol. Res.* **2021**, *9*, 454–469. [CrossRef]
81. Li, C.; Wang, G.; Sivasami, P.; Ramirez, R.N.; Zhang, Y.; Benoist, C.; Mathis, D. Interferon-α-producing plasmacytoid dendritic cells drive the loss of adipose tissue regulatory T cells during obesity. *Cell Metab.* **2021**, *33*, 1610–1623.e1615. [CrossRef] [PubMed]
82. Kim, S.; Lee, N.; Park, E.S.; Yun, H.; Ha, T.U.; Jeon, H.; Yu, J.; Choi, S.; Shin, B.; Yu, J.; et al. T-Cell Death Associated Gene 51 Is a Novel Negative Regulator of PPARγ That Inhibits PPARγ-RXRα Heterodimer Formation in Adipogenesis. *Mol. Cells* **2021**, *44*, 1–12. [CrossRef] [PubMed]
83. Van Herck, M.A.; Vonghia, L.; Kwanten, W.J.; Vanwolleghem, T.; Ebo, D.G.; Michielsen, P.P.; De Man, J.G.; Gama, L.; De Winter, B.Y.; Francque, S.M. Adoptive Cell Transfer of Regulatory T Cells Exacerbates Hepatic Steatosis in High-Fat High-Fructose Diet-Fed Mice. *Front. Immunol.* **2020**, *11*, 1711. [CrossRef]
84. Fontecha-Barriuso, M.; Martin-Sanchez, D.; Martinez-Moreno, J.M.; Monsalve, M.; Ramos, A.M.; Sanchez-Niño, M.D.; Ruiz-Ortega, M.; Ortiz, A.; Sanz, A.B. The Role of PGC-1α and Mitochondrial Biogenesis in Kidney Diseases. *Biomolecules* **2020**, *10*, 347. [CrossRef] [PubMed]
85. Fontecha-Barriuso, M.; Martín-Sánchez, D.; Martinez-Moreno, J.M.; Carrasco, S.; Ruiz-Andrés, O.; Monsalve, M.; Sanchez-Ramos, C.; Gómez, M.J.; Ruiz-Ortega, M.; Sánchez-Niño, M.D.; et al. PGC-1α deficiency causes spontaneous kidney inflammation and increases the severity of nephrotoxic AKI. *J. Pathol.* **2019**, *249*, 65–78. [CrossRef] [PubMed]
86. Bagattin, A.; Hugendubler, L.; Mueller, E. Transcriptional coactivator PGC-1alpha promotes peroxisomal remodeling and biogenesis. *Proc. Natl. Acad. Sci. USA* **2010**, *107*, 20376–20381. [CrossRef] [PubMed]
87. Huang, T.Y.; Zheng, D.; Houmard, J.A.; Brault, J.J.; Hickner, R.C.; Cortright, R.N. Overexpression of PGC-1α increases peroxisomal activity and mitochondrial fatty acid oxidation in human primary myotubes. *Am. J. Physiol. Endocrinol. Metab.* **2017**, *312*, e253–e263. [CrossRef]
88. Huang, T.Y.; Linden, M.A.; Fuller, S.E.; Goldsmith, F.R.; Simon, J.; Batdorf, H.M.; Scott, M.C.; Essajee, N.M.; Brown, J.M.; Noland, R.C. Combined effects of a ketogenic diet and exercise training alter mitochondrial and peroxisomal substrate oxidative capacity in skeletal muscle. *Am. J. Physiol. Endocrinol. Metab.* **2021**, *320*, e1053–e1067. [CrossRef]
89. Summermatter, S.; Baum, O.; Santos, G.; Hoppeler, H.; Handschin, C. Peroxisome proliferator-activated receptor {gamma} coactivator 1{alpha} (PGC-1{alpha}) promotes skeletal muscle lipid refueling in vivo by activating de novo lipogenesis and the pentose phosphate pathway. *J. Biol. Chem.* **2010**, *285*, 32793–32800. [CrossRef]
90. Held-Hoelker, E.; Klein, S.L.; Rings, F.; Salilew-Wondim, D.; Saeed-Zidane, M.; Neuhoff, C.; Tesfaye, D.; Schellander, K.; Hoelker, M. Cryosurvival of in vitro produced bovine embryos supplemented with l-Carnitine and concurrent reduction of fatty acids. *Theriogenology* **2017**, *96*, 145–152. [CrossRef]
91. Tan, X.; Zhu, T.; Zhang, L.; Fu, L.; Hu, Y.; Li, H.; Li, C.; Zhang, J.; Liang, B.; Liu, J. miR-669a-5p promotes adipogenic differentiation and induces browning in preadipocytes. *Adipocyte* **2022**, *11*, 120–132. [CrossRef] [PubMed]
92. Zhou, X.; Yan, Q.; Yang, H.; Ren, A.; He, Z.; Tan, Z. Maternal intake restriction programs the energy metabolism, clock circadian regulator and mTOR signals in the skeletal muscles of goat offspring probably via the protein kinase A-cAMP-responsive element-binding proteins pathway. *Anim. Nutr.* **2021**, *7*, 1303–1314. [CrossRef] [PubMed]

93. Distel, B.; Erdmann, R.; Gould, S.J.; Blobel, G.; Crane, D.I.; Cregg, J.M.; Dodt, G.; Fujiki, Y.; Goodman, J.M.; Just, W.W.; et al. A unified nomenclature for peroxisome biogenesis factors. *J. Cell Biol.* **1996**, *135*, 1–3. [CrossRef]
94. Barros-Barbosa, A.; Rodrigues, T.A.; Ferreira, M.J.; Pedrosa, A.G.; Teixeira, N.R.; Francisco, T.; Azevedo, J.E. The intrinsically disordered nature of the peroxisomal protein translocation machinery. *FEBS J.* **2019**, *286*, 24–38. [CrossRef] [PubMed]
95. Wang, W.; Xia, Z.J.; Farré, J.C.; Subramani, S. TRIM37, a novel E3 ligase for PEX5-mediated peroxisomal matrix protein import. *J. Cell Biol.* **2017**, *216*, 2843–2858. [CrossRef]
96. Klouwer, F.C.C.; Falkenberg, K.D.; Ofman, R.; Koster, J.; van Gent, D.; Ferdinandusse, S.; Wanders, R.J.A.; Waterham, H.R. Autophagy Inhibitors Do Not Restore Peroxisomal Functions in Cells with the Most Common Peroxisome Biogenesis Defect. *Front. Cell Dev. Biol.* **2021**, *9*, 661298. [CrossRef]
97. Lotz-Havla, A.S.; Woidy, M.; Guder, P.; Friedel, C.C.; Klingbeil, J.M.; Bulau, A.M.; Schultze, A.; Dahmen, I.; Noll-Puchta, H.; Kemp, S.; et al. iBRET Screen of the ABCD1 Peroxisomal Network and Mutation-Induced Network Perturbations. *J. Proteome Res.* **2021**, *20*, 4366–4380. [CrossRef]
98. Coppa, A.; Guha, S.; Fourcade, S.; Parameswaran, J.; Ruiz, M.; Moser, A.B.; Schlüter, A.; Murphy, M.P.; Lizcano, J.M.; Miranda-Vizuete, A.; et al. The peroxisomal fatty acid transporter ABCD1/PMP-4 is required in the C. elegans hypodermis for axonal maintenance: A worm model for adrenoleukodystrophy. *Free Radic. Biol. Med.* **2020**, *152*, 797–809. [CrossRef]
99. Zierfuss, B.; Weinhofer, I.; Kühl, J.S.; Köhler, W.; Bley, A.; Zauner, K.; Binder, J.; Martinović, K.; Seiser, C.; Hertzberg, C.; et al. Vorinostat in the acute neuroinflammatory form of X-linked adrenoleukodystrophy. *Ann. Clin. Transl. Neurol.* **2020**, *7*, 639–652. [CrossRef]
100. Violante, S.; Achetib, N.; van Roermund, C.W.T.; Hagen, J.; Dodatko, T.; Vaz, F.M.; Waterham, H.R.; Chen, H.; Baes, M.; Yu, C.; et al. Peroxisomes can oxidize medium- and long-chain fatty acids through a pathway involving ABCD3 and HSD17B4. *FASEB J.* **2019**, *33*, 4355–4364. [CrossRef]
101. Wang, J.; Ma, H.; Zhao, S.; Huang, J.; Yang, Y.; Tabashnik, B.E.; Wu, Y. Functional redundancy of two ABC transporter proteins in mediating toxicity of Bacillus thuringiensis to cotton bollworm. *PLoS Pathog.* **2020**, *16*, e1008427. [CrossRef] [PubMed]
102. Zhao, S.; Jiang, D.; Wang, F.; Yang, Y.; Tabashnik, B.E.; Wu, Y. Independent and Synergistic Effects of Knocking out Two ABC Transporter Genes on Resistance to Bacillus thuringiensis Toxins Cry1Ac and Cry1Fa in Diamondback Moth. *Toxins* **2020**, *13*, 9. [CrossRef]
103. Ahmad, S.M.; Bhat, B.; Bhat, S.A.; Yaseen, M.; Mir, S.; Raza, M.; Iquebal, M.A.; Shah, R.A.; Ganai, N.A. SNPs in Mammary Gland Epithelial Cells Unraveling Potential Difference in Milk Production Between Jersey and Kashmiri Cattle Using RNA Sequencing. *Front. Genet.* **2021**, *12*, 666015. [CrossRef] [PubMed]
104. Lopez, B.I.; Santiago, K.G.; Lee, D.; Ha, S.; Seo, K. RNA Sequencing (RNA-Seq) Based Transcriptome Analysis in Immune Response of Holstein Cattle to Killed Vaccine against Bovine Viral Diarrhea Virus Type I. *Animals* **2020**, *10*, 344. [CrossRef] [PubMed]
105. Xu, J.; Zhou, Y.; Yang, Y.; Lv, C.; Liu, X.; Wang, Y. Involvement of ABC-transporters and acyltransferase 1 in intracellular cholesterol-mediated autophagy in bovine alveolar macrophages in response to the Bacillus Calmette-Guerin (BCG) infection. *BMC Immunol.* **2020**, *21*, 26. [CrossRef]
106. Sales, A.D.; Duarte, A.B.G.; Rocha, R.M.P.; Brito, I.R.; Locatelli, Y.; Alves, B.G.; Alves, K.A.; Figueiredo, J.R.; Rodrigues, A.P.R. Transcriptional downregulation of ABC transporters is related to follicular degeneration after vitrification and in vitro culture of ovine ovarian tissue. *Theriogenology* **2022**, *177*, 127–132. [CrossRef]
107. Zhou, F.; Ding, M.; Gu, Y.; Fan, G.; Liu, C.; Li, Y.; Sun, R.; Wu, J.; Li, J.; Xue, X.; et al. Aurantio-Obtusin Attenuates Non-Alcoholic Fatty Liver Disease Through AMPK-Mediated Autophagy and Fatty Acid Oxidation Pathways. *Front. Pharmacol.* **2021**, *12*, 826628. [CrossRef]
108. Xie, Y.H.; Xiao, Y.; Huang, Q.; Hu, X.F.; Gong, Z.C.; Du, J. Role of the CTRP6/AMPK pathway in kidney fibrosis through the promotion of fatty acid oxidation. *Eur. J. Pharmacol.* **2021**, *892*, 173755. [CrossRef]
109. Zhang, X.; Li, L.; Li, Y.; Li, Z.; Zhai, W.; Sun, Q.; Yang, X.; Roth, M.; Lu, S. mTOR regulates PRMT1 expression and mitochondrial mass through STAT1 phosphorylation in hepatic cell. *Biochim. Biophys. Acta. Mol. Cell Res.* **2021**, *1868*, 119017. [CrossRef]
110. Liu, Z.; Liao, W.; Yin, X.; Zheng, X.; Li, Q.; Zhang, H.; Zheng, L.; Feng, X. Resveratrol-induced brown fat-like phenotype in 3T3-L1 adipocytes partly via mTOR pathway. *Food Nutr. Res.* **2020**, *64*, 3656. [CrossRef]
111. Wang, R.; Hu, W. Asprosin promotes β-cell apoptosis by inhibiting the autophagy of β-cell via AMPK-mTOR pathway. *J. Cell. Physiol.* **2021**, *236*, 215–221. [CrossRef] [PubMed]
112. Dai, H.; Sinclair, D.A.; Ellis, J.L.; Steegborn, C. Sirtuin activators and inhibitors: Promises, achievements, and challenges. *Pharmacol. Ther.* **2018**, *188*, 140–154. [CrossRef] [PubMed]
113. Curry, A.M.; White, D.S.; Donu, D.; Cen, Y. Human Sirtuin Regulators: The "Success" Stories. *Front. Physiol.* **2021**, *12*, 752117. [CrossRef] [PubMed]
114. Chen, X.F.; Tian, M.X.; Sun, R.Q.; Zhang, M.L.; Zhou, L.S.; Jin, L.; Chen, L.L.; Zhou, W.J.; Duan, K.L.; Chen, Y.J.; et al. SIRT5 inhibits peroxisomal ACOX1 to prevent oxidative damage and is downregulated in liver cancer. *EMBO Rep.* **2018**, *19*, e45124. [CrossRef]
115. Shumar, S.A.; Kerr, E.W.; Geldenhuys, W.J.; Montgomery, G.E.; Fagone, P.; Thirawatananond, P.; Saavedra, H.; Gabelli, S.B.; Leonardi, R. Nudt19 is a renal CoA diphosphohydrolase with biochemical and regulatory properties that are distinct from the hepatic Nudt7 isoform. *J. Biol. Chem.* **2018**, *293*, 4134–4148. [CrossRef]

116. Shumar, S.A.; Kerr, E.W.; Fagone, P.; Infante, A.M.; Leonardi, R. Overexpression of Nudt7 decreases bile acid levels and peroxisomal fatty acid oxidation in the liver. *J. Lipid Res.* **2019**, *60*, 1005–1019. [CrossRef]
117. Li, C.; Jiang, L.; Jin, Y.; Zhang, D.; Chen, J.; Qi, Y.; Fan, R.; Luo, J.; Xu, L.; Ma, W.; et al. Lipid metabolism disorders effects of 6:2 chlorinated polyfluorinated ether sulfonate through Hsa-miRNA-532-3p/Acyl-CoA oxidase 1(ACOX1) pathway. *Ecotoxicol. Environ. Saf.* **2021**, *228*, 113011. [CrossRef]
118. Zhang, F.; Xiong, Q.; Tao, H.; Liu, Y.; Zhang, N.; Li, X.F.; Suo, X.J.; Yang, Q.P.; Chen, M.X. ACOX1, regulated by C/EBPα and miR-25-3p, promotes bovine preadipocyte adipogenesis. *J. Mol. Endocrinol.* **2021**, *66*, 195–205. [CrossRef]
119. Wang, J.J.; Zhang, Y.T.; Tseng, Y.J.; Zhang, J. miR-222 targets ACOX1, promotes triglyceride accumulation in hepatocytes. *Hepatobiliary Pancreat. Dis. Int. HBPD INT* **2019**, *18*, 360–365. [CrossRef]
120. Lai, Y.H.; Liu, H.; Chiang, W.F.; Chen, T.W.; Chu, L.J.; Yu, J.S.; Chen, S.J.; Chen, H.C.; Tan, B.C. MiR-31-5p-ACOX1 Axis Enhances Tumorigenic Fitness in Oral Squamous Cell Carcinoma Via the Promigratory Prostaglandin E2. *Theranostics* **2018**, *8*, 486–504. [CrossRef]
121. Li, G.; Fu, S.; Chen, Y.; Jin, W.; Zhai, B.; Li, Y.; Sun, G.; Han, R.; Wang, Y.; Tian, Y.; et al. MicroRNA-15a Regulates the Differentiation of Intramuscular Preadipocytes by Targeting ACAA1, ACOX1 and SCP2 in Chickens. *Int. J. Mol. Sci.* **2019**, *20*, 63. [CrossRef] [PubMed]
122. Zhou, H.X.; Wang, L.Y.; Chen, S.; Wang, D.D.; Fang, Z. circ_0005379 inhibits the progression of oral squamous cell carcinoma by regulating the miR-17-5p/acyl-CoA oxidase 1 axis. *Hua Xi Kou Qiang Yi Xue Za Zhi = Huaxi Kouqiang Yixue Zazhi = West China J. Stomatol.* **2021**, *39*, 425–433. [CrossRef]
123. Lee, S.A.; Lee, J.; Kim, K.; Moon, H.; Min, C.; Moon, B.; Kim, D.; Yang, S.; Park, H.; Lee, G.; et al. The Peroxisomal Localization of Hsd17b4 Is Regulated by Its Interaction with Phosphatidylserine. *Mol. Cells* **2021**, *44*, 214–222. [CrossRef] [PubMed]
124. Zhang, Y.; Xu, Y.Y.; Yao, C.B.; Li, J.T.; Zhao, X.N.; Yang, H.B.; Zhang, M.; Yin, M.; Chen, J.; Lei, Q.Y. Acetylation targets HSD17B4 for degradation via the CMA pathway in response to estrone. *Autophagy* **2017**, *13*, 538–553. [CrossRef] [PubMed]
125. Huang, H.; Liu, R.; Huang, Y.; Feng, Y.; Fu, Y.; Chen, L.; Chen, Z.; Cai, Y.; Zhang, Y.; Chen, Y. Acetylation-mediated degradation of HSD17B4 regulates the progression of prostate cancer. *Aging* **2020**, *12*, 14699–14717. [CrossRef]
126. Wang, Y.; Li, X.; Cao, Y.; Xiao, C.; Liu, Y.; Jin, H.; Cao, Y. Effect of the ACAA1 Gene on Preadipocyte Differentiation in Sheep. *Front. Genet.* **2021**, *12*, 649140. [CrossRef]
127. Ranea-Robles, P.; Violante, S.; Argmann, C.; Dodatko, T.; Bhattacharya, D.; Chen, H.; Yu, C.; Friedman, S.L.; Puchowicz, M.; Houten, S.M. Murine deficiency of peroxisomal L-bifunctional protein (EHHADH) causes medium-chain 3-hydroxydicarboxylic aciduria and perturbs hepatic cholesterol homeostasis. *Cell. Mol. Life Sci. CMLS* **2021**, *78*, 5631–5646. [CrossRef]
128. Landau, Y.E.; Heimer, G.; Barel, O.; Shalva, N.; Marek-Yagel, D.; Veber, A.; Javasky, E.; Shilon, A.; Nissenkorn, A.; Ben-Zeev, B.; et al. Four patients with D-bifunctional protein (DBP) deficiency: Expanding the phenotypic spectrum of a highly variable disease. *Mol. Genet. Metab. Rep.* **2020**, *25*, 100631. [CrossRef]
129. Ferdinandusse, S.; Denis, S.; Hogenhout, E.M.; Koster, J.; van Roermund, C.W.T.; Ijlst, L.; Moser, A.B.; Wanders, R.J.; Waterham, H.R. Clinical, biochemical, and mutational spectrum of peroxisomal acyl-coenzyme A oxidase deficiency. *Hum. Mutat.* **2007**, *28*, 904–912. [CrossRef]
130. Turk, B.R.; Theda, C.; Fatemi, A.; Moser, A.B. X-linked adrenoleukodystrophy: Pathology, pathophysiology, diagnostic testing, newborn screening and therapies. *Int. J. Dev. Neurosci.* **2020**, *80*, 52–72. [CrossRef]
131. Sugiura, A.; Mattie, S.; Prudent, J.; McBride, H.M. Newly born peroxisomes are a hybrid of mitochondrial and ER-derived pre-peroxisomes. *Nature* **2017**, *542*, 251–254. [CrossRef] [PubMed]
132. Ding, L.; Sun, W.; Balaz, M.; He, A.; Klug, M.; Wieland, S.; Caiazzo, R.; Raverdy, V.; Pattou, F.; Lefebvre, P.; et al. Peroxisomal β-oxidation acts as a sensor for intracellular fatty acids and regulates lipolysis. *Nat. Metab.* **2021**, *3*, 1648–1661. [CrossRef] [PubMed]
133. Cui, W.; Liu, D.; Gu, W.; Chu, B. Peroxisome-driven ether-linked phospholipids biosynthesis is essential for ferroptosis. *Cell Death Differ.* **2021**, *28*, 2536–2551. [CrossRef] [PubMed]
134. MacLean, G.E.; Argyriou, C.; Di Pietro, E.; Sun, X.; Birjandian, S.; Saberian, P.; Hacia, J.G.; Braverman, N.E. Zellweger spectrum disorder patient-derived fibroblasts with the PEX1-Gly843Asp allele recover peroxisome functions in response to flavonoids. *J. Cell. Biochem.* **2019**, *120*, 3243–3258. [CrossRef]
135. Enns, G.M.; Ammous, Z.; Himes, R.W.; Nogueira, J.; Palle, S.; Sullivan, M.; Ramirez, C. Diagnostic challenges and disease management in patients with a mild Zellweger spectrum disorder phenotype. *Mol. Genet. Metab.* **2021**, *134*, 217–222. [CrossRef]
136. Wangler, M.F.; Hubert, L.; Donti, T.R.; Ventura, M.J.; Miller, M.J.; Braverman, N.; Gawron, K.; Bose, M.; Moser, A.B.; Jones, R.O.; et al. A metabolomic map of Zellweger spectrum disorders reveals novel disease biomarkers. *Genet. Med.* **2018**, *20*, 1274–1283. [CrossRef]
137. Klouwer, F.C.C.; Koot, B.G.P.; Berendse, K.; Kemper, E.M.; Ferdinandusse, S.; Koelfat, K.V.K.; Lenicek, M.; Vaz, F.M.; Engelen, M.; Jansen, P.L.M.; et al. The cholic acid extension study in Zellweger spectrum disorders: Results and implications for therapy. *J. Inherit. Metab. Dis.* **2019**, *42*, 303–312. [CrossRef]
138. Zeynelabidin, S.; Klouwer, F.C.C.; Meijers, J.C.M.; Suijker, M.H.; Engelen, M.; Poll-The, B.T.; van Ommen, C.H. Coagulopathy in Zellweger spectrum disorders: A role for vitamin K. *J. Inherit. Metab. Dis.* **2018**, *41*, 249–255. [CrossRef]
139. Anderson, J.N.; Ammous, Z.; Eroglu, Y.; Hernandez, E.; Heubi, J.; Himes, R.; Palle, S. Cholbam®and Zellweger spectrum disorders: Treatment implementation and management. *Orphanet J. Rare Dis.* **2021**, *16*, 388. [CrossRef]

140. Semenova, N.A.; Kurkina, M.V.; Marakhonov, A.V.; Dadali, E.L.; Taran, N.N.; Strokova, T.V. A novel mutation in the PEX26 gene in a family from Dagestan with members affected by Zellweger spectrum disorder. *Mol. Genet. Metab. Rep.* **2021**, *27*, 100754. [CrossRef]
141. Havali, C.; Dorum, S.; Akbaş, Y.; Görükmez, O.; Hirfanoglu, T. Two different missense mutations of PEX genes in two similar patients with severe Zellweger syndrome: An argument on the genotype-phenotype correlation. *J. Pediatric Endocrinol. Metab. JPEM* **2020**, *33*, 437–441. [CrossRef] [PubMed]
142. Lucaccioni, L.; Righi, B.; Cingolani, G.M.; Lugli, L.; Della Casa, E.; Torcetta, F.; Iughetti, L.; Berardi, A. Overwhelming sepsis in a neonate affected by Zellweger syndrome due to a compound heterozygosis in PEX 6 gene: A case report. *BMC Med. Genet.* **2020**, *21*, 229. [CrossRef] [PubMed]
143. Galvez-Ruiz, A.; Galindo-Ferreiro, A.; Alkatan, H. A clinical case of Zellweger syndrome in a patient with a previous history of ocular medulloepithelioma. *Saudi J. Ophthalmol.* **2018**, *32*, 241–245. [CrossRef] [PubMed]
144. Miyamoto, T.; Hosoba, K.; Itabashi, T.; Iwane, A.H.; Akutsu, S.N.; Ochiai, H.; Saito, Y.; Yamamoto, T.; Matsuura, S. Insufficiency of ciliary cholesterol in hereditary Zellweger syndrome. *EMBO J.* **2020**, *39*, e103499. [CrossRef]
145. Lipiński, P.; Stawiński, P.; Rydzanicz, M.; Wypchło, M.; Płoski, R.; Stradomska, T.J.; Jurkiewicz, E.; Ferdinandusse, S.; Wanders, R.J.A.; Vaz, F.M.; et al. Mild Zellweger syndrome due to functionally confirmed novel PEX1 variants. *J. Appl. Genet.* **2020**, *61*, 87–91. [CrossRef]
146. He, Y.; Lin, S.B.; Li, W.X.; Yang, L.; Zhang, R.; Chen, C.; Yuan, L. PEX26 gene genotype-phenotype correlation in neonates with Zellweger syndrome. *Transl. Pediatrics* **2021**, *10*, 1825–1833. [CrossRef]
147. Cheillan, D. Zellweger Syndrome Disorders: From Severe Neonatal Disease to Atypical Adult Presentation. *Adv. Exp. Med. Biol.* **2020**, *1299*, 71–80. [CrossRef]
148. Al-Essa, M.; Dhaunsi, G.S. Selective receptor-mediated impairment of growth factor activity in neonatal- and X-linked adrenoleukodystrophy patients. *J. Pediatric Endocrinol. Metab. JPEM* **2019**, *32*, 733–738. [CrossRef]
149. Farrell, D.F. Neonatal adrenoleukodystrophy: A clinical, pathologic, and biochemical study. *Pediatric Neurol.* **2012**, *47*, 330–336. [CrossRef]
150. Elghawy, O.; Zhang, A.Y.; Duong, R.; Wilson, W.G.; Shildkrot, E.Y. Ophthalmic Diagnosis and Novel Management of Infantile Refsum Disease with Combination Docosahexaenoic Acid and Cholic Acid. *Case Rep. Ophthalmol. Med.* **2021**, *2021*, 1345937. [CrossRef]
151. Dhaunsi, G.S.; Alsaeid, M.; Akhtar, S. Phytanic acid attenuates insulin-like growth factor-1 activity via nitric oxide-mediated γ-secretase activation in rat aortic smooth muscle cells: Possible implications for pathogenesis of infantile Refsum disease. *Pediatr. Res.* **2017**, *81*, 531–536. [CrossRef] [PubMed]
152. Warren, M.; Mierau, G.; Wartchow, E.P.; Shimada, H.; Yano, S. Histologic and ultrastructural features in early and advanced phases of Zellweger spectrum disorder (infantile Refsum disease). *Ultrastruct. Pathol.* **2018**, *42*, 220–227. [CrossRef] [PubMed]
153. Chen, K.; Zhang, N.; Shao, J.B.; Li, H.; Li, J.; Xi, J.M.; Xu, W.H.; Jiang, H. Allogeneic Hematopoietic Stem Cell Transplantation for PEX1-Related Zellweger Spectrum Disorder: A Case Report and Literature Review. *Front. Pediatrics* **2021**, *9*, 672187. [CrossRef] [PubMed]
154. Gao, F.J.; Hu, F.Y.; Xu, P.; Qi, Y.H.; Li, J.K.; Zhang, Y.J.; Chen, F.; Chang, Q.; Song, F.; Shen, S.M.; et al. Expanding the clinical and genetic spectrum of Heimler syndrome. *Orphanet J. Rare Dis.* **2019**, *14*, 290. [CrossRef]
155. Kim, Y.J.; Abe, Y.; Kim, Y.J.; Fujiki, Y.; Kim, J.W. Identification of a Homozygous PEX26 Mutation in a Heimler Syndrome Patient. *Genes* **2021**, *12*, 646. [CrossRef]
156. Daich Varela, M.; Jani, P.; Zein, W.M.; D'Souza, P.; Wolfe, L.; Chisholm, J.; Zalewski, C.; Adams, D.; Warner, B.M.; Huryn, L.A.; et al. The peroxisomal disorder spectrum and Heimler syndrome: Deep phenotyping and review of the literature. *Am. J. Med Genet. Part C Semin. Med. Genet.* **2020**, *184*, 618–630. [CrossRef]
157. Cordisco, A.; Pelo, E.; Di Tommaso, M.; Biagiotti, R. A new GNPAT variant of foetal rhizomelic chondrodysplasia punctata. *Mol. Genet. Genom. Med.* **2021**, *9*, e1733. [CrossRef]
158. Fallatah, W.; Schouten, M.; Yergeau, C.; Di Pietro, E.; Engelen, M.; Waterham, H.R.; Poll-The, B.T.; Braverman, N. Clinical, biochemical, and molecular characterization of mild (nonclassic) rhizomelic chondrodysplasia punctata. *J. Inherit. Metab. Dis.* **2021**, *44*, 1021–1038. [CrossRef]
159. Abousamra, O.; Kandula, V.; Duker, A.L.; Rogers, K.J.; Bober, M.B.; Mackenzie, W.G. Cervical Spine Deformities in Children With Rhizomelic Chondrodysplasia Punctata. *J. Pediatric Orthop.* **2019**, *39*, e680–e686. [CrossRef]
160. Duker, A.L.; Niiler, T.; Kinderman, D.; Schouten, M.; Poll-The, B.T.; Braverman, N.; Bober, M.B. Rhizomelic chondrodysplasia punctata morbidity and mortality, an update. *Am. J. Med. Genet. Part A* **2020**, *182*, 579–583. [CrossRef]
161. Braverman, N.E.; Steinberg, S.J.; Fallatah, W.; Duker, A.; Bober, M. Rhizomelic Chondrodysplasia Punctata Type 1. In *GeneReviews(®)*; Adam, M.P., Mirzaa, G.M., Pagon, R.A., Wallace, S.E., Bean, L.J.H., Gripp, K.W., Amemiya, A., Eds.; University of Washington: Seattle, WA, USA, 1993.
162. Lee, D.K.; Long, N.P.; Jung, J.; Kim, T.J.; Na, E.; Kang, Y.P.; Kwon, S.W.; Jang, J. Integrative lipidomic and transcriptomic analysis of X-linked adrenoleukodystrophy reveals distinct lipidome signatures between adrenomyeloneuropathy and childhood cerebral adrenoleukodystrophy. *Biochem. Biophys. Res. Commun.* **2019**, *508*, 563–569. [CrossRef] [PubMed]

163. Kanakis, G.; Kaltsas, G. Adrenal Insufficiency Due to X-Linked Adrenoleukodystrophy. In *Endotext*; Feingold, K.R., Anawalt, B., Boyce, A., Chrousos, G., de Herder, W.W., Dhatariya, K., Dungan, K., Hershman, J.M., Hofland, J., Kalra, S., et al., Eds.; MDText.com, Inc.: South Dartmouth, MA, USA, 2000.
164. Manor, J.; Chung, H.; Bhagwat, P.K.; Wangler, M.F. ABCD1 and X-linked adrenoleukodystrophy: A disease with a markedly variable phenotype showing conserved neurobiology in animal models. *J. Neurosci. Res.* **2021**, *99*, 3170–3181. [CrossRef] [PubMed]
165. Palakuzhiyil, S.V.; Christopher, R.; Chandra, S.R. Deciphering the modifiers for phenotypic variability of X-linked adrenoleukodystrophy. *World J. Biol. Chem.* **2020**, *11*, 99–111. [CrossRef] [PubMed]
166. Van de Stadt, S.I.W.; Huffnagel, I.C.; Turk, B.R.; van der Knaap, M.S.; Engelen, M. Imaging in X-Linked Adrenoleukodystrophy. *Neuropediatrics* **2021**, *52*, 252–260. [CrossRef] [PubMed]
167. Morita, A.; Enokizono, T.; Ohto, T.; Tanaka, M.; Watanabe, S.; Takada, Y.; Iwama, K.; Mizuguchi, T.; Matsumoto, N.; Morita, M.; et al. Novel ACOX1 mutations in two siblings with peroxisomal acyl-CoA oxidase deficiency. *Brain Dev.* **2021**, *43*, 475–481. [CrossRef]
168. Chung, H.L.; Wangler, M.F.; Marcogliese, P.C.; Jo, J.; Ravenscroft, T.A.; Zuo, Z.; Duraine, L.; Sadeghzadeh, S.; Li-Kroeger, D.; Schmidt, R.E.; et al. Loss- or Gain-of-Function Mutations in ACOX1 Cause Axonal Loss via Different Mechanisms. *Neuron* **2020**, *106*, 589–606.e586. [CrossRef]
169. Raas, Q.; Saih, F.E.; Gondcaille, C.; Trompier, D.; Hamon, Y.; Leoni, V.; Caccia, C.; Nasser, B.; Jadot, M.; Ménétrier, F.; et al. A microglial cell model for acyl-CoA oxidase 1 deficiency. *Biochim. Biophys. Acta. Mol. Cell Biol. Lipids* **2019**, *1864*, 567–576. [CrossRef]
170. Yang, S.M.; Cao, C.D.; Ding, Y.; Wang, M.J.; Yue, S.J. D-bifunctional protein deficiency caused by HSD17B4 gene mutation in a neonate. *Zhongguo Dang Dai Er Ke Za Zhi = Chin. J. Contemp. Pediatrics* **2021**, *23*, 1058–1063. [CrossRef]
171. Werner, K.M.; Cox, A.J.; Qian, E.; Jain, P.; Ji, W.; Tikhonova, I.; Castaldi, C.; Bilguvar, K.; Knight, J.; Ferdinandusse, S.; et al. D-bifunctional protein deficiency caused by splicing variants in a neonate with severe peroxisomal dysfunction and persistent hypoglycemia. *Am. J. Med. Genet. Part A* **2022**, *188*, 357–363. [CrossRef]
172. Veenvliet, A.R.J.; Garrelfs, M.R.; Udink ten Cate, F.E.A.; Ferdinandusse, S.; Denis, S.; Fuchs, S.A.; Schwantje, M.; Geurtzen, R.; van Wegberg, A.M.J.; Huigen, M.C.D.G.; et al. Neonatal Long-Chain 3-Ketoacyl-CoA Thiolase deficiency: Clinical-biochemical phenotype, sodium-D,L-3-hydroxybutyrate treatment experience and cardiac evaluation using speckle echocardiography. *Mol. Genet. Metab. Rep.* **2022**, *31*, 100873. [CrossRef]
173. Martines, A.M.F.; van Eunen, K.; Reijngoud, D.J.; Bakker, B.M. The promiscuous enzyme medium-chain 3-keto-acyl-CoA thiolase triggers a vicious cycle in fatty-acid beta-oxidation. *PLoS Comput. Biol.* **2017**, *13*, e1005461. [CrossRef] [PubMed]
174. Alsalamah, A.K.; Khan, A.O. Asymptomatic retinal dysfunction in alpha-methylacyl-CoA racemase deficiency. *Mol. Vis.* **2021**, *27*, 396–402. [PubMed]
175. Gündüz, M.; Ünal, Ö.; Küçükçongar-Yavaş, A.; Kasapkara, Ç. Alpha methyl acyl CoA racemase deficiency: Diagnosis with isolated elevated liver enzymes. *Turk. J. Pediatrics* **2019**, *61*, 289–291. [CrossRef]
176. Herzog, K.; van Lenthe, H.; Wanders, R.J.A.; Vaz, F.M.; Waterham, H.R.; Ferdinandusse, S. Identification and diagnostic value of phytanoyl- and pristanoyl-carnitine in plasma from patients with peroxisomal disorders. *Mol. Genet. Metab.* **2017**, *121*, 279–282. [CrossRef] [PubMed]
177. Tanti, M.J.; Maguire, M.J.; Warren, D.J.; Bamford, J. Late onset AMACR deficiency with metabolic stroke-like episodes and seizures. *BMJ Case Rep.* **2022**, *15*, e247964. [CrossRef]
178. Boursier, G.; Rittore, C.; Milhavet, F.; Cuisset, L.; Touitou, I. Mevalonate Kinase-Associated Diseases: Hunting for Phenotype-Genotype Correlation. *J. Clin. Med.* **2021**, *10*, 1552. [CrossRef]
179. Bekkering, S.; Arts, R.J.W.; Novakovic, B.; Kourtzelis, I.; van der Heijden, C.; Li, Y.; Popa, C.D.; Ter Horst, R.; van Tuijl, J.; Netea-Maier, R.T.; et al. Metabolic Induction of Trained Immunity through the Mevalonate Pathway. *Cell* **2018**, *172*, 135–146.e139. [CrossRef]
180. Bader-Meunier, B.; Martins, A.L.; Charbit-Henrion, F.; Meinzer, U.; Belot, A.; Cuisset, L.; Faye, A.; Georgin-Lavialle, S.; Quartier, P.; Remy-Piccolo, V.; et al. Mevalonate Kinase Deficiency: A Cause of Severe Very-Early-Onset Inflammatory Bowel Disease. *Inflamm. Bowel Dis.* **2021**, *27*, 1853–1857. [CrossRef]
181. Politiek, F.A.; Waterham, H.R. Compromised Protein Prenylation as Pathogenic Mechanism in Mevalonate Kinase Deficiency. *Front. Immunol.* **2021**, *12*, 724991. [CrossRef]
182. Jeyaratnam, J.; Frenkel, J. Management of Mevalonate Kinase Deficiency: A Pediatric Perspective. *Front. Immunol.* **2020**, *11*, 1150. [CrossRef]
183. Leandro, J.; Bender, A.; Dodatko, T.; Argmann, C.; Yu, C.; Houten, S.M. Glutaric aciduria type 3 is a naturally occurring biochemical trait in inbred mice of 129 substrains. *Mol. Genet. Metab.* **2021**, *132*, 139–145. [CrossRef] [PubMed]
184. Dorum, S.; Havalı, C.; Görükmez, Ö.; Görükmez, O. Two patients with glutaric aciduria type 3: A novel mutation and brain magnetic resonance imaging findings. *Turk. J. Pediatrics* **2020**, *62*, 657–662. [CrossRef] [PubMed]
185. Waters, P.J.; Kitzler, T.M.; Feigenbaum, A.; Geraghty, M.T.; Al-Dirbashi, O.; Bherer, P.; Auray-Blais, C.; Gravel, S.; McIntosh, N.; Siriwardena, K.; et al. Glutaric Aciduria Type 3: Three Unrelated Canadian Cases, with Different Routes of Ascertainment. *JIMD Rep.* **2018**, *39*, 89–96. [CrossRef]
186. Singh, R.; Singh, S. Redox-dependent catalase mimetic cerium oxide-based nanozyme protect human hepatic cells from 3-AT induced acatalasemia. *Colloids Surf. B Biointerfaces* **2019**, *175*, 625–635. [CrossRef] [PubMed]

187. Ogino, N.; Nagaoka, K.; Tomizuka, K.; Matsuura-Harada, Y.; Eitoku, M.; Suganuma, N.; Ogino, K. Compromised glutathione synthesis results in high susceptibility to acetaminophen hepatotoxicity in acatalasemic mice. *Food Chem. Toxicol.* **2021**, *156*, 112509. [CrossRef]
188. Zarrouk, A.; Nury, T.; El Hajj, H.I.; Gondcaille, C.; Andreoletti, P.; Moreau, T.; Cherkaoui-Malki, M.; Berger, J.; Hammami, M.; Lizard, G.; et al. Potential Involvement of Peroxisome in Multiple Sclerosis and Alzheimer's Disease: Peroxisome and Neurodegeneration. *Adv. Exp. Med. Biol.* **2020**, *1299*, 91–104. [CrossRef]

*agriculture*

MDPI

*Review*

# Progress of m$^6$A Methylation in Lipid Metabolism in Humans and Animals

**Zimeng Xin [1], Tianying Zhang [2], Qinyue Lu [1], Zhangping Yang [1,3] and Zhi Chen [1,3,*]**

1 College of Animal Science and Technology, Yangzhou University, Yangzhou 225009, China
2 Shaanxi Key Laboratory of Brain Disorders & Institute of Basic and Translational Medicine, Xi'an Medical University, No.1 Xinwang Road, Xi'an 710021, China
3 Joint International Research Laboratory of Agriculture & Agri-Product Safety, Ministry of Education, Yangzhou University, Yangzhou 225009, China
* Correspondence: zhichen@yzu.edu.cn; Tel.: +86-0514-87979269

**Abstract:** N6-methyladenosine (m$^6$A) methylation is a type of methylation modification discovered on RNA molecules, mainly on mRNAs, as well as on other RNAs. Similar to DNA methylation, m$^6$A methylation regulates the post-transcriptional expression level of genes without altering their base sequences. It modulates gene expression mainly by affecting the binding of mRNAs to reader proteins, thereby regulating variable splicing, translation efficiency, and stability of mRNAs. Early in the research, the study of m$^6$A-related biological functions was greatly hindered due to the lack of effective detection methods. As second-generation sequencing and bioinformatics develop, several methods have been available to detect and predict m$^6$A methylation sites in recent years. Moreover, m$^6$A methylation is also closely related to the development of lipid metabolism, as shown in current studies. Combined with recent research, this paper reviews the concept, detection, and prediction means of m$^6$A methylation, especially the relationship between m$^6$A and lipid metabolism, providing a new clue to enrich the molecular mechanism of lipid metabolism.

**Keywords:** m$^6$A methylation; lipid metabolism; molecular mechanism; detection tools

**Citation:** Xin, Z.; Zhang, T.; Lu, Q.; Yang, Z.; Chen, Z. Progress of m$^6$A Methylation in Lipid Metabolism in Humans and Animals. *Agriculture* **2022**, *12*, 1683. https://doi.org/10.3390/agriculture12101683

Academic Editor: Linhai Wang

Received: 4 September 2022
Accepted: 9 October 2022
Published: 13 October 2022



## 1. m$^6$A Methylation of RNA

### 1.1. Concepts Related to m$^6$A Methylation

Early in the 1970s, Desrosiers first proposed RNA-N6-methyladenosine (m$^6$A) methylation as a new type of epigenetic modification [1], that is, a methyl substituent is inserted into the N atom at position 6 of adenosine. This only occurs in a small fraction of fragments containing DRACH (D = A/G/U, R = A/G, H = A/C/U) motif sequences, mainly in the CDS region and 3′UTR region of mRNAs [2]. Numerous studies have shown that m6A methylation plays a crucial role in nearly every stage of RNA's life cycle, including maturation and degradation. Additionally, it contributes to mRNA processing, splicing, export from the nucleus to the cytoplasm, translation, and decay. M$^6$A methylation modifications are reversible and dynamic in nature, and mainly involve three types of regulators: methyltransferase (writer), demethylase (eraser), and reading protein (reader). Among them, the writer catalyzes the m$^6$A modification of adenosine on mRNAs, the eraser demethylates bases that have undergone m$^6$A modification, and the reader identifies bases that have undergone m$^6$A modification so that they can participate in downstream regulatory pathways, such as RNA translational degradation [3].

#### 1.1.1. Methyltransferase-Mediated m$^6$A Methylation Modification

Methyltransferase, or "writer," mainly catalyzes the m$^6$A modification of mRNA bases, and its members include methyltransferase-like 3/14/5/16 (METTL3/14/5/16), Wilms tumor 1-associating protein (WTAP), KIAA1492, and Zinc Finger CCHC-Type Containing

4 (ZCCHC4). METTL3, the first discovered core methyltransferase subunit, plays an important role in m$^6$A methylation. In addition, it acts as a releasing agent in the cytoplasm, promoting the translation of m$^6$A-modified mRNAs, a process independent of its methyltransferase activity [4]. WTAP and KIAA1492 assist in m$^6$A methylation modifications of mRNAs. WTAP helps localize the METTL3/METTL14 complex by interacting with them to bind to the optimal substrates. KIAA1492 recruits methyltransferase components (METTL3, METTL14, and WTAP) and catalyzes them to specific RNA regions [5]. Notably, METTL5, METTL16, and ZCCHC4 all perform independent catalytic functions. For instance, METTL5 can catalyze m$^6$A modification of some structural RNAs, such as 18S rRNA, 28srRNA, and snRNA. In addition to binding U6 snRNA, ncRNAs, and pre-mRNAs [6], METTL16 can directly catalyze the m6A modification of mRNAs in the nucleus, especially nascent mRNAs, such as MALAT1, XIST, and MAT2A [7]. Among the key independent RNA methylation enzymes, ZCCHC4 targets ribosomal subunits and plays an important role in translation as well as cell proliferation. In addition, other members of the writer's family, including Vir-like m6A methyltransferase-associated (VIRMA), RNA binding motif protein 15 (RBM15), and RNA binding motif protein 15B (RBM15B), have been shown to be capable of methylating modifications of adenosine (Figure 1) [8].

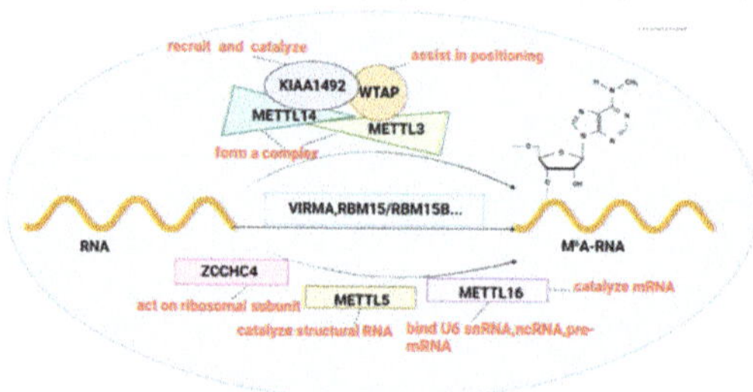


**Figure 1.** Mechanism of action and functions of methyltransferases.

Methyltransferase is also responsible for the methylation of m$^6$A. METTL3 and METTL14 usually form a complex, which is then combined with WTAP to locate. KIAA1492 is another auxiliary catalytic enzyme. ZCCHC4, METTL5, and METTL16 play independent catalytic functions on ribosomal subunits, structural RNA, and snRNA, respectively.

Studies have shown that methyltransferase-mediated m$^6$A modifications, particularly METTL3 and METTL14, are associated with the occurrence of a variety of cancers, including gastric, prostate, breast, colon, pancreatic, kidney, mesothelioma, sarcoma, and leukemia. Among them, gastric cancer (GC) is the fifth most prevalent type of cancer and the third leading cause of cancer-related death worldwide [9]. A recent study showed that the H3K27ac modification of METTL3 in the promoter region is regulated by p300, which in turn induces transcriptional activation of METTL3 and catalyzes the m$^6$A modification of GC-associated HDGF genes. This enables IGF2BP3 to directly bind the m$^6$A site on HDGF, maintaining gene stability and accelerating the progression of GC [4]. Furthermore, METTL3 promotes GC angiogenesis and progression through the m$^6$A-YTHDF2-dependent pathway to reduce the stability of ADAMTS9 mRNA by activating the PI3K/AKT signaling pathway [10].

There is strong evidence that METTL3 also promotes breast cancer cell proliferation and is associated with poor clinical outcomes in breast cancer patients when METTL3 is highly expressed. METTL3 interacts through the hepatitis B virus X protein (HBXIP), which can suppress METTL3 expression by affecting the binding of METTL3's 3′UTR and miRNA-let-7h, thus promoting cell proliferation. Meanwhile, METTL3 promotes the expression of HBXIP through m$^6$A modification, forming a positive feedback regulatory mechanism among HBXI, let-7h, and METTL3, to promote the proliferation of breast cancer cells [11]. METTL3 has also been suggested as a new potential target for metformin in the treatment of breast cancer. This is because metformin reduces m$^6$A modifications of METTL3 by targeting miR-483-3p, thereby upregulating the expression of p21. In turn, it inhibits the proliferation of breast cancer cells [12]. Wu et al. revealed that the specific binding of Lnc942 to METTL14 protein enhances METTL14 activity. By upregulating the m$^6$A methylation levels of downstream target genes C-X-C Motif Chemokine Receptor 4 (CXCR4) and CYP1B1, the mRNA stability and expression of these genes are enhanced, which ultimately promotes the proliferation and progression of breast cancer cells [13].

As for renal cell carcinoma, METTL3 increases the level of m$^6$A methylation in the 5′UTR of ABCD1, directly enhancing the expression of ABCD1 protein level and promoting the occurrence of renal tumorigenesis. Interestingly, the mRNA levels of ABCD1 do not alter during this process [14]. Chen et al. demonstrated that METTL14 regulates the stability of TRAF1 mRNA through m$^6$A-IGF2BP2, which in turn increases TRAF1 expression levels and induces apoptosis and angiogenesis in cancer cells [15].

METTL3 and METTL14 have also been shown to play potential roles in colorectal cancer (CRC) progression. According to Chen et al., METTL3 catalyzes m$^6$A methylation of BHLHE41, thus promoting expression at the protein level and forming the m6A-BHLHE41-CXCL1/CXCR2 pathway to suppress anti-tumor immunity to promote cancer [16]. In addition, Yu et al. demonstrated that METTL3 can directly bind to the m$^6$A site of the PLAU RNA 3′UTR, form the MAPK/ERK pathway, and upregulate PLAU mRNA through m$^6$A modification, thereby promoting angiogenesis and metastasis of colorectal cancer [17]. METTL3 also serves as a functional clinical oncogene in an m$^6$A-dependent manner to promote cancer progression [18]. Additionally, based on the study of Chen et al., METTL14-dependent m$^6$A methylation regulates DGCR8 processing of pri-miR-375, accelerates maturation of miR-375, and establishes the METTL14-miR-375-YAP1 pathway, thereby inhibiting CRC progression [19].

There are several other methyltransferases associated with cancer. Gong's group revealed that lncRNA ZNF582-AS1 can inhibit the m$^6$A modification level of mitochondrial MT-RNR1 (DNA12SrRNA) by reducing the expression level of methyltransferase A8K0B9 protein. Accordingly, MT-RNR1 expression is suppressed, leading to mitochondrial dysfunction and inhibiting the growth of renal cell carcinoma, cells, and invasive metastasis [20].

#### 1.1.2. Demethylase-Mediated Methylation Modification

There are only two known demethylases, α-ketone glutarate (α-KG)-dependent dioxygenase and fat mass and obesity-related proteins (FTO) and AlkB Homolog 5 (ALKBH5). Their main function is to remove methylation modifications from m6A that have already occurred (Figure 2). FTO is highly expressed in metabolically active organs such as the brain and fat and affects internal m6A and its modification in the 5′ end cap structure [9]. An increasing number of studies have confirmed that FTO plays an important regulatory role in lipid metabolism. A typical example is that FTO can mediate RNA demethylation, inhibit hepatocyte mitochondrial function, and promote hepatic lipid accumulation [21]. Apart from that, FTO-dependent m$^6$A demethylation promotes adipocyte differentiation and participates in the differentiation and transformation of adipose tissue by downregulating levels of miR-130 and miR-55, thereby upregulating C/EBPβ expression in preadipocytes [22]. Particularly, FTO can directly upregulate C/EBPβ levels to inhibit the differentiation of adipose tissue into brown adipose tissue [23]. According to Lee et al., FTO-dependent demethylation promotes lipogenesis by targeting miR-130 to upregulate the level of peroxisome proliferator-activated receptor γ (PPARγ) [24]. A number of other processes, including lipid metabolism in macrophages and skeletal muscle, are also regulated by FTO (Figure 2) [25]. Based on existing research, we speculate that FTO-mediated m$^6$A modification is closely related to lipid metabolism and is expected to be an important molecular mechanism regulating lipid metabolism. At this stage, however, comprehensive and thorough studies are needed to fully elucidate the specific regulatory mechanisms by which FTO regulates various processes of lipid metabolism. Besides, it is substantiated that FTO-mediated demethylation could be a potential regulatory mechanism in cancer. FTO was found by Niu et al. to be highly expressed in breast cancer and its expression is negatively associated with prognosis [26]. As reported by RUAN et al., under hypoxic conditions, ubiquitination of FTO mediated protein degradation is accelerated and its protein levels are inhibited. Meanwhile, FTO reduces mRNA stability by targeting MTA1, thus inhibiting CRC metastasis [27].

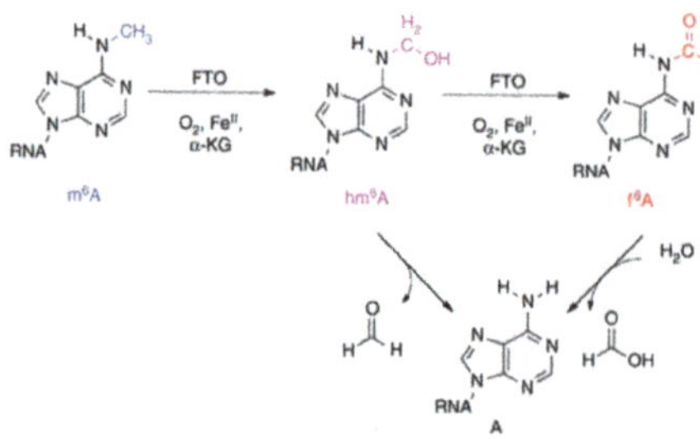


**Figure 2.** Mechanism of action of demethylase FTO [28].

Another identified m$^6$A demethylase, ALKBH5, can regulate mRNA export and RNA metabolism [29]. Numerous studies have shown that ALKBH5 is essential for metastasis of cancer cells. In gastric cancer, ALKBH5 downregulates the expression of PKMYT1 in an m$^6$A-dependent manner, and IGF2BP3 helps stabilize PKMYT1 mRNA by recognizing its m$^6$A modification site, forming the ALKBH5-PKMYT1-IGF2BP3 mechanism of gastric cancer metastasis [30]. Additionally, it promotes invasion and metastasis of GC by reducing the methylation of lncRNA NEAT1 [31]. In breast cancer, ALKBH5 upregulates the expression of NANOG by regulating the m$^6$A site on NANOG, thus promoting metastasis of breast cancer [32]. Furthermore, ALKBH5 was found by TAN et al. to stabilize AURKB RNA in an m$^6$A-dependent manner, thereby exerting an oncogenic effect in renal cell carcinoma [33]. However, studies related to the regulation of lipid metabolism by ALKBH5 have been inadequately explored.

#### 1.1.3. Reading Protein-Mediated m$^6$A Methylation Modifications

The m$^6$A reader proteins are composed of YTH functional domain family proteins (YTHDF1-3), YTH functional domain-containing proteins (YTHDC1-2), insulin-like growth factor 2 mRNA-binding proteins (IGF2PBs), and heterogeneous nuclear ribonucleoproteins (HNRNPs, including HNRNPA2/B1, HNRNPC, and HNRNPG). In the cytoplasm, YTHDF1/YTHDF3 can promote mRNA translation after recruiting translation initiation factors by recognizing m$^6$A, while YTHDF2 promotes mRNA degradation by binding m$^6$A [34]. Additionally, YTHDC1 can regulate the mRNA splicing process by recruiting splicing factor 3 (SRSF13) and blocking the binding of splicing factor 10 (SRSF10) to the nuclear site, which in turn facilitates mRNA translocation from the nucleus to the cytoplasm [35].

YTHDC2, the only protein in the YTH structural domain family that contains a decapping enzyme, regulates the stability of m$^6$A mRNAs by recognizing modifications of methylated m$^6$A and recruiting RNA degradation factors. Moreover, it links m$^6$A-modified mRNAs with ribosomes, facilitating their efficient translation. Unlike YTHDC2, after obtaining m$^6$A modification, HNRNPA2/B1, rather than acting as a direct "reader" of the m$^6$A modification, opens the secondary structure of RNA and promotes recognition of the recognition proteins. In addition, HNRNPA2/B1 recognizes the m$^6$A modification site of pri-miRNAs and accelerates their processing by recruiting the microprocessor complex Drosha-DGCR8 [35]. It is noteworthy that unlike other reader proteins, this process occurs in the nucleus (Figure 3).

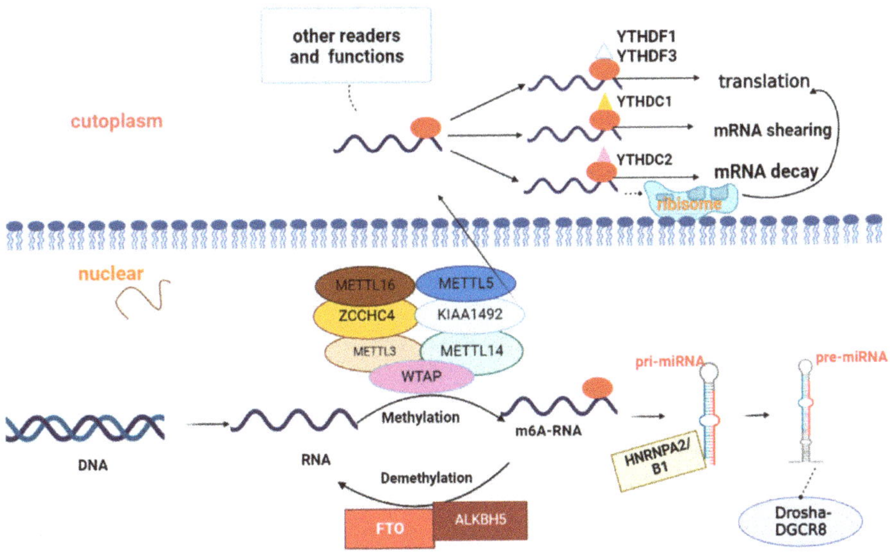


**Figure 3.** Mechanism of action of m$^6$A. The m$^6$A methylation by methyltransferase catalytic modification is mainly through the formation of a complex through METTL3, METTL14, and WTAP, and is independently catalyzed by KIAA1429 cofactors, METTL5, METTL16, and ZCCHC4. The m$^6$A modification is removed by the demethylases FTO or ALKBH5. Nuclei m$^6$A readers promote miRNA processing and export. Cytoplasmic readers are involved in RNA translation, splicing, and degradation processes.

It is found that YT521-B homologous structural domain family protein 1 (YTHDF1), an m$^6$A-modified reading protein, plays an essential role in transcription and translation, immune escape, epithelial mesenchymal transition (EMT), and chemoresistance in tumors [36]. In addition to contributing to gastric carcinogenesis by promoting the translation of USP14 and FZD7 proteins in an m$^6$A-dependent manner [37], YTHDF1 has also been testified to promote breast cancer metastasis by recognizing and binding m$^6$A-modified Forkhead Box M1 (FOXM1) mRNAs, thereby accelerating the translation process of FOXM1. In contrast, overexpression of FOXM1 in breast cancer cells could partially reverse the tumor suppressive effect caused by YTHDF1 silencing [38]. In addition, Han et al. found that YTHDF1 recognizes m$^6$A-modified FZD9 and promotes its translation, leading to aberrant activation of Wnt/β-catenin signaling, enhancing the tumorigenicity and stem cell-like activity of CRC [39].

Taken together, m$^6$A methylation regulates the expression of proto-oncogenes or oncogenes, thus affecting cancer development, metastasis, and invasion. The specific process is to first identify the m$^6$A regulators that trigger abnormal methylation levels in cancer and verify their roles in cancer cells. The following steps are combined with the methylation site prediction method to identify target genes and explore their positive or negative regulatory effects on target mRNAs before the final formation of the m$^6$A-regulator-target gene axis. Another conclusion is that there is an interaction between non-coding small molecule RNAs and m$^6$A methylation modifications in cancer. Taking miRNAs as an example, m$^6$A can regulate their processing and maturation to perform cancer-promoting or cancer-suppressing functions. In turn, miRNAs can target m$^6$A regulators to achieve regulation of methylation levels. Moreover, m$^6$A-modified reader proteins (YTH proteins) have been shown to directly regulate lncRNAs' molecular functions. Similar to miRNAs, lncRNAs can also directly affect m$^6$A levels. Except for this, circRNAs can also serve as targets for m$^6$A methylation modification, although their feedback on m$^6$A modification is less studied. Based on the property that circRNAs act as miRNA sponges in cancer, Ma et al. proposed that circRNAs could be indirectly involved in the regulation of $m^6A$ methylation levels, but this still remains to be tested [40]. Previous studies proved that regulators or inhibitors of $m^6A$ modification may provide potential therapeutic strategies for cancers.

As a nonsteroidal anti-inflammatory drug, Meclofenamic acid (MA) acts as an FTO inhibitor by competing with the FTO binding site, while MA2, an ethyl ester derivative of MA, increases $m^6A$ modification and inhibits tumor progression [41]. SPI1 is a hematopoietic transcription factor that directly inhibits METTL14 expression in malignant hematopoietic cells and may be a potential therapeutic target for acute myeloid leukemia [42]. As an enzyme-specific inhibitor of METTL3, STM2457 shows better oncogenic effects as well as less toxic side effects, revealing promise in the clinical treatment of acute myeloid leukemia [43]. The Wnt signaling program has been shown to promote cancer metastasis. Recent studies have found that in cancer cells with low FTO levels, $m^6A$ was enriched in mRNAs belonging to the Wnt signalling program. And through clinical trials, they found that cancer cells or tumours with reduced FTO levels and enhanced Wnt activity were more sensitive to Wnt inhibitors [44]. Collectively, $m^6A$ methylation is expected to be a potential target for cancer diagnosis, metastasis identification, and therapy. Meanwhile, the design of small molecules based on $m^6A$ regulators also provides new perspectives for cancer intervention, radiotherapy, chemotherapy, and other related fields. Therefore, there is an urgent need for researchers to thoroughly explore the relationship between $m^6A$ and cancer, and to improve related strategies to serve cancer diagnosis and clinical treatment.

### 1.2. Detection of $m^6A$ Methylation

Direct detection of $m^6A$ methylation bases is difficult because $m^6A$ methylation base pairing remains constant and cannot be distinguished from regular bases by reverse transcription. Several tools have been developed successively to predict m6A methylation sites, including the following:

Scilab v1.1SRAMP, software at http://www.cuilab.cn/sramp/ (accessed on 20 February 2016) [45];

Scilab v1.1iRNA-Methyl, software at http://lin.uestc.edu.cn/server/ (accessed on 24 August 2015) [46];

Scilab v1.11 iRNAm$^5$C-PseDNC, software at http://www.jci-bioinfo.cn/iRNAm5CPseDNC (accessed on 20 June 2017) [47];

Scilab v1.1m$^6$A MRFS, software at http://server.malab.cn/M6AMRFS/ (accessed on 25 October 2018) [48];

Scilab v1.1m$^6$APred-EL, software at http://server.malab.cn/M6APred-EL/ (accessed on 7 September 2018) [49];

Scilab v1.1Deep m$^6$A Seq, eq, software at https://github.com/rreybeyb/ (accessed on 31 December 2018) [50];

On the other hand, commonly used techniques for detecting $m^6A$ include methylated RNA immunoprecipitation sequencing (MeRIP-seq), $m^6A$ individual-nucleotide-resolution crosslinking and immunoprecipitation (miCLIP-seq), liquid chromatography-tandem mass spectrometry (LC-MS/MS), etc.

#### 1.2.1. Introduction of MeRIP-Seq

As the most commonly used method for $m^6A$ methylation detection, MeRIP-seq identifies $m^6A$ methylation levels at the whole transcriptome level. RNA-specific enrichment and interruption are first achieved by RNAs fragmentation reagents (Thermo, Waltham, MA, USA). After enriching the enriched mRNAs with $m^6A$ antibodies, their fragments are purified, sequentially reverse transcribed, and PCR amplified to construct a high-throughput sequencing library (IP). Furthermore, a normal transcriptome library (input) is constructed separately as a control to reflect the abundance of the base RNAs. Finally, the two sequencing libraries are sequenced together to obtain an m$^6$A map of the whole transcriptome following biological analysis (Figure 4) [51].

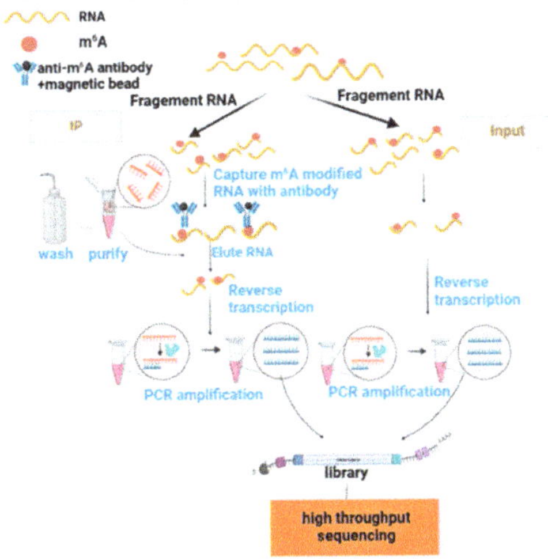


**Figure 4.** Specific process of MeRIP-seq. This figure introduces the operation process of MeRIP seq.

This technique has been extensively used to study m$^6$A modifications because it is easy, fast, and relatively inexpensive. It enables the qualitative analysis of mRNA regions undergoing hypermethylation. Wang et al. detected 9085 m$^6$A methylation peaks by performing MeRIP-seq analysis on the transcriptome of velvet goat skin tissue. The transcriptome and differential peaks were then combined to screen out 19 differentially expressed genes that contain RNA methylation modifications associated with velvet growth [52]. In addition, MeRIP-seq has been widely used as an assay for investigating the mechanisms related to cancer regulation by measuring m$^6$A methylation modifications. Yang et al. identified m$^6$A modifications in the EOC transcriptome of endometrial ovarian cancer by MeRIP-seq for the first time and described methylated m$^6$A to modify differentially expressed genes. This provides a new direction for the underlying molecular mechanisms and signaling pathways of EOC development [53]. During a comprehensive analysis of mRNA m$^6$A modifications in human colorectal cancer, Li et al. conducted a combined analysis by MeRIP-seq and RNA-seq to predict RNA-binding proteins and identify methylation-related genes FMR1, IGF2BP2, and IGF2BP3 that may be involved in the development of CRC [54]. At this stage, MeRIP-seq can only identify regions of m$^6$A hypermethylation but cannot realize single-base resolution sequencing.

#### 1.2.2. Introduction of miCLIP

MiCLIP induces mutations during reverse transcription by crosslinking single nucleotides that detect methylated bases at the antibody binding sites. As another widely applied m$^6$A detection technology, like MeRIP-seq, it performs immunoprecipitation using UV crosslinking after completing RNA fragmentation and binding of m$^6$A methylation-modified mRNA fragments using antibody immunomagnetic beads. Next, reverse transcription is performed after digestion with K protein. M$^6$A modifications on the RNA result in highly specific mutations or truncations on the corresponding cDNAs. Finally, co-sequencing of the library is achieved to perform single methyl site detection (Figure 5). Furthermore, miCLIP enables high-resolution detection of individual m$^6$A residues and m$^6$A clustering analysis of the entire RNAs. Using miCLIP, combined with RIP-seq, Liu et al. identified a total of eight m$^6$A sites in the genome at single-base resolution, further verifying that SARS-CoV-2 genomic RNA can dynamically modify m$^6$A, making it a negative regulatory RNA in human and monkey cells [54]. This approach was also employed by Zhao and Liu's team to address and compare the differences in mRNA $m^6A$ modifications in temozolomide-sensitive and drug-resistant glioblastoma GBM tissues [55]. They also probed into the interaction mechanism of m6A modifications and histone modifications in regulating drug resistance in glioblastoma.

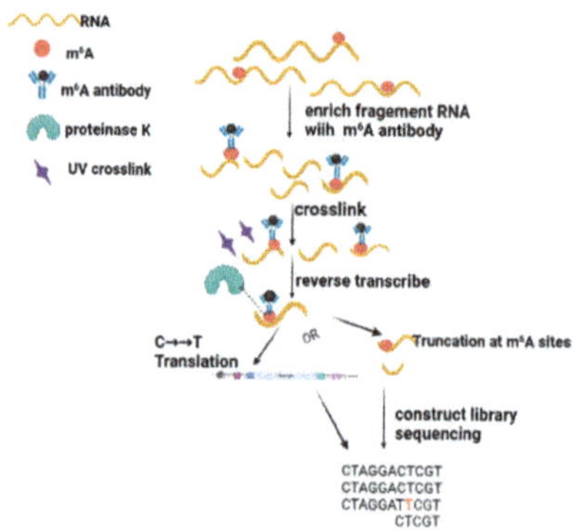


**Figure 5.** Specific process of miCLIP. This figure introduces the operation process of miCLIP.

1.2.3. Introduction of LC-MS/MS

LC-MS/MS enables the separation of sample components based on multiple migration rates because various substances interact differently with the stationary and mobile phases in the chromatograph. Specifically, after the extraction of total RNA using the NA extraction reagent TRIzol, oligodT magnetic beads are used to enrich mRNA, or the RNA removal kit is used to obtain RNA including mRNA, lncRNA, and circRNA. In the following step, nuclease P1 is used to digest RNA from single strands to individual bases. After several hours, samples are incubated with alkaline phosphatase and ammonium bicarbonate. Ultimately, following the injection of the sample into the liquid chromatograph, the overall methylation of $m^6A$ on the mRNA can be calculated using mass spectrometry tandem analysis based on the ratio of $m^6A$ to total adenine (Figure 6) [56].

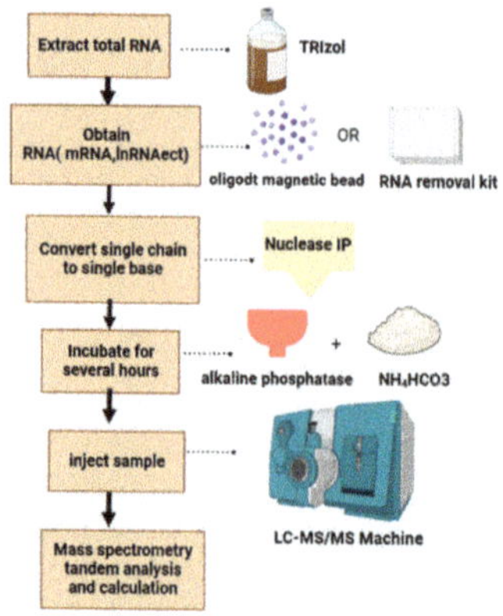


**Figure 6.** Specific process of LC-MS/MS. This figure introduces the operation process of miCLIP.

Compared with the previous two methods, LC-MS/MS is able to detect the overall $m^6A$ level of RNA but is more suitable for smaller-scale gene detection and cannot pinpoint methylation sites. ZHANG et al. determined the levels of m$^6$A and A in RNAs by electrospray ionization (ESI) in positive ion multiple reaction monitoring (MRM) mode. They successfully applied this method to detect the level of m$^6$A in RNAs of mouse spleen T cells under different treatment conditions. As reported by SHEN et al., LC-MS/MS analysis of m$^6$A methylation glycolysis in CRC patients demonstrated that levels of m$^6$A methylation are significantly higher in CRC patients who ingested higher levels of fluorodeoxyglucose [18].

In addition to conventional methods, some improved assays are also included. One is m$^6$A-label-seq developed by SHU et al., a method to label RNA m$^6$A sites throughout transcriptome by cellular self-metabolism. Using this method, labeled sites can undergo chemical treatment-induced reverse transcription base mutations, which in turn enables single-base resolution determination [57]. The other is the m$^6$A-SEAl technique created by Wang et al. This is the first technology to achieve m$^6$A chemical labeling using FTO enzyme and has been successfully applied to high-throughput sequencing. Due to its high sensitivity and specificity, it has been validated to be suitable for a small number of m$^6$A samples [58].

## 2. m$^6$A Methylation in Lipid Metabolism

### *2.1. m$^6$A Methylation Involved in the Regulation of Human Lipid Metabolism*

Lipids are essential components of biological membranes and structural cellular units. Mainly used for energy storage and metabolism, they also play a pivotal role in a variety of cellular activities as signaling molecules. In vivo, lipid metabolism refers to the process of digestion, absorption, synthesis, and decomposition of fat with the help of various related enzymes, including the biosynthesis of saturated fatty acids, the extension of fatty acid carbon chains, and the production of unsaturated fatty acids. This process requires the participation of many enzymes, among which acetyl coenzyme A carboxylase (ACC) and fatty acid synthase (FAS) are the key synthases involved in fat synthesis. DGAT1 is mainly responsible for the synthesis of TAG during fat absorption and storage, while adipose triglyceride lipase (ATGL) and hormone-sensitive lipase (HSL) are mainly involved in fat hydrolysis. In addition, stearoyl coenzyme A desaturase 1 (SCD1) catalyzes the conversion of saturated fatty acids into monounsaturated fatty acids, and peroxisome proliferator-activated receptor (PPAR) and sterol regulatory element binding protein-1 (SREBP-1) regulate lipid metabolism.

Several recent studies have indicated that RNA m$^6$A methylation modifications are closely associated with lipid metabolism and human lipid disorders, and m$^6$A can regulate lipid metabolism-related gene expression. As reported by Liu et al., the zinc finger protein 217 (Zfp217) gene upregulates m$^6$A levels of cell cyclin D1 by increasing the expression of m$^6$A methyltransferase METTL3. As a result, the knockout of Zfp217 prevents the amplification process of mitotic clones and induces inhibition of adipogenesis [59]. As Yang et al. proved, overexpression of METTL3/14 upregulates levels of m$^6$A modification, enhances mRNA stability of ACLY and SCD1, and engenders increasing expression levels of ACLY and SCD1 proteins, thereby promoting FA de novo synthesis and lipid accumulation [60]. According to PENG et al., on the other hand, a novel mechanism is revealed by which m$^6$A mRNA methylation modification affects the lipid metabolism of hepatocytes through the regulation of autophagy. Additionally, YTHDF1 binds to the mRNA of the key autophagy gene rubicon mainly through the m$^6$A motif and inhibits its degradation. When rubicon is expressed at high levels, autophagic vesicles are less likely to fuse with lysosomes, leading to reduced degradation of lipid droplets and increased lipid droplet accumulation in hepatocytes via the lysosomal pathway (Figure 7) [61]. Regarding hepatic lipid metabolism, Li et al. revealed that the METTL3-mediated elevated m$^6$A levels on a high-fat diet exacerbate the disorder of hepatic lipid metabolism, manifested by increased subcutaneous fat, hepatic steatosis, and increased total cholesterol in serum. Further studies showed that M6A methylation achieves the regulation of lipid metabolism by affecting the stability of genes related to lipid metabolism, especially Lpin1 mRNA. Interestingly, in the HFD group, METTL3 is mainly involved in fatty acid catabolism as well as oxidation-related genes. In contrast, METTL3-targeted genes are mainly enriched in the sterol synthesis pathway under normal diet conditions. Therefore, researchers propose that METTL3 can simultaneously participate in the bidirectional regulation of lipid metabolism [62]. Another methyltransferase, WTAP, is also shown to be involved in the regulation of lipid metabolism. The deficiency of WTAP induces the expression and secretion of IGFBP1, thus enhancing lipolysis in eWAT and increasing serum FFA, leading to hepatic steatosis. Furthermore, the hepatic WTAP deficiency increases expression of both CD36 and CCL2, thereby enhancing the hepatic uptake capacity of FFA [63].

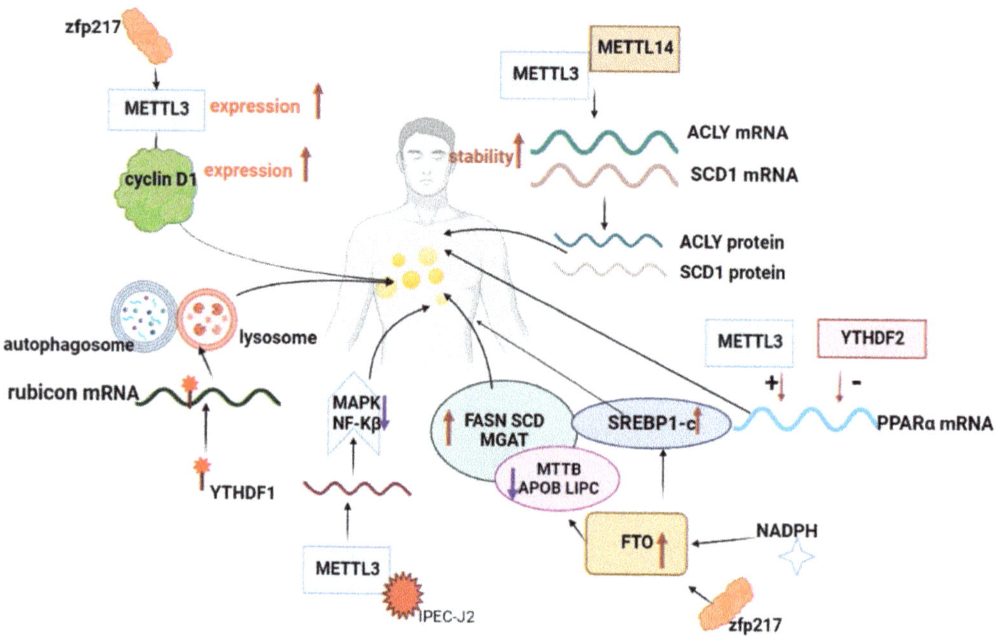


**Figure 7.** Regulation mechanism of human lipid metabolism by m$^6$A.

FTO functions as a major demethyltransferase that directly regulates lipid metabolism and participates in it by removing m$^6$A methylation modifications. Zfp217 regulates m$^6$A mRNA methylation by activating transcription of the m$^6$A demethylase FTO, and couples gene transcription to m$^6$A mRNA modification to promote adipogenesis [64]. FTO was found by Kang et al. to mediate the reduction in m$^6$A methylation levels and enhance the expression of lipid metabolism-related genes FASN, SCD, and MGAT. Meanwhile, it also inhibits lipid transport-related genes MTTP, APOB, and LIPC, which are involved in fatty acid synthesis as well as triglyceride accumulation [65]. As reported by Zhou et al., FTO-dependent m$^6$A demethylation promotes the expression of key SREBP-1c genes to indirectly increase SREBP1-c expression and promote the formation of hepatocytes and lipid droplets [25]. A subsequent study by Wang et al. further verified that NAPDH binds and enhances FTO expression to promote lipid synthesis in 3T3 cells through m$^6$A demethylation [66]. Adipogenesis is also regulated by FTO-mediated mRNA demethylation, which impacts selective splicing of mRNAs during mitotic clonal amplification (Figure 7) [67]. Sun provided novel insights into the epigenetic mechanism of FTO's function in hepatic lipid synthesis. Specifically, knockdown of FTO significantly increases m$^6$A levels in FASN mRNA and promotes mRNA decay, thereby decreasing FASN mRNA expression. The protein expression of acetyl-CoA carboxylase and ATP-citrate lyase is further reduced, ultimately inhibiting de novo lipogenesis in HepG2 cells [68]. Another important target gene for the function of $m^6A$ methylation on hepatic lipid metabolism was demonstrated by Zhong et al. METTL3 promotes the expression of PPARα by positively regulating the $m^6A$ modification of PPARα RNA, while YTHDF2 plays the opposite role. By combining the actions of two regulators, PPARα mRNA remains stable, thus regulating lipid metabolism [69]. Similarly, carboxylesterase 2(CES2), which is involved in the regulation of hepatic lipid metabolism, has also been shown to be regulated by $m^6A$ modification. Dual knockdown of methyltransferases METTL3 and METTL14 upregulates CES2 expression and then reduces lipid accumulation. However, knockdown of demethyltransferases FTO or ALKBH5 downregulates CES2 expression and achieves lipid increase in HepaRG and HepG2 cells. Furthermore, the reader protein YTHDC2 reduces CES2 mRNA stability by recognizing $m^6A$ in the 5′-UTR of CES2 [70].

In addition, $m^6A$ plays a role in long-chain fatty acid uptake, just as Zong et al. revealed. Deletion of methyltransferase METTL3 decreases Traf6 expression in the lipopolysaccharide-mediated IPEC-J2 inflammatory response, thereby inhibiting the downstream NF-kB and MAPK signaling pathways and promoting the uptake of long-chain fatty acids (Figure 7) [71]. Xu et al. found that METTL3 increases levels of caprylic, gamma-linolenic, arachidonic, and docosapentaenoic acid by upregulating the expression of the lipid metabolism regulator CYP4F40, thereby reducing free fatty acid FFA-induced steatosis in hepatocytes. In addition, recent studies have shown that $m^6A$ in 18S rRNAs can also promote fatty acid metabolism. Specifically, METTL5-mediated alterations in 18S rRNA $m^6A$ modification results in impaired assembly of the 80S ribosome and inhibiting translation of mRNAs involved in fatty acid metabolism. In particular, acyl-CoA synthetase long chain family member 4 (ACSL4), a target gene of METTL5, also enhances the regulatory function of METLT5 in promoting lipid metabolism [72].

The figure shows the regulation of partial $m^6A$ on human lipid metabolism. The $m^6A$ regulatory factor targets the mRNA of lipid metabolism regulation genes, and the levels of positive or negative regulation genes to achieve the regulation of lipid metabolism.

As shown in the above studies, $m^6A$ modification plays a dual role in human lipid metabolism. On the one hand, the discovery of $m^6A$ modification is a new direction to regulate lipid metabolism at the molecular level. $M^6A$ methylation acts on genes related to lipid metabolism regulation, affects the translation, splicing, transport, and other processes of mRNAs, and realizes the regulation of expression level. On the other hand, $m^6A$ levels can be regulated as in Zfp217, thus affecting the role in lipid metabolism. An increasing number of studies have demonstrated that FTO mediates $m^6A$ modifications in the regulation of human lipid metabolism, while FTO deficiency triggers the upregulation of $m^6A$ modification levels. Ultimately, this leads to browning and thermogenesis in white adipocytes, providing a potential target to combat obesity and metabolic diseases [66]. Nonetheless, there are relatively few studies examining the role of other $m^6A$ methylation regulators in lipid metabolism and their specific mechanisms of action.

### 2.2. *$m^6A$ Methylation Involved in the Regulation of Animal Lipid Metabolism*

There is evidence that hypermethylation is primarily involved in cellular responses to peptide hormone stimulation and adipogenesis-related pathways, especially fat metabolism of related tissue in pigs and poultry. In sows subjected to heat stress, methylation regulators FTO, WTAP, YTHDF2, and METTL14 upregulate the expression levels of HSP70 protein and genes, such as ACCAA and FASN that are involved in hepatic fatty acid biosynthesis and fatty acid metabolism. As for abdominal fat, METTL3, METTL4, WTAP, and FTO up-regulate the levels of HSP27 protein as well as fat metabolism-related genes, such as SCD36, but down-regulate the levels of ATGL and CPT1A [73]. Wang's team revealed that $m^6A$ participates in the molecular mechanism of porcine fat deposition through post-transcriptional regulation of the JAK2-STAT3-C/EBPβ signaling pathway. The absence of FTO promotes the recognition and degradation of JAK2 by YTHDF2, which in turn inhibits the expression of JAK2 and the phosphorylation of STAT3, leading to increased difficulty in translocation to the nucleus, thus inhibiting the transcription factor C/EBPβ that is necessary for early adipogenesis. It is also validated that in porcine precursor adipocytes, $m^6A$ modifications at the locus of the autophagy-associated protein ATG5 and ATG7 are enhanced by interfering with FTO that is recognized and degraded by YTHDF2, resulting in a reduction in protein expression that inhibits autophagosome formation and autophagic flow occurrence. This ultimately hinders the fat deposition process (Figure 8) [74]. Regarding the process of intramuscular fat formation in pigs, Jiang et al. sequenced $m^6A$ in the longest dorsal muscles (LDMs) of Changbai and Jinhua pigs and concluded that $m^6A$ modifications may play a regulatory role in intramuscular adipogenesis [75].

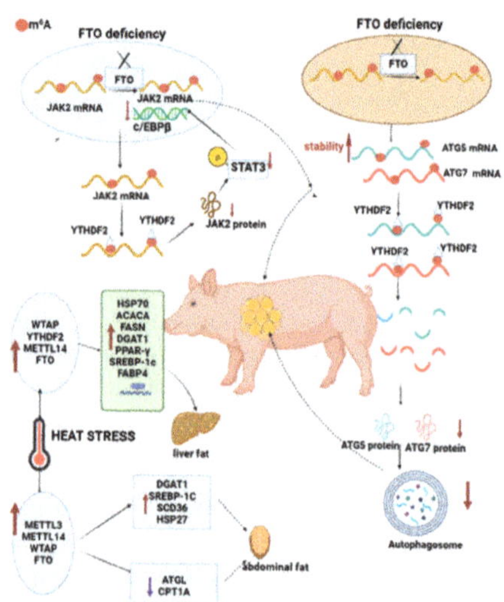


**Figure 8.** Regulation mechanism of pig lipid metabolism by $m^6A$. The figure shows the part of the regulation of $m^6A$ on pig lipid metabolism. FTO plays an important role in the early stages of porcine adipogenesis, as FTO deletion inhibits JAK2 m6A demethylation modification and thus increases JAK2 mRNA stability YTHDF2 is in-volved in mediating m6A-dependent JAK2 mRNA stability, thereby reducing JAK2 expression and consequently STAT3 phosphorylation levels, making translocation to the nucleus more difficult and thus inhibits the transcription of C/EBPβ, a transcription factor essential for early adipogenesis.In porcine precursor adipocytes, knockdown of the demethylase FTO increased the level of m6A modification of the autophagy-related genes ATG5 and ATG7 and thus enhanced mRNA stability, which was then recognized and degraded by YTHDF2, resulting in a significant decrease in ATG5 and ATG7 protein expression, inhibiting autophagosome formation and autophagic flow, and ul-timately fat deposition.

An increasing number of studies have suggested that $m^6A$ modifications are crucial for adipogenesis and lipid metabolism in poultry [76]. A study by Cheng et al. analyzed the $m^6A$ methylationome of chicken abdominal adipose tissue and found that hypermethylation of ACSL1 and FASN mRNAs is associated with increased mRNA stability, thereby promoting TG formation. Through $m^6A$ methylation, adipocyte differentiation-associated LPIN1 maintains energy metabolism homeostasis by increasing mRNA expression levels. Besides, researchers also suggested that hypomethylation of LRP4 may promote LRP4 mRNA levels by reducing YTHDF2-mediated mRNA decay, thereby increasing adipocyte size. This conjecture needs to be further tested [77]. According to Hu et al., glucocorticoid (GR)-mediated transactivation of FTO and m6A demethylation at the mRNA level of adipogenic genes SREBP-1, FAS, and SCD facilitate activation of chicken liver adipogenic genes and TG accumulation in primary chicken hepatocyte cells treated with OA/DEX [78]. Zhang et al. have proposed another ALKBH5 as a novel regulator of chicken preadipocyte proliferation and differentiation. ALKBH5 was found to promote chicken preadipocyte differentiation by directly or indirectly enhancing the expression of PPAR $\gamma$, FABP, and FAS. The deficiency is that the specific mechanism of action is not clearly explained [79]. N6-methyladenosine demethylase ALKBH5 is a novel regulator of the proliferation and differentiation of chicken preadipocytes. IGF2BP1 is also reported to play a significant role in chicken lipid metabolism. In chicken adipocyte proliferation, cell cycle-related genes, such as CDK1, CCNB3, and so forth, can be induced to promote cell proliferation. Moreover, its overexpression reduces cell population arrest in the G1 phase but increases cell population in the S phase. During adipogenesis, upregulating IGF2BP1 expression-related genes (ACSL5, CPT1APCK1, PPAR$\alpha$, FTO, etc.) could promote chicken adipocyte differentiation, fatty acid metabolism, and lipid droplet accumulation (Figure 9) [80].

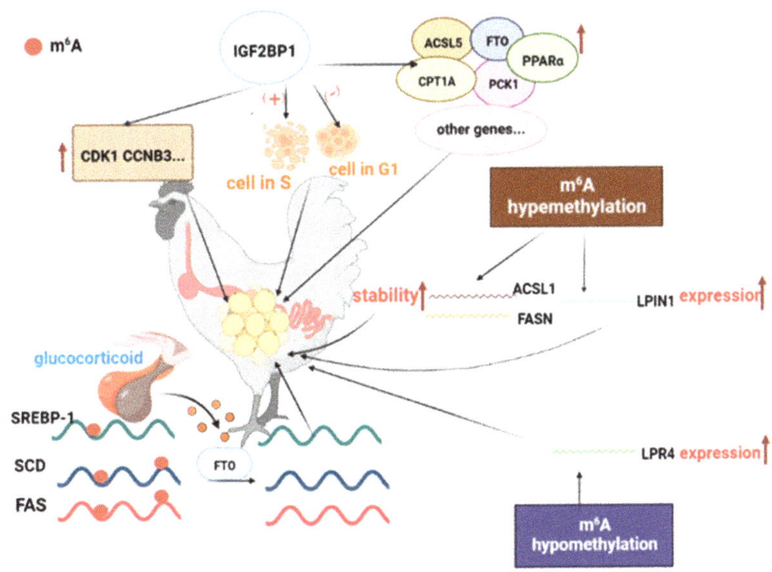


**Figure 9.** Regulation mechanism of poultry lipid metabolism by m$^6$A. The figure shows the part of regulation of m$^6$A on poultry lipid metabolism.

Taken together, m$^6$A modification is responsible for regulating lipid metabolism in muscle, viscera, bone, and other body tissues of animals. Research in this field also provides new ideas for improving the growth and breeding of livestock as well as the control of diseases. In addition, recent studies have shown that rumen-protected methionine and lysine in low-protein diets enhance FADS and ACC expression through m$^6$A methylation modifications, affecting polyunsaturated fatty acid synthesis in lamb liver instead of muscle [81]. In exploring the whole transcriptome profile of yak m6A and its potential biological functions in adipocyte differentiation, Zhang et al. presented that several m$^6$A differentially expressed genes, such as KLF9 and FOXO1, are involved in yak adipocyte differentiation through activation of Foxo and Hippo signaling pathways [82].

### 2.3. m$^6$A Methylation Involved in the Regulation of Milk Fat Metabolism

As economic life continues to develop, people's demands and quality requirements for dairy products increase. In addition to affecting flavor and nutritional value, milk fat becomes a major economic trait in ruminant reproduction. The process of milk fat regulation is influenced by multiple factors, such as genes and regulatory factors, among which SREBP-1 is one of the important nuclear transcription factors in mammals. In cow mammary tissue, a key function of SREBP-1 is to promote the ab initio synthesis of fatty acids and regulate the composition and content of beneficial fatty acids in milk by activating the expression of key genes and enzymes during milk fat synthesis and secretion. A lactation model established by Wang et al. revealed that METTL3 could promote translation to protein formation by positively regulating the methylation of SREBP1 mRNA, thereby accelerating milk fat synthesis [83]. When exploring the biological functions of m$^6$A methylation in S. aureus-stimulated bovine mammary epithelial cells, Li et al. spotted that m$^6$A hypermethylated genes are significantly associated with fatty acid degradation and adipocytokine signaling pathways, while m$^6$A modification alters the mRNA levels of genes related to milk fat metabolism [84]. It is still necessary to validate the specific mechanism of action. More importantly, this study offers a new proposal for m$^6$A modification to regulate milk fat metabolism.

Currently, m$^6$A modifications have been found to be involved in the regulation of milk fat metabolism, such as SREBP1, which has been shown to be a target factor for m$^6$A methylation. Additionally, genes and signaling pathways related to milk fat metabolism have also been shown to be closely associated with m$^6$A methylation. Furthermore, m$^6$A metabolism-related genes and signaling pathways have also been shown to be closely related to m$^6$A methylation. Excitingly, the mechanism of m$^6$A methylation in the human mammary gland, particularly in breast cancer, has been well established and is expected to provide a reference for research on mammary gland development and milk fat metabolism in ruminants [85].

## 3. Summary and Outlook

At present, a growing number of researchers are focusing on studying m$^6$A methylation modification, which is leading the public to a deeper understanding of m$^6$A methylation modification. Moreover, m$^6$A methylation modification is a type of epistatic modification that can perform multiple functions, including both molecular and biological functions. At the molecular level, in addition to regulating RNA splicing, translation, transport, and stability, m$^6$A methylation also affects the cleavage, transport, stability, and degradation of non-coding small molecules, such as miRNAs, lncRNAs, and circRNAs. Because of various factors, such as technological limitations, their specific mechanisms of action in ncRNAs, especially with circRNAs molecules, have not yet been clearly elucidated.

In recent years, technologies for m$^6$A detection and methylation site prediction have emerged as a result of the advancement of second-generation sequencing and bioinformatics. In addition to the traditional techniques for detection, such as MeRIP-seq and miCLIPSRAMP, an increasing number of new techniques are being developed. However, new techniques are not as efficient, accurate, convenient, or inexpensive as traditional techniques. Therefore, it is necessary to conduct further research to enhance their applicability and convenience.

Nowadays, m$^6$A is one of the major research hotspots in the field of cancer research, and its regulatory mechanism on cancer is becoming more and more refined, making it a potential target for cancer identification and treatment in the future. In addition, m$^6$A modifications play a regulatory role in lipid metabolism. However, the specific regulatory mechanisms remain unclear due to limited studies, especially in ruminant milk lipid metabolism. The existing studies on the function of m$^6$A in regulating lipid metabolism in organisms mainly focus on the demethyl transferase FTO. It remains to be determined whether other m$^6$A regulators are involved in the regulatory mechanisms and what the specific regulatory mechanisms are. Aside from this, exosomes, as intercellular communication carriers, also play a role in lipid metabolism by interfering with the synthetic transport and degradation processes. According to investigators, melatonin reduces adipocyte-derived exosome resistin levels via m$^6$A RNA demethylation in adipocytes, further reducing hepatic steatosis [86]. In light of this, the authors infer that exosome and m$^6$A might provide new approaches to the regulation of lipid metabolism.

**Author Contributions:** Writing—original draft preparation, Z.X. and Z.C.; writing—review and editing, T.Z., Q.L. and Z.Y. All authors have read and agreed to the published version of the manuscript.

**Funding:** Jiangsu Agricultural Science and Technology Independent Innovation Fund (CX(21)3119) Yangzhou University "Blue Project" and Scientific Research Program Funded by the Education Department of Shaanxi Provincial Government (Program No. 21JP110).

**Institutional Review Board Statement:** Not applicable.

**Informed Consent Statement:** Not applicable.

**Data Availability Statement:** Not applicable.

**Conflicts of Interest:** The authors declare no conflict of interest.

## References

1. Desrosiers, R.; Friderici, K.; Rottman, F. Identification of Methylated Nucleosides in Messenger RNA from Novikoff Hepatoma Cells. *Proc. Natl. Acad. Sci. USA* **1974**, *71*, 3971–3975. [CrossRef]
2. Zhang, H.; Shi, X.; Huang, T.; Zhao, X.; Chen, W.; Gu, N.; Zhang, R. Dynamic landscape and evolution of m6A methylation in human. *Nucleic Acids Res.* **2020**, *48*, 6251–6264. [CrossRef]
3. Maity, A.; Das, B. N6-methyladenosine modification in mRNA: Machinery, function and implications for health and diseases. *FEBS J.* **2016**, *283*, 1607–1630. [CrossRef] [PubMed]
4. Wang, Q.; Chen, C.; Ding, Q.; Zhao, Y.; Wang, Z.; Chen, J.; Jiang, Z.; Zhang, Y.; Xu, G.; Zhang, J.J.G. METTL3-mediated m6A modification of HDGF mRNA promotes gastric cancer progression and has prognostic significance. *Gut* **2019**, *69*, 1193–1205. [CrossRef] [PubMed]
5. Jiang, X.; Liu, B.; Nie, Z.; Duan, L.; Xiong, Q.; Jin, Z.; Yang, C.; Chen, Y. The role of m6A modification in the biological functions and diseases. *Signal Transduct. Target. Ther.* **2021**, *6*, 74. [CrossRef]
6. Warda, A.S.; Kretschmer, J.; Hackert, P.; Lenz, C.; Urlaub, H.; Höbartner, C.; Sloan, K.E.; Bohnsack, M.T. Human METTL16 is a N6-methyladenosine (m6A) methyltransferase that targets pre-mRNAs and various non-coding RNAs. *EMBO Rep.* **2017**, *18*, 2004. [CrossRef]
7. Deng, X.; Su, R.; Weng, H.; Huang, H.; Li, Z.; Chen, J.J.C.R. RNA N 6 -methyladenosine modification in cancers: Current status and perspectives. *Cell Res.* **2018**, *28*, 507–517. [CrossRef]
8. Yue, Y.; Liu, J.; Cui, X.; Cao, J.; Luo, G.; Zhang, Z.; Cheng, T.; Gao, M.; Shu, X.; Ma, H.; et al. VIRMA mediates preferential m(6)A mRNA methylation in 3′UTR and near stop codon and associates with alternative polyadenylation. *Cell Discov.* **2018**, *4*, 10. [CrossRef]
9. Smyth, E.C.; Nilsson, M.; Grabsch, H.I.; van Grieken, N.C.; Lordick, F. Gastric cancer. *Lancet* **2020**, *396*, 635–648. [CrossRef]
10. Wang, N.; Huo, X.; Zhang, B.; Chen, X.; Zhao, S.; Shi, X.; Xu, H.; Wei, X. METTL3-Mediated ADAMTS9 Suppression Facilitates Angiogenesis and Carcinogenesis in Gastric Cancer. *Front. Oncol.* **2022**, *12*, 861807. [CrossRef]
11. Cai, X.L.; Wang, X.; Cao, C.; Gao, Y.E.; Zhang, S.Q.; Yang, Z.; Liu, Y.X.; Zhang, X.D.; Zhang, W.Y.; Ye, L.H. HBXIP-elevated methyltransferase METTL3 promotes the progression of breast cancer via inhibiting tumor suppressor let-7g. *Cancer Lett.* **2018**, *415*, 11–19. [CrossRef]
12. Cheng, L.; Zhang, X.; Huang, Y.Z.; Zhu, Y.L.; Xu, L.Y.; Li, Z.; Dai, X.Y.; Shi, L.; Zhou, X.J.; Wei, J.F.; et al. Metformin exhibits antiproliferation activity in breast cancer via miR-483-3p/METTL3/m(6)A/p21 pathway. *Oncogenesis* **2021**, *10*, 7. [CrossRef]
13. Sun, T.; Wu, Z.; Wang, X.; Wang, Y.; Hu, X.; Qin, W.; Lu, S.; Xu, D.; Wu, Y.; Chen, Q.; et al. LNC942 promoting METTL14-mediated m6A methylation in breast cancer cell proliferation and progression. *Oncogene* **2020**, *39*, 5358–5372. [CrossRef]
14. Shi, Y.; Dou, Y.; Zhang, J.; Qi, J.; Xin, Z.; Zhang, M.; Xiao, Y.; Ci, W. The RNA N6-Methyladenosine Methyltransferase METTL3 Promotes the Progression of Kidney Cancer via N6-Methyladenosine-Dependent Translational Enhancement of ABCD1. *Front. Cell Dev. Biol.* **2021**, *9*, 737498. [CrossRef]
15. Chen, Y.; Lu, Z.; Qi, C.; Yu, C.; Li, Y.; Huan, W.; Wang, R.; Luo, W.; Shen, D.; Ding, L.; et al. N(6)-methyladenosine-modified TRAF1 promotes sunitinib resistance by regulating apoptosis and angiogenesis in a METTL14-dependent manner in renal cell carcinoma. *Mol. Cancer* **2022**, *21*, 111. [CrossRef]
16. Chen, H.; Pan, Y.; Zhou, Q.; Liang, C.; Wong, C.C.; Zhou, Y.; Huang, D.; Liu, W.; Zhai, J.; Gou, H.; et al. METTL3 inhibits anti-tumor immunity by targeting m(6)A-BHLHE41-CXCL1/CXCR2 axis to promote colorectal cancer. *Gastroenterology* **2022**, *163*, 891–907. [CrossRef]
17. Yu, T.; Liu, J.; Wang, Y.; Chen, W.; Liu, Z.; Zhu, L.; Zhu, W. METTL3 promotes colorectal cancer metastasis by stabilizing PLAU mRNA in an m6A-dependent manner. *Biochem. Biophys. Res. Commun.* **2022**, *614*, 9–16. [CrossRef]
18. Shen, C.; Xuan, B.; Yan, T.; Ma, Y.; Xu, P.; Tian, X.; Zhang, X.; Cao, Y.; Ma, D.; Zhu, X.; et al. m6A-dependent glycolysis enhances colorectal cancer progression. *Mol. Cancer* **2020**, *19*, 72. [CrossRef]
19. Chen, X.; Xu, M.; Xu, X.; Zeng, K.; Liu, X.; Sun, L.; Pan, B.; He, B.; Pan, Y.; Sun, H. METTL14 Suppresses CRC Progression via Regulating N6-Methyladenosine-Dependent Primary miR-375 Processing—ScienceDirect. *Mol. Ther.* **2020**, *28*, 599–612. [CrossRef]

20. Zhao, W.; Quansah, E.; Yuan, M.; Li, P.; Yi, C.; Cai, X.; Zhu, J. Next-generation sequencing analysis reveals segmental patterns of microRNA expression in yak epididymis. *Reprod. Fertil. Dev.* **2020**, *32*, 1067–1083. [CrossRef]
21. He, P.C.; He, C. m6A RNA methylation: From mechanisms to therapeutic potential. *EMBO J.* **2021**, *40*, e105977. [CrossRef]
22. Berulava, T.; Rahmann, S.; Rademacher, K.; Klein-Hitpass, L.; Horsthemke, B. N6-adenosine methylation in MiRNAs. *PLoS ONE* **2015**, *10*, e0118438. [CrossRef]
23. Kajimura, S.; Seale, P.; Kubota, K.; Lunsford, E.; Frangioni, J.V.; Gygi, S.P.; Spiegelman, B.M. Initiation of myoblast to brown fat switch by a PRDM16-C/EBP-beta transcriptional complex. *Nature* **2009**, *460*, 1154–1158. [CrossRef]
24. Lee, E.K.; Lee, M.J.; Abdelmohsen, K.; Kim, W.; Kim, M.M.; Srikantan, S.; Martindale, J.L.; Hutchison, E.R.; Kim, H.H.; Marasa, B.S.; et al. miR-130 Suppresses Adipogenesis by Inhibiting Peroxisome Proliferator-Activated Receptor gamma Expression. *Mol. Cell Biol.* **2011**, *31*, 626–638. [CrossRef]
25. Yang, Z.; Yu, G.L.; Zhu, X.; Peng, T.H.; Lv, Y.C. Critical roles of FTO-mediated mRNA m6A demethylation in regulating adipogenesis and lipid metabolism: Implications in lipid metabolic disorders. *Genes Dis.* **2022**, *9*, 51–61. [CrossRef]
26. Niu, Y.; Lin, Z.; Wan, A.; Chen, H.; Liang, H.; Sun, L.; Wang, Y.; Li, X.; Xiong, X.-f.; Wei, B.; et al. RNA N6-methyladenosine demethylase FTO promotes breast tumor progression through inhibiting BNIP3. *Mol. Cancer* **2019**, *18*, 46. [CrossRef]
27. Ruan, D.-Y.; Li, T.; Wang, Y.-N.; Meng, Q.; Li, Y.; Yu, K.; Wang, M.; Lin, J.-F.; Luo, L.-Z.; Wang, D.-S.; et al. FTO downregulation mediated by hypoxia facilitates colorectal cancer metastasis. *Oncogene* **2021**, *40*, 5168–5181. [CrossRef]
28. Fu, Y.; Jia, G.; Pang, X.; Wang, R.N.; Wang, X.; Li, C.J.; Smemo, S.; Dai, Q.; Bailey, K.A.; Nobrega, M.A.; et al. FTO-mediated formation of N6-hydroxymethyladenosine and N6-formyladenosine in mammalian RNA. *Nat. Commun.* **2013**, *4*, 1798. [CrossRef]
29. Chen, Y.; Zhao, Y.; Chen, J.; Peng, C.; Zhang, Y.; Tong, R.; Cheng, Q.; Yang, B.; Feng, X.; Lu, Y.; et al. ALKBH5 suppresses malignancy of hepatocellular carcinoma via m6A-guided epigenetic inhibition of LYPD1. *Mol. Cancer* **2020**, *19*, 123. [CrossRef]
30. Hu, Y.; Gong, C.; Li, Z.; Liu, J.; Chen, Y.; Huang, Y.; Luo, Q.; Wang, S.; Hou, Y.; Yang, S.; et al. Demethylase ALKBH5 suppresses invasion of gastric cancer via PKMYT1 m6A modification. *Mol Cancer* **2022**, *21*, 34. [CrossRef]
31. Zhang, J.; Guo, S.; Piao, H.Y.; Wang, Y.; Wu, Y.; Meng, X.Y.; Yang, D.; Zheng, Z.C.; Zhao, Y. ALKBH5 promotes invasion and metastasis of gastric cancer by decreasing methylation of the lncRNA NEAT1. *J. Physiol. Biochem.* **2019**, *75*, 379–389. [CrossRef]
32. Zhang, C.Z.; Samanta, D.; Lu, H.Q.; Bullen, J.W.; Zhang, H.M.; Chen, I.; He, X.S.; Semenza, G.L. Hypoxia induces the breast cancer stem cell phenotype by HIF-dependent and ALKBH5-mediated m(6)A-demethylation of NANOG mRNA. *Proc. Natl. Acad. Sci. USA* **2016**, *113*, E2047–E2056. [CrossRef]
33. Tan, L.; Tang, Y.; Li, H.; Li, P.; Ye, Y.; Cen, J.; Gui, C.; Luo, J.; Cao, J.; Wei, J. N6-Methyladenosine Modification of LncRNA DUXAP9 Promotes Renal Cancer Cells Proliferation and Motility by Activating the PI3K/AKT Signaling Pathway. *Front. Oncol.* **2021**, *11*, 641833. [CrossRef]
34. Shi, H.; Wang, X.; Lu, Z.; Zhao, B.S.; Ma, H.; Hsu, P.J.; Liu, C.; He, C. YTHDF3 facilitates translation and decay of N6-methyladenosine-modified RNA. *Cell Res.* **2017**, *27*, 315–328. [CrossRef]
35. Hsu, P.J.; Zhu, Y.F.; Ma, H.H.; Guo, Y.H.; Shi, X.D.; Liu, Y.Y.; Qi, M.J.; Lu, Z.K.; Shi, H.L.; Wang, J.Y.; et al. Ythdc2 is an N-6-methyladenosine binding protein that regulates mammalian spermatogenesis. *Cell Res.* **2017**, *27*, 1115–1127. [CrossRef]
36. Luo, X.; Cao, M.; Gao, F.; He, X. YTHDF1 promotes hepatocellular carcinoma progression via activating PI3K/AKT/mTOR signaling pathway and inducing epithelial-mesenchymal transition. *Exp. Hematol. Oncol.* **2021**, *10*, 35. [CrossRef]
37. Chen, D.; Cheung, H.; Lau, H.C.; Yu, J.; Wong, C.C. N(6)-Methyladenosine RNA-Binding Protein YTHDF1 in Gastrointestinal Cancers: Function, Molecular Mechanism and Clinical Implication. *Cancers* **2022**, *14*, 3489. [CrossRef]
38. Chen, H.; Yu, Y.; Yang, M.; Huang, H.; Ma, S.; Hu, J.; Xi, Z.; Guo, H.; Yao, G.; Yang, L.; et al. YTHDF1 promotes breast cancer progression by facilitating FOXM1 translation in an m6A-dependent manner. *Cell Biosci.* **2022**, *12*, 19. [CrossRef]
39. Han, B.; Yan, S.J.; Wei, S.S.; Xiang, J.; Liu, K.L.; Chen, Z.H.; Bai, R.P.; Sheng, J.H.; Xu, Z.P.; Gao, X.W. YTHDF1-mediated translation amplifies Wnt-driven intestinal stemness. *EMBO Rep.* **2020**, *21*, e49229. [CrossRef]
40. Ma, S.; Chen, C.; Ji, X.; Liu, J.; Zhou, Q.; Wang, G.; Yuan, W.; Kan, Q.; Sun, Z. The interplay between m6A RNA methylation and noncoding RNA in cancer. *J. Hematol. Oncol.* **2019**, *12*, 121. [CrossRef]
41. Huang, Y.; Yan, J.L.; Li, Q.; Li, J.F.; Gong, S.Z.; Zhou, H.; Gan, J.H.; Jiang, H.L.; Jia, G.F.; Luo, C.; et al. Meclofenamic acid selectively inhibits FTO demethylation of m(6)A over ALKBH5. *Nucleic Acids Res.* **2015**, *43*, 373–384. [CrossRef] [PubMed]
42. Weng, H.; Huang, H.; Wu, H.; Qin, X.; Zhao, B.S.; Dong, L.; Shi, H.; Skibbe, J.; Shen, C.; Hu, C.; et al. METTL14 Inhibits Hematopoiet-ic Stem/Progenitor Differentiation and Promotes Leukemogenesis via mRNA m(6)A Modification. *Cell Stem Cell* **2018**, *22*, 191–205.e9. [CrossRef] [PubMed]
43. Yanko-va, E.; Blackaby, W.; Albertella, M.; Rak, J.; De Braekeleer, E.; Tsagkogeorga, G.; Pilka, E.S.; Aspris, D.; Leggate, D.; Hendrick, A.G.; et al. Small-molecule inhibition of METTL3 as a strategy against myeloid leukaemia. *Nature* **2021**, *593*, 597–601. [CrossRef] [PubMed]
44. Jesch-ke, J.; Collignon, E.; Al Wardi, C.; Krayem, M.; Bizet, M.; Jia, Y.; Garaud, S.; Wimana, Z.; Calonne, E.; Hassabi, B.; et al. Downregula-tion of the FTO m(6)A RNA demethylase promotes EMT-mediated progression of epithelial tumors and sensitivity to Wnt inhibitors. *Nat. Cancer* **2021**, *2*, 611–628. [CrossRef] [PubMed]
45. Zhou, Y.; Zeng, P.; Li, Y.H.; Zhang, Z.; Cui, Q. SRAMP: Prediction of mammalian N6-methyladenosine (m6A) sites based on sequence-derived features. *Nucleic Acids Res.* **2016**, *44*, e91. [CrossRef] [PubMed]
46. Chen, W.; Feng, P.; Ding, H.; Lin, H.; Chou, K.-C. iRNA-Methyl: Identifying N6-methyladenosine sites using pseudo nucleotide composition. *Anal. Biochem.* **2015**, *490*, 26–33. [CrossRef]

47. Qiu, W.R.; Jiang, S.Y.; Xu, Z.C.; Xiao, X.; Chou, K.C. iRNAm5C-PseDNC: Identifying RNA 5-methylcytosine sites by incorporating physical-chemical properties into pseudo dinucleotide composition. *Oncotarget* **2017**, *8*, 41178–41188. [CrossRef]
48. Qiang, X.; Chen, H.; Ye, X.; Su, R.; Wei, L. M6AMRFS: Robust Prediction of N6-Methyladenosine Sites with Sequence-Based Features in Multiple Species. *Front. Genet.* **2018**, *9*, 495. [CrossRef]
49. Wei, L.; Chen, H.; Su, R. M6APred-EL: A Sequence-Based Predictor for Identifying N6-methyladenosine Sites Using Ensemble Learning. Molecular therapy. *Nucleic Acids* **2018**, *12*, 635–644. [CrossRef]
50. Zhang, Y.; Hamada, M. DeepM6ASeq: Prediction and characterization of m6A-containing sequences using deep learning. *BMC Bioinform.* **2018**, *19*, 524. [CrossRef] [PubMed]
51. Tao, X.; Chen, J.; Jiang, Y.; Wei, Y.; Chen, Y.; Xu, H.; Zhu, L.; Tang, G.; Li, M.; Jiang, A.; et al. Transcriptome-wide N (6) -methyladeno-si-ne methylome profiling of porcine muscle and adipose tissues reveals a potential mechanism for transcriptional regulation and differential methylation pattern. *BMC Genom.* **2017**, *18*, 336. [CrossRef] [PubMed]
52. Wang, Y.; Zheng, Y.; Guo, D.; Zhang, X.; Guo, S.; Hui, T.; Yue, C.; Sun, J.; Guo, S.; Bai, Z.; et al. m6A Methylation Analysis of Differentially Expressed Genes in Skin Tissues of Coarse and Fine Type Liaoning Cashmere Goats. *Front. Genet.* **2019**, *10*, 1318. [CrossRef]
53. Yang, L.; Chen, X.; Qian, X.; Zhang, J.J.; Wu, M.J.; Yu, A.J. Comprehensive Analysis of the Transcriptome-Wide m6A Methylome in Endometrioid Ovarian Cancer. *Front. Oncol.* **2022**, *12*, 583. [CrossRef] [PubMed]
54. Li, N.; Guo, Q.; Zhang, Q.; Chen, B.J.; Li, X.A.; Zhou, Y. Comprehensive Analysis of Differentially Expressed Profiles of mRNA N6-Methyladenosine in Colorectal Cancer. *Front. Cell Dev. Biol.* **2022**, *9*, 760912. [CrossRef] [PubMed]
55. Liu, J.; Xu, Y.P.; Li, K.; Ye, Q.; Zhou, H.Y.; Sun, H.; Li, X.; Yu, L.; Deng, Y.Q.; Li, R.T.; et al. The m(6)A methylome of SARS-CoV-2 in host cells. *Cell Res.* **2021**, *31*, 404–414. [CrossRef] [PubMed]
56. Ma-thur, L.; Jung, S.; Jang, C.; Lee, G. Quantitative analysis of m(6)A RNA modification by LC-MS. *STAR Protoc.* **2021**, *2*, 100724. [CrossRef] [PubMed]
57. Shu, X.; Cao, J.; Liu, J. m(6)A-label-seq: A metabolic labeling protocol to detect transcriptome-wide mRNA N (6)-methyladenosine (m(6)A) at base resolution. *STAR Protoc.* **2022**, *3*, 101096. [CrossRef] [PubMed]
58. Wang, Y.; Xiao, Y.; Dong, S.; Yu, Q.; Jia, G. Antibody-free enzyme-assisted chemical approach for detection of N6-methyladenosine. *Nat. Chem. Biol.* **2020**, *16*, 896–903. [CrossRef]
59. Liu, Q.; Zhao, Y.; Wu, R.; Jiang, Q.; Cai, M.; Bi, Z.; Liu, Y.; Yao, Y.; Feng, J.; Wang, Y.; et al. ZFP217 regulates adipogenesis by controlling mitotic clonal expansion in a METTL3-m6A dependent manner. *RNA Biol.* **2019**, *16*, 1785–1793. [CrossRef] [PubMed]
60. Yang, Y.; Cai, J.; Yang, X.; Wang, K.; Sun, K.; Yang, Z.; Zhang, L.; Yang, L.; Gu, C.; Huang, X.; et al. Dysregulated m6A modifica-tion promotes lipogenesis and development of non-alcoholic fatty liver disease and hepatocellular carcinoma. *Mol. Ther.* **2022**, *30*, 2342–2353. [CrossRef]
61. Peng, Z.S.; Gong, Y.Y.; Wang, X.J.; He, W.M.; Wu, L.T.; Zhang, L.Y.; Xiong, L.; Huang, Y.R.; Su, L.; Shi, P.J.; et al. METTL3-m(6)A-Rubicon axis inhibits autophagy in nonalcoholic fatty liver disease. *Mol. Ther.* **2022**, *30*, 932–946. [CrossRef] [PubMed]
62. Li, Y.; Zhang, Q.; Cui, G.; Zhao, F.; Tian, X.; Sun, B.F.; Yang, Y.; Li, W. m(6)A Regulates Liver Metabolic Disorders and Hepatogenous Diabetes. *Genom. Proteom. Bioinform.* **2020**, *18*, 371–383. [CrossRef] [PubMed]
63. Li, X.; Ding, K.; Li, X.; Yuan, B.; Wang, Y.; Yao, Z.; Wang, S.; Huang, H.; Xu, B.; Xie, L.; et al. Deficiency of WTAP in hepatocytes induces lipoatrophy and non-alcoholic steatohepatitis (NASH). *Nat. Commun.* **2022**, *13*, 4549. [CrossRef] [PubMed]
64. Song, T.; Yang, Y.; Wei, H.; Xie, X.; Lu, J.; Zeng, Q.; Peng, J.; Zhou, Y.; Jiang, S.; Peng, J. Zfp217 mediates m6A mRNA methyla-tion to orchestrate transcriptional and post-transcriptional regulation to promote adipogenic differentiation. *Nucleic Acids Res.* **2019**, *47*, 6130–6144. [CrossRef] [PubMed]
65. Merkestein, M.; Laber, S.; McMurray, F.; Andrew, D.; Sachse, G.; Sanderson, J.; Li, M.; Usher, S.; Sellayah, D.; Ashcroft, F.M.; et al. FTO influences adipogenesis by regulating mitotic clonal expansion. *Nat. Commun.* **2015**, *6*, 6792. [CrossRef]
66. Wang, L.; Song, C.; Wang, N.; Li, S.; Liu, Q.; Sun, Z.; Wang, K.; Yu, S.-C.; Yang, Q. NADP modulates RNA m6A methylation and adipogenesis via enhancing FTO activity. *Nat. Chem. Biol.* **2020**, *16*, 1394–1402. [CrossRef] [PubMed]
67. Liao, X.; Liu, J.; Chen, Y.; Liu, Y.; Chen, W.; Zeng, B.; Liu, Y.; Luo, Y.; Huang, C.; Guo, G.; et al. Metformin combats obesity by targeting FTO in an m6A-YTHDF2-dependent manner. *J. Drug Target.* **2022**, *30*, 983–991. [CrossRef] [PubMed]
68. Sun, D.; Zhao, T.; Zhang, Q.; Wu, M.; Zhang, Z. Fat mass and obesity-associated protein regulates lipogenesis via m6A modification in fatty acid synthase mRNA. *Cell Biol. Int.* **2021**, *45*, 334–344. [CrossRef]
69. Zhong, X.; Yu, J.; Frazier, K.; Weng, X.; Li, Y.; Cham, C.M.; Dolan, K.; Zhu, X.; Hubert, N.; Tao, Y.; et al. Circadian Clock Regulation of Hepatic Lipid Metabolism by Modulation of m6A mRNA Methylation. *Cell Rep.* **2018**, *25*, 1816–1828.e4. [CrossRef] [PubMed]
70. Takemoto, S.; Nakano, M.; Fukami, T.; Nakajima, M. m(6)A modification impacts hepatic drug and lipid metabolism properties by regulating carboxylesterase 2. *Biochem. Pharmacol.* **2021**, *193*, 114766. [CrossRef] [PubMed]
71. Zong, X.; Zhao, J.; Wang, H.; Lu, Z.; Wang, F.; Du, H.; Wang, Y. Mettl3 Deficiency Sustains Long-Chain Fatty Acid Absorption through Suppressing Traf6-Dependent Inflammation Response. *J. Immunol.* **2019**, *202*, 567–578. [CrossRef] [PubMed]
72. Peng, H.; Chen, B.; Wei, W.; Guo, S.; Han, H.; Yang, C.; Ma, J.; Wang, L.; Peng, S.; Kuang, M.; et al. N6-methyladenosine (m6A) in 18S rRNA promotes fatty acid metabolism and oncogenic transformation. *Nat. Metab.* **2022**, *4*, 1041–1054. [CrossRef] [PubMed]
73. Heng, J.H.; Wu, Z.H.; Tian, M.; Chen, J.M.; Song, H.Q.; Chen, F.; Guan, W.T.; Zhang, S.H. Excessive BCAA regulates fat metabolism partially through the modification of m(6)A RNA methylation in weanling piglets. *Nutr. Metab.* **2020**, *17*, 10. [CrossRef]

74. Wang, X.; Wu, R.; Liu, Y.; Zhao, Y.; Bi, Z.; Yao, Y.; Liu, Q.; Shi, H.; Wang, F.; Wang, Y. m(6)A mRNA methylation controls autophagy and adipogenesis by targeting Atg5 and Atg7. *Autophagy* **2020**, *16*, 1221–1235. [CrossRef]
75. Jiang, Q.; Sun, B.F.; Liu, Q.; Cai, M.; Wu, R.F.; Wang, F.Q.; Yao, Y.X.; Wang, Y.Z.; Wang, X.X. MTCH2 promotes adipogenesis in intramuscular preadipocytes via an m(6)A-YTHDF1-dependent mechanism. *FASEB J.* **2019**, *33*, 8690–8691. [CrossRef]
76. Guo, F.; Zhang, Y.; Ma, J.; Yu, Y.; Wang, Q.; Gao, P.; Wang, L.; Xu, Z.; Wei, X.; Jing, M. m(6)A mRNA Methylation Was Associat-ed with Gene Expression and Lipid Metabolism in Liver of Broilers under Lipopolysaccharide Stimulation. *Front. Genet.* **2022**, *13*, 818357. [CrossRef]
77. Cheng, B.; Leng, L.; Li, Z.; Wang, W.; Jing, Y.; Li, Y.; Wang, N.; Li, H.; Wang, S. Profiling of RNA N (6)-Methyladenosine Methyla-tion Reveals the Critical Role of m(6)A in Chicken Adipose Deposition. *Front. Cell Dev. Biol.* **2021**, *9*, 590468. [CrossRef]
78. Hu, Y.; Feng, Y.; Zhang, L.; Jia, Y.; Cai, D.; Qian, S.B.; Du, M.; Zhao, R. GR-mediated FTO transactivation induces lipid accumulation in hepatocytes via demethylation of m(6)A on lipogenic mRNAs. *RNA Biol.* **2020**, *17*, 930–942. [CrossRef]
79. Zhang, Q.; Cheng, B.; Jiang, H.; Zhang, H.; Li, H. N6-methyladenosine demethylase ALKBH5: A novel regulator of prolifera-tion and differentiation of chicken preadipocytes. *Acta Biochim. Biophys. Sin.* **2022**, *54*, 55. [CrossRef]
80. Chen, J.; Ren, X.; Li, L.; Lu, S.; Chen, T.; Tan, L.; Liu, M.; Luo, Q.; Liang, S.; Nie, Q.; et al. Integrative Analyses of mRNA Expres-si-on Profile Reveal the Involvement of IGF2BP1 in Chicken Adipogenesis. *Int. J. Mol. Sci.* **2019**, *20*, 2923. [CrossRef]
81. Gebeyew, K.; Yang, C.; Mi, H.; Cheng, Y.; Zhang, T.; Hu, F.; Yan, Q.; He, Z.; Tang, S.; Tan, Z. Lipid metabolism and m6A RNA methylation are altered in lambs supplemented rumen-protected methionine and lysine in a low-protein diet. *J. Anim. Sci. Biotechnol.* **2022**, *13*, 85. [CrossRef] [PubMed]
82. Zhang, Y.; Liang, C.; Wu, X.; Pei, J.; Guo, X.; Chu, M.; Ding, X.; Bao, P.; Kalwar, Q.; Yan, P. Integrated Study of Transcriptome-wide m6A Methylome Reveals Novel Insights into the Character and Function of m6A Methylation during Yak Adipocyte Differentiation. *Front. Cell Dev. Biol.* **2021**, *9*, 689067. [CrossRef] [PubMed]
83. Wang, L.; Qi, H.; Li, D.; Liu, L.; Chen, D.; Gao, X. METTL3 is a key regulator of milk synthesis in mammary epithelial cells. *Cell Biol. Int.* **2022**, *46*, 359–369. [CrossRef] [PubMed]
84. Li, T.; Lin, C.; Zhu, Y.; Xu, H.; Yin, Y.; Wang, C.; Tang, X.; Song, T.; Guo, A.; Chen, Y.; et al. Transcriptome Profiling of m(6)A mRNA Modification in Bovine Mammary Epithelial Cells Treated with *Escherichia coli*. *Int. J. Mol. Sci.* **2021**, *22*, 6254. [CrossRef] [PubMed]
85. Zhao, W.; Ahmed, S.; Ahmed, S.; Yangliu, Y.; Wang, H.; Cai, X. Analysis of long non-coding RNAs in epididymis of cattleyak associated with male infertility. *Theriogenology* **2021**, *160*, 61–71. [CrossRef]
86. Rong, B.; Feng, R.; Liu, C.; Wu, Q.; Sun, C. Reduced delivery of epididymal adipocyte-derived exo-somal resistin is essential for melatonin ameliorating hepatic steatosis in mice. *J. Pineal Res.* **2019**, *66*, e12561. [CrossRef]

agriculture

MDPI

Review

# Genetic Load of Mutations Causing Inherited Diseases and Its Classification in Dairy Cattle Bred in the Russian Federation

**Saida N. Marzanova [1], Davud A. Devrishov [1], Irina S. Turbina [2], Nurbiy S. Marzanov [3], Darren K. Griffin [4] and Michael N. Romanov [4,*]**

1 K.I. Skryabin Moscow State Academy of Veterinary Medicine and Biotechnology, 109472 Moscow, Russia
2 Head Center for the Reproduction of Farm Animals, 142143 Bykovo, Moscow Oblast, Russia
3 L.K. Ernst Federal Research Center for Animal Husbandry, 142132 Dubrovitsy, Moscow Oblast, Russia
4 School of Biosciences, University of Kent, Canterbury CT2 7NJ, UK
* Correspondence: m.romanov@kent.ac.uk

**Abstract:** This review addresses the concept of genetic load from the point of view of molecular genetics, development and efforts in selective breeding. As typical examples, the assessment of animals in the Holstein breed and its high-blooded crossbreeds is considered for mutations that cause three inherited diseases: bovine leukocyte adhesion deficiency (*CD18* locus), complex vertebral malformation (*SLC35A3* locus), and brachyspina (*FANCI* locus). The reasons for their occurrence and accumulation in the breeding herds of the black-pied genealogical root are discussed. These include an intense artificial-selection of bulls and cows in highly productive herds and the intensive sale (within and between countries) of breeding material (animals, semen, embryos) from a small population of sires from countries with a high level of dairy-cattle breeding development. There is a founder effect when the source of mutant-allele spread is a prominent sire. For example, the greatest contribution to the spread of mutant alleles $CD18^{G}$, $SLC35A3^{T}$ and $FANCI^{BY}$ was made by the descendants of three closely related bulls. A genogeographic generalization of the mutation occurrence in the world and Russia is provided for these hereditary-disease loci and, includes a total of 31 countries where these mutations were detected. The genetic-load classification for these and other mutations is given. The mutations are inherited both recessively ($CD18^{G}$, $SLC35A3^{T}$, $FANCI^{BY}$) and codominantly ($CSN3^{A}$, $CSN3^{C}$, $CSN3^{E}$, $CSN2^{A1}$, $CSN2^{B}$). Genetic load is classified into the following types: mutational, segregation, substitutional, and immigration. For each of these, examples are given that explain their occurrence. Overall, it can be concluded that the phenomenon of genetic load in industrial herds of dairy cattle requires special attention when creating healthy livestock and obtaining high-quality dairy products.

**Keywords:** hereditary diseases; breeds; dairy cattle; mutations; alleles; genetic-load classification

**Citation:** Marzanova, S.N.; Devrishov, D.A.; Turbina, I.S.; Marzanov, N.S.; Griffin, D.K.; Romanov, M.N. Genetic Load of Mutations Causing Inherited Diseases and Its Classification in Dairy Cattle Bred in the Russian Federation. *Agriculture* **2023**, *13*, 299. https://doi.org/10.3390/agriculture13020299

Academic Editors: Zhi Chen and Cong Li

Received: 23 December 2022
Revised: 19 January 2023
Accepted: 23 January 2023
Published: 26 January 2023



## 1. Introduction

Genetic load is a term used to refer to the sum of unfavorable lethal and sublethal mutations in the genome of individuals within a population that reduces their viability or increases the risk of death. The concept was first proposed by the English population geneticist J.B.S. Haldane [1].

Geneticists who proposed the genetic-load theory expanded on the concept of an ideal genotype that conforms to ideal fitness. Currently, a drawback of this approach is apparent. That is, it is somewhat meaningless to proceed from such an assumption when referring to various types of farm animals, where we deal with extremely high requirements for adaptive plasticity. Nevertheless, the term "genetic load" here may be convenient for designating the burden of the gene pool due to a rather poor-quality heredity. Its artificial accumulation in the gene pool reduces the overall fitness of a particular species or leads to a low quality in the products obtained from it [2,3]. In this regard, it is important to study the prevalence (i.e., genogeography) of mutations that cause certain diseases and disorders in animals.

The purpose of this review is to summarize the data on the distribution of mutations associated with a number of inherited diseases and classify the genetic load, mainly in populations of dairy cattle. As representative examples of hereditary diseases, we have focused in this review on bovine leukocyte adhesion deficiency (BLAD; *CD18*$^{G}$ mutation), complex vertebral malformation (CVM; *SLC35A3*$^{T}$ mutation), and brachyspina or short spine lethal syndrome (BY; *FANCI*$^{BY}$ mutation), as well as abnormalities associated with mutations of beta- and kappa-caseins (*CSN2* and *CSN3* loci, respectively). These genetic syndromes are diagnosed on the basis of certain developed methods, according to which many patents have also been obtained (e.g., [4–8]). The method for diagnosing four alleles of beta-casein (*CSN2*$^{A1}$, *CSN2*$^{A2}$, *CSN2*$^{A3}$, *CSN2*$^{B}$) has come into practice thanks to the respective developments by Lien et al. [9] and Dinç [10], with subsequent modifications. An understanding of the distribution of the three mutations (*CD18*$^{G}$, *SLC35A3*$^{T}$ and *FANCI*$^{BY}$) diagnosed in Holstein cattle can be obtained from the analysis of various literature sources, as will be discussed in this review below.

## 2. What Does Genetic Load Mean?

As summarized by Bertorelle et al. [11], the genetic load is part of the hereditary variability of a population that occurs as a result of natural or artificial selection and determines the appearance of less adapted individuals that undergo selective death. Along with diminishing a population's biodiversity, the genetic load is linked to a decline in the selection value of individuals. A population's incapacity to adjust to a given set of environmental factors is measured by the genetic load. The term "genetic load" refers to the accumulation of lethal and sublethal harmful mutations that significantly reduce an individual's viability or result in their death when the mutation enters a homozygous state. Thus, in a stricter sense, the genetic load is accepted in population genetics as an expression of a lowering of the selective value of a population compared to that which the population had in the past [12,13]. In dairy cattle, mutations causing hereditary diseases are often breed-specific, since they were introduced worldwide from a limited number of genealogical roots, e.g., through bulls that were the founders of the Holstein breed. Mutant alleles and genotypes at the casein loci can also be breed-specific (e.g., [14]).

## 3. Mutations Leading to Disorders in Cattle Breeds

Over 559 genetically determined morphological and functional disorders have been identified in cattle [15]. In Russia, among the six dairy breeds that are exploited in many regions, the largest number of such inherited deviations was recorded in Holstein cattle followed by the Friesian, Black Pied, Simmental, Brown Swiss and Ayrshire breeds (Figure 1).

The difference between the Holstein breed and others is associated with the peculiarities of its breeding and reproduction. Holstein cattle are known to have a limited number of sire lines and related groups (e.g., Osbornedale Ivanhoe, A.B.C. Reflection Sovereign, Montwick Chieftain, etc.). In addition, the formation of Holstein populations in the United States occurred with the intensive use of a small number of bulls. Therefore, in the pedigrees of almost all animals of the Holstein breed in the 7th–10th lines of ancestors, there are genes of at least one of the 20 founding bulls. Thus, in formal outbreeding, it is actually difficult to avoid the selection of pairs in the pedigrees of which there is no "blood" of these founders or their descendants. On the one hand, such a breeding system with intensive selection contributes to the consolidation of the breed, but on the other hand, it increases the likelihood of accumulation and transition to the homozygous state of a complex of mutant genes that cause various disorders [3,16].

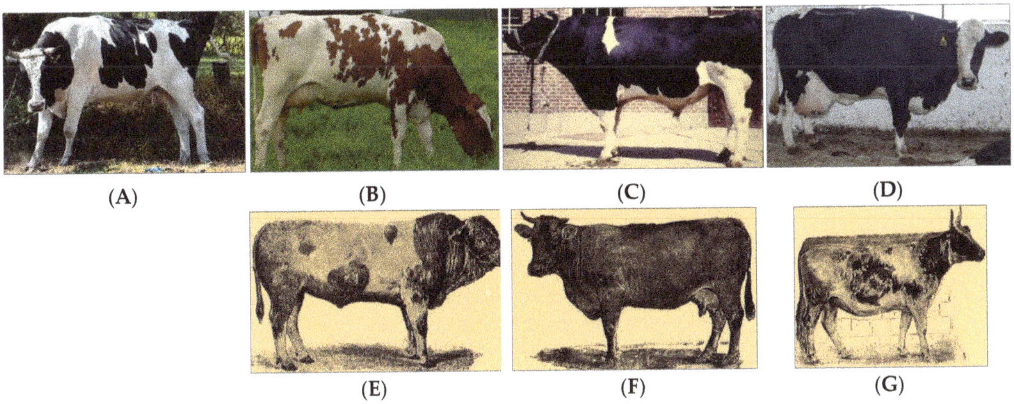


**Figure 1.** Some notable dairy cattle breeds that can be carriers of deleterious mutations. (**A**), Holstein Friesian bull with the black-pied coat color; (**B**), Holstein Friesian heifer with the red-pied coat color; (**C**), German Black Pied bull; (**D**), Russian Black Pied cow; (**E**), Simmental bull; (**F**), Brown Swiss cow; (**G**), Ayrshire cattle (Image sources: (**A**), https://commons.wikimedia.org/wiki/File:Tethered_bull_Holstein_Mexico_p1.jpg (accessed on 19 January 2022), by Cvmontuy, 2020, Creative Commons Attribution-Share Alike 4.0 International license, Category: Holstein Friesian cattle, cropped; (**B**), https://commons.wikimedia.org/wiki/File:Red_Holstein.jpg (accessed on 19 January 2022), by MGA73bot2, 2011, Creative Commons Attribution-Share Alike 3.0 Unported license, Category: Red Holstein, cropped; (**C**), https://commons.wikimedia.org/wiki/File:%20Schwarzbunte_%3D_%E4%B8%96%E7%95%8C%E3%81%AE%E7%89%9B_%E3%83%89%E3%82%A4%E3%83%84%E9%BB%92%E7%99%BD%E6%96%91%E7%89%9B%EF%BC%88%E9%9B%84%EF%BC%89_(36567900171).jpg (accessed on 19 January 2022), by Tomasina, 2019, Creative Commons Attribution 2.0 Generic license, Category: Deutsches Schwarzbuntes Niederungsrind, cropped; (**D**), https://ru.wikipedia.org/wiki/%D0%9A%D0%BE%D1%80%D0%BE%D0%B2%D0%B0_%D1%87%D1%91%D1%80%D0%BD%D0%BE-%D0%BF%D1%91%D1%81%D1%82%D1%80%D0%BE%D0%B9_%D0%BF%D0%BE%D1%80%D0%BE%D0%B4%D1%8B.jpg (accessed on 19 January 2022), by Nicolas-a, 2018, Creative Commons Attribution 3.0 license, cropped; (**E**), https://commons.wikimedia.org/wiki/File:Brockhaus_and_Efron_Encyclopedic_Dictionary_b59_264-2.jpg (accessed on 19 January 2022), by ButkoBot, 2009, public domain, Category: Mammals illustrations from Brockhaus and Efron Encyclopedic Dictionary, cropped; (**F**), https://commons.wikimedia.org/wiki/File:Brockhaus_and_Efron_Encyclopedic_Dictionary_b59_264-1.jpg (accessed on 19 January 2022), by ButkoBot, 2009, public domain, Category: Mammals illustrations from Brockhaus and Efron Encyclopedic Dictionary, cropped; (**G**), https://commons.wikimedia.org/wiki/File:Brockhaus_and_Efron_Encyclopedic_Dictionary_b59_264-1.jpg (accessed on 19 January 2022), by ButkoBot, 2009, public domain, Category: Mammals illustrations from Brockhaus and Efron Encyclopedic Dictionary, cropped).

The described disorders are an example of the direct transfer of mutant genes from one breed or from one country to another. Recently, however, the opposite situation has developed: when using Holstein bulls to "improve" populations of black-pied cattle, recessive mutations causing BLAD, CVM and BY were transferred to their gene pool, along with the introduction of beneficial traits [17–19].

Many of the above mutations, which represent a genetic load that is characteristic only for the Holstein breed of black-pied and red-pied varieties and their crosses, probably arose recently and in a particular breed, while some other mutations are of an earlier origin. First of all, this concerns the BLAD, CVM and BY diseases that have become widespread in populations of the Holstein cattle, due to the intensive use of the descendants of the Osbornedale Ivanhoe 1189870 bull, including his son Penstate Ivanhoe Star 1441440 and especially his grandson Carlin-M Ivanhoe Bell 1667366, in reproduction [20,21].

As a result of the culling of bulls carrying inherited-disorder mutations in the 1990s and 2000s, numbers have been reduced to a minimum or have been eradicated. In recent years, the emergence of carrier bulls has been associated with the transmission of mutations from non-certified bull-breeding mothers [22].

## 4. Mutation Occurrence in the World and Russia

Genogeographic analysis of the distribution of the most well-known mutations ($CD18^G$, $SLC35A3^T$, $FANCI^{BY}$, $CSN3^A$, $CSN3^C$, $CSN3^E$, $CSN2^{A1}$, $CSN2^B$) was carried out broadly in the Holstein breed of both black-pied and red-pied types [4,21–26]. It was reported that the average frequency of heterozygous $CD18^{A/G}$ carriers in the early studies on Holsteins in the USA and other countries was approximately 20% [27].

According to our generalization (Table 1), the mutant alleles $CD18^G$, $SLC35A3^T$ and $FANCI^{BY}$ were identified in most of the 31 countries where the Holstein breed or its high-blooded crosses were bred. As for the $SLC35A3^T$ allele, it was discovered in a smaller number of countries, due to the past long-term lack of access to diagnostic methods. Even fewer studies have been conducted on the $FANCI^{BY}$ allele, for the same reason [19]. In Table 1, some of the main world regions and countries are listed for which the occurrence of the alleles $CD18^G$, $SLC35A3^T$ and $FANCI^{BY}$ has been reported.

**Table 1.** Geography of dispersion of mutant alleles $CD18^G$, $SLC35A3^T$ and $FANCI^{BY}$.

| Region | Country | References |
|---|---|---|
| North America | US USA | [24,28–36] * |
| | CA Canada | [37,38] * |
| Latin America and Caribbean | AR Argentina | [39] |
| | BR Brazil | [40] |
| Europe | AT Austria | [41] |
| | BE Belgium | [37,42,43] * |
| | GB United Kingdom | [44] |
| | HU Hungary | [45,46] |
| | DK Denmark | [47–54] * |
| | IT Italy | [55,56] * |
| | ES Spain | [57] |
| | MK North Macedonia | [58] |
| | NL The Netherlands | [53,59–61] * |
| | PL Poland | [62–64] * |
| | RU Russia | [19,20,27,65–69] * |
| | RO Romania | [70] |
| | FR France | [23,71–73] * |
| | DE Germany | [74–82] * |
| | UA Ukraine | [83–85] |
| | CH Switzerland | [86] |
| | CZ Czech Republic | [87] |
| Southeast Asia and the Pacific Basin | CN China | [88–92] * |
| | TW Taiwan | [22] * |
| | JP Japan | [93–96] |
| | NZ New Zealand<br>AU Australia | [97] |
| Middle East | IR Iran | [98–100] |
| | TR Turkey | [101] |
| South Asia | IN India | [102,103] |
| | PK Pakistan | [104] |
| Africa | ZA South Africa | [105] |

* Reports of the diagnosed $FANCI^{BY}$ allele in breeding animals. Otherwise, references for reports of the established mutant alleles $CD18^G$ and $SLC35A3^T$ are provided.

Since the Holstein breed is widely spread worldwide in over 150 countries [106], the range of mutations is most likely much broader. Based on the data obtained, it should be noted that the mutant alleles came to certain countries through the purchase of breeding material from the United States, Canada, and some European countries (Denmark, Germany, the Netherlands, France, etc.), i.e., through the acquisition of sires, semen and embryos from these countries. In this regard, the livestock-breeding communities of the USA, Canada, Germany, Belgium, the Netherlands, Russian Federation and other countries with the developed dairy-cattle breeding decided to implement the mandatory tests of breeding material for the presence of mutant alleles [107]. The imported materials are entered into the catalogs of Holstein bulls and their high-blooded crosses, along with recently discovered haplotypes. The mutant alleles $CD18^{G}$, $SLC35A3^{T}$ and $FANCI^{BY}$ are also considered to be breed-specific traits for the Holstein cattle. One of the reasons for this situation is the strict selection of sires that are carriers of genes for extraordinary milk production, while the occurrence of the $CD18^{G}$, $SLC35A3^{T}$ and $FANCI^{BY}$ alleles are a side effect of high milk-yield, fertility or milk quality [22,34,108].

Thus, based on the above observations, it can be stated that the struggle to cleanse the breeding stock from mutations has been undertaken in those countries where elite animals of the Holstein breed or synthetic populations produced by its crossing with local cattle breeds are used [69].

In the Russian Federation, an assessment based on genetic studies was made with respect to dispersion of the $CD18^{G}$, $SLC35A3^{T}$ and $FANCI^{BY}$ mutations. Since detailed results of the occurrence of $CD18^{G}$ and $SLC35A3^{T}$ alleles in Russia were reported in our own previous works (e.g., [4]), here, we can limit ourselves to a general characterization of the state of the $FANCI^{BY}$ mutation spread in the regions of the Russian Federation, covered by the respective studies. These latter involved animals from the Bryansk Oblast, Ivanovo Oblast, Lipetsk Oblast, Moscow Oblast and the Republic of Karelia. In particular, the $FANCI^{BY}$ allele was found in fifteen breeding sires and five breeding cows of the Holstein breed maintained by the joint-stock company "Head Center for Animal Reproduction" (JSC GTsV). The $FANCI^{BY}$ allele carriers were the following bulls of the Holstein breed, both black-pied and red-pied: Avanti 76845, Barkhat 38, Braslet 106759921, Garus 10917481, Kankan 11033687, Kapral 1400, Laur 10990032, Leroy 10990031, Opal 11007858, Pikul 106894920, Ramos 96286, Flint 1223, Floks 1448, Shtabel' 1780, and Etiket 7754. They were culled from the JSC GTsV breeding stock. Currently, there are no carriers of the $CD18^{G}$, $SLC35A3^{T}$ and $FANCI^{BY}$ alleles in this breeding center [19]. It should be noted that sires carrying the three alleles were purchased by the JSC GTsV before the development of methods for diagnosing these mutations in animals.

In the course of our investigations [19,21], it was found that the main way for a mutant allele to enter the Russian Federation is the acquisition of breeding material without checking for the carriage of mutations. Obviously, the Holstein breed had a hidden segregation load of mutations from the Dutch cattle from which it originated. Indeed, a number of anomalies are found both in the Dutch Black Pied, Friesian, and Holstein cattle [16,27,109].

## 5. Classification of Genetic Load

Each type of genetic load in farm animals correlates with a certain type of natural or artificial selection in them [110]. In animal husbandry, four types of genetic load are generally distinguished: mutation, immigration, segregation, and substitution (Figure 2).

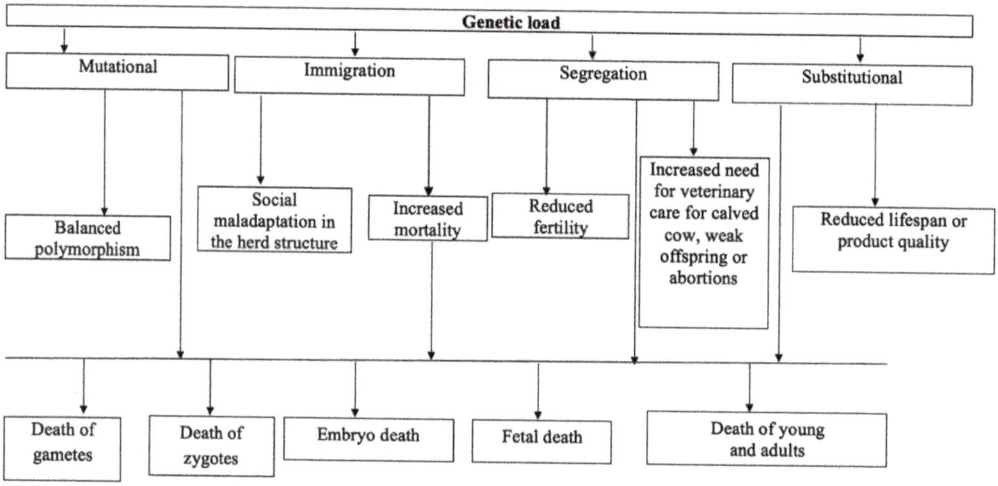


**Figure 2.** Classification of genetic load in dairy cattle and forms of its manifestation.

### 5.1. Mutational Genetic-Load

Most often, this type of genetic load includes autosomal-recessive mutations. A side effect of a mutation is the weakening of a population, due to the accumulation of various forms of unwanted alleles. Stabilizing natural selection either removes deleterious mutations from a population or, conversely, preserves them. In the process of artificial selection, which is widely used in breeding dairy cattle, when screening sires for mutant alleles such as $CD18^{G}$, $SLC35A3^{T}$ and $FANCI^{BY}$ causing BLAD, CVM and BY, respectively, and as well as reducing the cheese suitability of milk ($CSN3^{C}$, $CSN3^{E}$) or milk quality ($CSN2^{A1}$, $CSN2^{B}$) in daughters, most often such sires are culled from the user part of the breeders. Often, they are used under strict control, due to their high prepotency, selecting female non-carriers of one or another mutation for mating. This scheme is implemented by a breeder when a bull-reproducing group of cows is created and there is a need to obtain bulls desirable for reproduction. It also happens, on the contrary, that mutations are preserved due to associations with certain economically important traits (milk yield, high fat-content in milk or high fertility), most often in a heterozygous state and due to the impossibility of their presence in a homozygote. Thus, the mutation in this case is a trail of an important economic or biological trait. At the same time, there are more cows than sires in herds, so they are a kind of biological reserve, i.e., hidden carriers of the above mutant alleles. Marzanova [18] reported that, when creating a bull-reproducing group at one breeder site, a genetic test of the selected cows was performed for carriers of unwanted mutations. Of the 34 selected cows, 17 turned out to be carriers of the $SLC35A3^{T}$ allele causing CVM; they turned out to be the daughters of two bulls carrying this mutation. It was also noted that under environmental conditions, natural selection removes carriers of harmful mutations from the animal population, since they are weaker than healthy animals [18].

### 5.2. Immigration Genetic-Load

One more genetic-load type is immigration load when, due to the influx of genes from other populations or breeds, an improved breed is saturated with mutations, along with useful gene variants. The immigration load is created by the inclusion of alien gene alleles in a given gene pool, which in the new genotypic environment lead to lower fitness. Striking examples of this phenomenon are missense mutations ($CD18^{G}$ and $SLC35A3^{T}$) and deletions ($FANCI^{BY}$) in the Holstein breed, which cause the respective hereditary disorders in representatives of the black-pied genealogical root. They were introduced into the

Holstein breed from the aforementioned three famous sires [4]. There is another type of mutation: a missense mutation or deletion, which simultaneously have a codominant type of inheritance. Here, it is also necessary to undertake the entire course of genetic sanitation proposed for the purification of breeding herds in dairy breeds from recessive mutations. These mutations include abnormal allele variants of the beta-casein locus belonging to the A1 family [111], and there are only five alleles of this type: $CSN2^{A1}$, $CSN2^{B}$, $CSN2^{C}$, $CSN2^{F}$, and $CSN2^{G}$. However, the most remarkable representatives of this family are the $CSN2^{A1}$ and $CSN2^{B}$ alleles, which are most often discovered in herds. When a population finds itself in an extreme situation, it reacts in its own way through a change in the allelotype, first in individual animals and then in the entire population or breed which are dependent on their outstanding representatives used by humans, i.e., there is a founder effect in this case [4,111].

### *5.3. Segregation Genetic-Load*

Another type of genetic load is characteristic of populations that take advantage of heterozygotes. In this case, less-adapted homozygous individuals resulted from mating two heterozygotes are removed from a herd. By purposeful selection of heterozygous animals, researchers from the Veterinary Institute of Hannover (Germany) [17] obtained 50 homozygous calves for the BLAD syndrome [75,76]. Homozygous calves ($CD18^{G/G}$) with BLAD-syndrome fell ill in the first months after birth, and died within 2 months (50%) and 12 months (100%) of life. It was also reported that the frequency of the $CD18^{G}$ allele causing BLAD was as high as 24% in 2000, and the mutation rate of $SLC35A3^{T}$, the trigger of CVM, ranged from 9 to 16% between 2001 and 2007 in the German Holstein population [17]. The course of BLAD disease in calves was chronic. Animals significantly lagged behind in growth and development, lost weight, despite having a good appetite, and were very susceptible to various infections. Lichen was often observed in calves. At the same time, there were fever attacks, and constant disturbances in the gastrointestinal-tract functioning, as well as signs of the respiratory-tract inflammation. In most cases, the surface of the oral cavity was inflamed due to gingivitis, in calves. The treatment attempt was unsatisfactory, time consuming and ultimately unsuccessful. A similar case in relation to CVM was examined by Danish scientists [52]. They showed that in the homozygous form, a fetus for the mutant allele $SLC35A3^{T}$ was aborted or a stillborn calf was born; accordingly, the calving cow fell ill for a long time [52]. BY syndrome has long been confused with the course of CVM disease. With BY, we also deal only with a recessive homozygote ($FANCI^{BY/BY}$), as in the case of BLAD ($CD18^{G/G}$) and CVM ($SLC35A3^{T/T}$). However, the mutant allele in the homozygotes causes fetal death in the womb, abortion before day 40 or, rarely, stillbirth. Both of the latter pathological features are special characteristics in BY [19,50,52].

### *5.4. Substitutional Genetic-Load*

This type is manifested when the old allele is replaced by a new one. It conforms to the driving form of natural selection and transitional polymorphism. A distinct example is the substitution of the $CSN2^{A2}$ allele for $CSN2^{A1}$ and $CSN2^{A1}$ for $CSN2^{B}$ at the beta-casein locus during the evolution of cattle domestication. It is believed that the emergence of mutant alleles of the A1 family is more associated with domestication processes and the creation of high-milk cattle breeds [112]. Table 2 presents the data obtained for the diagnosed genotypes and alleles at the beta-casein locus in four dairy-cattle breeds of the Russian Federation. A comparative analysis of these data demonstrated a difference in the occurrence of three genotypes and two alleles in representatives of dairy breeds. In particular, the heterozygous genotype ($CSN2^{A1/A2}$) was most often detected in the Black Pied, Holstein and Yaroslavl breeds, while the Bestuzhev breed was homozygous for the $CSN2^{A1}$ mutant allele. This breed also had a high incidence of the heterozygous genotype ($n$ = 24). It was also found that the most common mutant $CSN2^{A1}$ allele occurred in the Bestuzhev breed (0.67), with a lower frequency in the Black Pied (0.56) and Yaroslavl (0.52) breeds. As can be seen from Table 2, the mutant allele was least detected in the Holstein breed, although its frequency of occurrence was also high (0.42). Disturbance of the genetic equilibrium at the beta-casein locus was not found in any of the studied breeds of dairy cattle [3,113,114].

**Table 2.** Polymorphism of the beta-casein locus in four different Russian dairy-cattle breeds ($n$ = 177) [3,113,114].

| Breed | $n$ | Gt | Genotype Frequency | | | Allele Frequency | | $\chi^2$ | df | $p$ |
|---|---|---|---|---|---|---|---|---|---|---|
| | | | $CSN2^{A1/A1}$ | $CSN2^{A1/A2}$ | $CSN2^{A2/A2}$ | $CSN2^{A1}$ | $CSN2^{A2}$ | | | |
| Russian Black Pied | 50 | O<br>E | 15<br>15.68 | 26<br>24.64 | 9<br>9.68 | 0.56 | 0.44 | 0.145 | 1 | >0.05 |
| Holstein Friesian | 30 | O<br>E | 4<br>5.21 | 17<br>14.58 | 9<br>10.21 | 0.42 | 0.58 | 0.82 | 1 | >0.05 |
| Yaroslavl | 30 | O<br>E | 9<br>8.0093 | 13<br>14.9833 | 8<br>7.0074 | 0.52 | 0.48 | 0.53 | 1 | >0.05 |
| Bestuzhev | 67 | O<br>E | 35<br>39 | 24<br>21 | 8<br>7 | 0.67 | 0.33 | 0.98 | 1 | >0.05 |

Abbreviations: $n$, number of animals examined; Gt, genotype; O, observed number of genotypes; E, expected number of genotypes; $\chi^2$, chi-squared test statistic; df, number of degrees of freedom; $p > 0.05$, genetic equilibrium in breeds is not disturbed.

It was also established that the formation of the allelotype in herds of cows for the $CSN2^{A1}$ and $CSN2^{A2}$ alleles was influenced by such factors as the genetic genealogy of a sire, the founder effect, and the drift of the mutant allele. Moreover, the drift of the mutant $CSN2^{A1}$ allele, both within one country and between countries, was due to artificial selection [3,113,114]. The main reason for this phenomenon is strict selection and widespread use of a small number of elite bulls, artificial insemination of a large number of cows, and multiple ovulation and embryo transfer (MOET). The dispersion of the mutant $CSN2^{A1}$ allele in Russia occurs through the purchase of breeding material (animals, semen, and embryos). The $CSN2^{A1}$ allele was found to be common where carrier bulls were used. Cows can also be suppliers of the mutant allele, but to a lesser extent. They serve more as a reserve, being homozygotes ($CSN2^{A1}/CSN2^{A1}$) or heterozygotes ($CSN2^{A1}/CSN2^{A2}$) in herds. The mutant $CSN2^{A1}$ allele is the codominant factor, and it should be noted that this is a new phenomenon in the diagnosis of abnormal alleles in cattle breeding. Previously identified mutant-alleles were found only as recessive factors [115,116].

## 6. Integrating Genomic Approaches to Improve Dairy Cattle

Over the last few decades, dairy-cattle breeding has been transformed by the implementation of genomic technologies [117]. High contributions from foreign sires (potentially with deleterious mutations) are almost always found in contemporary dairy-breeding programs. While lowering the projected returns from investment to increase the accuracy of genomic prediction in a home country, having a foreign supply of genetic material with a high rate of genetic advance significantly contributes to the advantages of domestic genetic-progress [118].

Genomic evaluations are recognized by producers as reliable predicators of a bull's ultimate daughter-based appraisal. The traditional evaluation approach has been improved by the incorporation of genomics and DNA-marker technology. This has resulted in a reduction in generation gaps, an improvement in selection accuracy, a drop in progeny-testing expenses, and the detection of recessive lethal and semi-lethal mutations [117]. Crossing between inbred lines significantly enhanced homozygosis, which contributed to the cumulative negative impacts of inbreeding, such as a loss in reproductive efficiency. Therefore, there is a higher risk of suboptimal outcomes from errors in the selection of candidates with high genetic-merit based only on low-heritability phenotypic features for empirical-conventional models of dairy-production systems. Due to the drastic drop in genetic gains, this lengthens generational intervals and raises costs. The recent significant advancement in genomic prediction increases the precision in choosing the best candidates [119]. Progeny testing of the top young males has been crucial to breeding programs' success, since it allows researchers to correctly determine each individual's genetic values and, consequently, breeding potential. Gains in the accuracy of projecting breeding-values for young animals without their own performance have been made possible by the incorporation of extensive genomic information into statistical algorithms used to make selection decisions, known as genomic selection [120].

Genome-wide approaches and tools will play an increasing role in the creation of a dairy sector that is strong, long-lasting, sustainable, and that prioritizes animal welfare (meeting the basic needs regarding animal health and promoting positive welfare and environmental efficiency in animal production) and productive effectiveness [120,121]. Precision management on contemporary dairy farms is facilitated by genomic selection-derived outcomes, and emerging genome-editing technologies will open up new perspectives on the future of dairy-cattle breeding [119]. Genomic applications relevant to Russian dairy-cattle breeding have also been implemented, and include genomic selection [122,123], genomic evaluation of the breeding value [124,125], identification of selection signatures [126], and genome editing [127].

## 7. Concluding Remarks

The genetic load is considered as part of the inherited variation of a population, which determines the appearance of less-adapted individuals that undergo selective death as a result of natural selection. In the 20th century, the intensive use of the world's breed gene-pool and reproduction biotechnologies (artificial insemination, embryo transplantation, cloning) facilitated the significant increase in the genetic potential of animal milk production by obtaining highly productive offspring, i.e., true-breed leaders [128].

At the same time, commercial breed-stocks are increasingly showing signs of genetic erosion, i.e., the accumulation of a harmful recessive-mutation load. Hereby, the reproductive ability and fertility, the viability of newborns and young animals, and the duration of the economic use of animals decrease, which negatively affects the profitability of production [129–132]. However, recently, the situation has changed in the reverse direction: while using Holstein bulls to "upgrade" populations of black-pied cattle, recessive genes producing BLAD, CVM, and BY were introduced to their gene pool along with the transfer of advantageous traits [4,17,133].

Many of the detected mutations which represent a genetic-load characteristic only of Holsteins, probably arose recently and in this breed alone. This applies to mutations that cause the BLAD, CVM and BY syndromes. They have become widespread in a number of countries of the world, including Russia, where populations of Holstein cattle are intensively used in the reproduction of the offspring of single bulls [109,134,135]. This has stimulated a further genogeographic analysis of the spread of known mutations ($CD18^{G}$, $SLC35A3^{T}$ and $FANCI^{BY}$) in the Holstein breed of both black-pied and red-pied varieties [3,4,23,25,26].

Because of the intensive use of the global gene-pool, as well as artificial insemination, embryo transplantation and cloning, it has become feasible to significantly increase the genetic potential of animal productivity by obtaining offspring of sires that are leaders of their breed. On the other hand, due to the fact that populations are increasingly showing signs of genetic erosion [3,25,45,136], social maladaptation of an animal in the herd exists, and manifests itself in a violation of the interaction of the individual with the external environment. This is characterized by the inability of such an individual to exercise a positive role in specific microsocial-conditions corresponding to the animal's capabilities. In this case, it manifests itself in the animal in the form of a reduced live-weight, frequent illnesses, decreased reproductive-ability, fertility, viability of newborns and young animals, declined resistance, and lower duration of economic use of animals, which ultimately negatively affects the profitability of livestock production.

Currently, if the *CD18*$^{G}$, *SLC35A3*$^{T}$ and *FANCI*$^{BY}$ mutations are detected, it is most often in cows that serve as a kind of reserve, since they often represent a source of inheritance of various mutations. Therefore, when conducting a genetic and genealogical analysis, it is often impossible to determine how a given cow received a mutant allele. As the data of various studies and pedigrees of Holstein bulls or high-blooded Holsteinized carriers show, the analysis of the occurrence of mutant alleles *CD18*$^{G}$, *SLC35A3*$^{T}$ and *FANCI*$^{BY}$ in the world continues to be relevant for developing healthy livestock intended for producing high-quality dairy products. Modern genomic technologies, including genomic selection and gene editing, will be instrumental for further genetic-progress and animal welfare in the dairy-breeding and production sectors.

**Author Contributions:** Conceptualization, S.N.M., D.A.D., I.S.T. and N.S.M.; resources, I.S.T.; writing—original draft preparation, S.N.M., D.A.D., I.S.T. and N.S.M.; writing—review and editing, N.S.M., D.K.G. and M.N.R.; visualization, N.S.M. and M.N.R.; supervision, N.S.M.; project administration, N.S.M. All authors have read and agreed to the published version of the manuscript.

**Funding:** This review received no external funding.

**Conflicts of Interest:** The authors declare no conflict of interest.

## References

1. Haldane, J.B.S. The cost of natural selection. *J. Genet.* **1957**, *55*, 511–524. [CrossRef]
2. Marzanov, N.S.; Devrishov, D.A.; Marzanova, S.N. Genetic Burden of Mutations in Cattle Breeds. In *Materials of the International Scientific and Practical Conference on Molecular Genetic Technologies for Analysis of Gene Expression Related to Animal Productivity and Disease Resistance, Moscow, Russia, 21–22 November 2019*; Kochish, I.I., Romanov, M.N., Surai, P.F., Nikonov, I.N., Selina, M.V., Eds.; Sel'skokhozyaistvennye tekhnologii: Moscow, Russia, 2019; pp. 124–130. Available online: https://elibrary.ru/item.asp?id=41444945 (accessed on 19 January 2022).
3. Marzanov, N.S.; Devrishov, D.A.; Abylkasymov, D.A.; Marzanova, S.N.; Konovalova, N.V.; Libet, I.S.; Novikova, L.F.; Popov, A.N.; Turbina, I.S.; Marzanova, L.K. Occurrence of β-CN$^{A2}$ and β-CN$^{A1}$ Alleles at the Beta-casein Locus in Cattle. In *Increasing the Competitiveness of Animal Husbandry and Staffing Tasks, Proceedings of the 25th International Scientific and Practical Conference, Podolsk, Bykovo, Moscow Oblast, Russia, 24–25 June 2019*; Pyzhov, A.P., Zhukov, V.F., Ponomarev, N.V., Pyzhova, E.A., Eds.; Russian Academy of Livestock Management: Podolsk, Moscow Oblast, Russia, 2019; pp. 134–143. Available online: https://elibrary.ru/item.asp?id=38589459 (accessed on 19 January 2022).
4. Marzanova, S.N.; Nagorniy, V.A.; Devrishov, D.A.; Alekseev, Y.I.; Konovalova, N.V.; Tokhov, M.K.; Eskin, G.V.; Turbina, I.S.; Lukashina, A.A.; Marzanov, N.S. Founder effect and genogeography of CV and BL mutations in Black and White cattle. *Russ. Agric. Sci.* **2016**, *42*, 90–93. [CrossRef]
5. Marzanova, S.N.; Devrishov, D.A.; Alekseev, J.I.; Konovalova, N.V.; Marzanov, N.S. Set of Sequence of Primers and Allele-specific Probes for Simultaneous Gene Diagnostics of Two Mutant Alleles Causing CVM and BLAD in Cattle. Proprietor: LLC "Agrovet". Patent for Invention RU 2 577 990 C1. Federal Service for Intellectual Property, Moscow, Russia. Applied 11 February 2015. Published 20.03.2016b. Bull. No. 8. Available online: https://patents.google.com/patent/RU2577990C1/en (accessed on 19 January 2022).
6. Marzanova, S.N.; Devrishov, D.A.; Alekseev, Y.I.; Konovalova, N.V.; Marzanov, N.S. Set of Primers Sequence and Allele-Specific Probes for Simultaneous Four Mutant Kappa-Casein Alleles Gene Diagnostics in Bovine Cattle. Proprietor: LLC "Agrovet". Patent for Invention RU 2 646 140 C1. Federal Service for Intellectual Property, Moscow, Russia. Applied 24 March 2017. Published 01.03.2018a. Bull. No. 7. Available online: https://patents.google.com/patent/RU2646140C1/en (accessed on 19 January 2022).
7. Marzanova, S.N.; Devrishov, D.A.; Alekseev, Y.I.; Konovalova, N.V.; Marzanov, N.S. Method for Genetic Diagnosis of Mutant Allele That Causes Short Spine Syndrome or Brachyspine Syndrome in Cattle and a Test System for Implementation Thereof. Proprietor: LLC "Agrovet". Patent for Invention RU 2 664 454 C2. Federal Service for Intellectual Property, Moscow, Russia. Applied 4 April 2016. Published 17.08.2018b. Bull. No. 23. Available online: https://patents.google.com/patent/RU2664454C2/en (accessed on 19 January 2022).
8. Marzanova, S.N.; Devrishov, D.A.; Alekseev, Y.I.; Konovalova, N.V.; Marzanov, N.S. Method for Simultaneous Genodiagnostic of Four Mutant Alleles of Kappa-Casein in Cattle and Test System for Implementation Thereof. Proprietor: LLC "Agrovet". Patent for Invention RU 2 691 995 C2. Federal Service for Intellectual Property, Moscow, Russia. Applied 2 March 2017. Published 19.06.2019. Bull. No. 17. Available online: https://patents.google.com/patent/RU2691995C2/en (accessed on 19 January 2022).
9. Lien, S.; Aleström, P.; Klungland, H.; Rogne, S. Detection of multiple β-casein (CASB) alleles by amplification created restriction sites (ACRS). *Anim. Genet.* **1992**, *23*, 333–338. [CrossRef] [PubMed]
10. Dinç, H. Genotyping of Beta-casein, Kappa-casein and Beta-lactoglobulin Genes in Turkish Native Cattle Breeds and Efforts to Delineate BCM-7 on Human PBMC. Ph.D. Thesis, Graduate School of Natural and Applied Sciences, Middle East Technical University, Ankara, Turkey. Available online: https://open.metu.edu.tr/handle/11511/18963 (accessed on 19 January 2022).

11. Bertorelle, G.; Raffini, F.; Bosse, M.; Bortoluzzi, C.; Iannucci, A.; Trucchi, E.; Morales, H.E.; van Oosterhout, C. Genetic load: Genomic estimates and applications in non-model animals. *Nat. Rev. Genet.* **2022**, *23*, 492–503. [CrossRef] [PubMed]
12. Altukhov, Y.P. *Genetic Processes in Populations*, 3rd ed.; Akademkniga: Moscow, Russia, 2003; ISBN 5-94628-083-X. Available online: https://search.rsl.ru/en/search#q=ISBN%205-94628-083-X (accessed on 19 January 2022).
13. Mutovin, G.R. *Clinical Genetics. Genomics and Proteomics of Hereditary Pathology: Textbook*, 3rd ed.; GEOTAR-Media: Moscow, Russia, 2010; ISBN 978-5-9704-1152-0. Available online: https://search.rsl.ru/en/search#q=ISBN%20978-5-9704-1152-0 (accessed on 19 January 2022).
14. Farrell, H.M., Jr.; Jimenez-Flores, R.; Bleck, G.T.; Brown, E.M.; Butler, J.E.; Creamer, L.K.; Hicks, C.L.; Hollar, C.M.; Ng-Kwai-Hang, K.F.; Swaisgood, H.E. Nomenclature of the proteins of cows' milk—Sixth revision. *J. Dairy Sci.* **2004**, *87*, 1641–1674. [CrossRef] [PubMed]
15. Online Mendelian Inheritance in Animals—OMIA. Sydney School of Veterinary Science. Available online: https://omia.org/ (accessed on 19 January 2022).
16. Zhigachev, A.I.; Ernst, L.K.; Bogachev, A.S. About accumulation of mutation weight in the cattle breeds as result of intensive reproduction technology and improvement on target determinants. *Sel'skokhozyaistvennaya Biol. (Agr. Biol.)* **2008**, *43*, 25–32. Available online: https://www.elibrary.ru/item.asp?id=11709790 (accessed on 19 January 2022).
17. Schütz, E.; Scharfenstein, M.; Brenig, B. Implication of complex vertebral malformation and bovine leukocyte adhesion deficiency DNA-based testing on disease frequency in the Holstein population. *J. Dairy Sci.* **2008**, *91*, 4854–4859. [CrossRef]
18. Marzanova, S.N. Development of Gene Diagnostics of the Complex of Spinal Anomalies [CVM] and Immunodeficiency [BLAD] in Black-and-White Holstein Cattle. Ph.D. Thesis, K.I. Skryabin Moscow State Academy of Veterinary Medicine and Biotechnology, Moscow, Russia, 2012. Available online: https://search.rsl.ru/ru/record/01005529649 (accessed on 19 January 2022).
19. Marzanova, S.N.; Devrishov, D.A.; Turbina, I.S.; Alekseev, Y.I.; Konovalova, N.V.; Marzanov, N.S. Genetic diagnostics of $FANCI^{BY}$ mutation in representatives of the Holstein breed and its crosses. *Biotekhnologiya* **2021**, *37*, 16–25. [CrossRef]
20. Turbina, I.S. Characterization of Sires by Various Genetic Markers. Ph.D. Thesis, Russian Academy of Livestock Management, Moscow, Russia, 2006.
21. Marzanova, S.N.; Devrishov, D.A.; Turbina, I.S.; Marzanov, N.S. DNA Technology and estimation of drift of mutant alleles in populations of the Holstein breed and its crosses. *Russ. J. Genet.* **2022**, *58*, 876–879. [CrossRef]
22. Chun, H.C.; Yi, M.C.; Kuo, H.L. Genotype screening of bovine brachyspina in Taiwan Holstein cows. *Am. J. Anim. Vet. Sci.* **2020**, *15*, 206–210. [CrossRef]
23. Boichard, D.; Amigue, Y. Investigation on possible linkage between the BLAD gene and a QTL for production or type traits in Holstein cattle. In *Book of Abstracts, Proceedings of the 46th Annual Meeting of the European Association for Animal Production (EAAP), Prague, Czech Republic, 4–7 September 1995*; van Arendonk, J.A.M., Ed.; Wageningen Pers.: Wageningen, The Netherlands, 1995; p. 27. Available online: https://hal.inrae.fr/hal-02772606 (accessed on 19 January 2022).
24. Casas, E.; Kehrli, M.E., Jr. A review of selected genes with known effects on performance and health of cattle. *Front. Vet. Sci.* **2016**, *3*, 113. [CrossRef] [PubMed]
25. Epishko, O.A.; Pestis, B.K.; Tanana, L.A.; Kuzmina, T.I.; Cheburanova, E.S.; Shevchenko, M.Y.; Petrova, A.P.; Glinskaya, H.A.; Trahimchik, R.V. Definition of recessive mutations of BLAD, CVM and BS in population of cattle of the lactic direction of Republic of Belarus. In *Agriculture—Problems and Prospects: Collection of Scientific Papers—Zootechnics*; Pestis, V.K., Ed.; Grodno State Agrarian University: Grodno, Belarus, 2017; Volume 37, pp. 44–51. Available online: https://www.elibrary.ru/item.asp?id=32443408 (accessed on 19 January 2022)ISBN 978-985-537-108-4.
26. Ivanov, V.A.; Marzanov, N.S.; Eliseeva, L.I.; Tadzhiyev, K.P.; Marzanova, S.N. Genotypes of cattle breeds and quality of milk. *Problemy Biologii Productivnykh Zhivotnykh (Probl. Product. Anim. Biol.)* **2017**, 3, 48–65. Available online: https://www.elibrary.ru/item.asp?id=30058078 (accessed on 19 January 2022).
27. Ignatiev, V.M. Screening of the Leukocyte Adhesion Gene (BLAD-Syndrome) in Black-and-White Bovine Animals. Ph.D. Thesis, All-Russia Research Institute for Animal Husbandry, Dubrovitsy, Moscow Oblast, Russia, 1998. Available online: https://elibrary.ru/item.asp?id=19145898 (accessed on 19 January 2022).
28. Hagemoser, W.A.; Roth, J.A.; Lofstedt, J.; Fagerland, J.A. Granulocytopathy in a Holstein heifer. *J. Am. Vet. Med. Assoc.* **1983**, *183*, 1093–1094. Available online: https://pubmed.ncbi.nlm.nih.gov/6643217/ (accessed on 19 January 2022). [PubMed]
29. Kehrli, M.E., Jr.; Schmalstieg, F.C.; Anderson, D.C.; Van der Maaten, M.J.; Hughes, B.J.; Ackermann, M.R.; Wilhelmsen, C.L.; Brown, G.B.; Stevens, M.G.; Whetstone, C.A. Molecular definition of the bovine granulocytopathy syndrome: Identification of deficiency of the Mac-1 (CD11b/CD18) glycoprotein. *Am. J. Vet. Res.* **1990**, *51*, 1826–1836. Available online: https://pubmed.ncbi.nlm.nih.gov/1978618/ (accessed on 19 January 2022). [PubMed]
30. Kehrli, M.E., Jr.; Shuster, D.E.; Ackermann, M.R. Leukocyte adhesion deficiency among Holstein cattle. *Cornell Vet.* **1992**, *82*, 103–109. Available online: https://pubmed.ncbi.nlm.nih.gov/1623723/ (accessed on 19 January 2022).
31. Ackermann, M.R.; Kehrli, M.E., Jr.; Morfitt, D.C. Ventral dermatitis and vasculitis in a calf with bovine leukocyte adhesion deficiency. *J. Am. Vet. Med. Assoc.* **1993**, *202*, 413–415. Available online: https://pubmed.ncbi.nlm.nih.gov/8440633/ (accessed on 19 January 2022).
32. Gilbert, R.O.; Rebhun, W.C.; Kim, C.A.; Kehrli, M.E., Jr.; Shuster, D.E.; Ackermann, M.R. Clinical manifestations of leukocyte adhesion deficiency in cattle: 14 cases (1977–1991). *J. Am. Vet. Med. Assoc.* **1993**, *202*, 445–449. Available online: https://pubmed.ncbi.nlm.nih.gov/8095042/ (accessed on 19 January 2022).

33. Duncan, R.B., Jr.; Carrig, C.B.; Agerholm, J.S.; Bendixen, C. Complex vertebral malformation in a Holstein calf: Report of a case in the USA. *J. Vet. Diagn. Investig.* **2001**, *13*, 333–336. [CrossRef]
34. VanRaden, P.M.; Olson, K.M.; Null, D.J.; Hutchison, J.L. Harmful recessive effects on fertility detected by absence of homozygous haplotypes. *J. Dairy Sci.* **2011**, *94*, 6153–6161. [CrossRef]
35. Holstein Association USA, Inc. Brattleboro, VT, USA. Available online: https://www.holsteinusa.com/ (accessed on 19 January 2022).
36. Select Sires Inc. Plain City, OH, USA. Available online: https://www.selectsires.com/ (accessed on 19 January 2022).
37. Anon. VDS-Beschluß zu BLAD. *Ostfr. Landvolk.* **1992**, *107*, 28.
38. Agerholm, J.S.; DeLay, J.; Hicks, B.; Fredholm, M. First confirmed case of the bovine brachyspina syndrome in Canada. *Can. Vet. J.* **2010**, *51*, 1349–1350. Available online: https://pubmed.ncbi.nlm.nih.gov/21358926/ (accessed on 19 January 2022).
39. Poli, M.A.; Dewey, R.; Semorile, L.; Lozano, M.E.; Albariño, C.G.; Romanowski, V.; Grau, O. PCR screening for carriers of bovine leukocyte adhesion deficiency (BLAD) and uridine monophosphate synthase (DUMPS) in Argentine Holstein cattle. *J. Vet. Med. Ser. A* **1996**, *43*, 163–168. [CrossRef] [PubMed]
40. Ribeiro, L.A.; Baron, E.E.; Martinez, M.L.; Coutinho, L.L. PCR screening and allele frequency estimation of bovine leukocyte adhesion deficiency in Holstein and Gir cattle in Brazil. *Genet. Mol. Biol.* **2000**, *23*, 831–834. [CrossRef]
41. Schilcher, F.; Hotter, H.; Tammen, I.; Schuh, M. Occurrence of bovine leukocyte adhesion deficiency (BLAD) of Holstein Friesian in Austria. *Wien. Tierarztl. Monatsschr.* **1995**, *82*, 207–212. Available online: http://a.xueshu.baidu.com/usercenter/paper/show?paperid=52ca18b04dbff9c033612a462f966bfd (accessed on 19 January 2022).
42. Cox, E.; Kuczka, A.; Tammen, I.; Schwenger, B. Leukocyten adhesien deficientie bij rund en hond: Een erfelijke stoornis van het immuunstelsel. *Vlaams Diergeneeskd. Tijdschr.* **1993**, *62*, 71–79. Available online: https://library.wur.nl/WebQuery/titel/830081 (accessed on 19 January 2022).
43. Georges, M.; Coppieters, W.; Charlier, C.; Agerholm, J.S.; Fredholm, M. A Genetic Marker Test for Brachyspina and Fertility in Cattle. Patent for Invention EP2310528B1. International Patent Publ. No. WO2010/012690, 20 November 2013. Available online: https://patents.google.com/patent/EP2310528B1/en (accessed on 19 January 2022).
44. Andrews, A.H.; Fishwick, J.; Waters, R.J. Bovine leukocyte adhesion deficiency (BLAD) in a one year old Holstein Friesian bull—The first report in the United Kingdom. *Br. Vet. J.* **1996**, *152*, 347–351. [CrossRef]
45. Zsolnai, A.; Fésüs, L. Simultaneous analysis of bovine κ-casein and BLAD alleles by multiplex PCR followed by parallel digestion with two restriction enzymes. *Anim. Genet.* **1996**, *27*, 207–209. [CrossRef]
46. Fésüs, L.; Zsolnai, A.; Anton, I.; Bárány, I.; Bozó, S. BLAD genotypes and cow production traits in Hungarian Holsteins. *J. Anim. Breed. Genet.* **1999**, *116*, 169–174. [CrossRef]
47. Agerholm, J.S.; Houe, H.; Jørgensen, C.B.; Basse, A. Bovine leukocyte adhesion deficiency in Danish Holstein-Friesian cattle. II. Patho-anatomical description of affected calves. *Acta Vet. Scand.* **1993**, *34*, 237–243. [CrossRef]
48. Agerholm, J.S.; Basse, A.; Christensen, K. Investigations on the occurrence of hereditary diseases in the Danish cattle population 1989–1991. *Acta Vet. Scand.* **1993**, *34*, 245–253. [CrossRef]
49. Jørgensen, C.B.; Agerholm, J.S.; Pedersen, J.; Thomsen, P.D. Bovine leukocyte adhesion deficiency in Danish Holstein-Friesian cattle. I. PCR screening and allele frequency estimation. *Acta Vet. Scand.* **1993**, *34*, 231–236. [CrossRef]
50. Agerholm, J.S.; Bendixen, C.; Andersen, O.; Arnbjerg, J. Complex vertebral malformation in Holstein calves. *J. Vet. Diagn. Investig.* **2001**, *13*, 283–289. [CrossRef]
51. Agerholm, J.S.; Andersen, O.; Almskou, M.B.; Bendixen, C.; Arnbjerg, J.; Aamand, G.P.; Nielsen, U.S.; Panitz, F.; Petersen, A.H. Evaluation of the inheritance of the complex vertebral malformation syndrome by breeding studies. *Acta Vet. Scand.* **2004**, *45*, 133–137. [CrossRef]
52. Agerholm, J.S.; McEvoy, F.; Arnbjerg, J. Brachyspina syndrome in a Holstein calf. *J. Vet. Diagn. Investig.* **2006**, *18*, 418–422. [CrossRef] [PubMed]
53. Agerholm, J.S.; Peperkamp, K. Familial occurrence of Danish and Dutch cases of the bovine brachyspina syndrome. *BMC Vet. Res.* **2007**, *3*, 8. [CrossRef] [PubMed]
54. Charlier, C.; Agerholm, J.S.; Coppieters, W.; Karlskov-Mortensen, P.; Li, W.; de Jong, G.; Fasquelle, C.; Karim, L.; Cirera, S.; Cambisano, N.; et al. A deletion in the bovine FANCI gene compromises fertility by causing fetal death and brachyspina. *PLoS ONE* **2012**, *7*, e43085. [CrossRef] [PubMed]
55. Stöber, M.; Pohlenz, J.; Leibold, W.; Simon, D.; Kuczka, A.; Schwenger, B.; Kuhlmann, E.; Tammen, L.; Piturru, P. Deficienza di adesione dei leucociti nei bovini (BLAD) un difetto ereditario da porre sotto vigilanza. *Prax. Vet.* **1992**, *8*, 5–7.
56. Testoni, S.; Diana, A.; Olzi, E.; Gentile, A. Brachyspina syndrome in two Holstein calves. *Vet. J.* **2008**, *177*, 144–146. [CrossRef]
57. Viana, J.L.; Fernández, A.; Iglesias, A.; Pernas, G.S. Diagnóstico y control de las principales enfermedades genéticas (citrulinemia, DUMPS y BLAD) descritas en ganado Holstein-Frisón. *Med. Vet.* **1998**, *15*, 538–544. Available online: https://dialnet.unirioja.es/servlet/articulo?codigo=4410498 (accessed on 19 January 2022).
58. Adamov, N.; Mitrov, D.; Esmerov, I.; Dovc, P. Detection of recessive mutations (BLAD and CVM) in Holstein-Friesian cattle population in Republic of Macedonia. *Maced. Vet. Rev.* **2014**, *37*, 61–68. [CrossRef]
59. Müller, K.; Bernadina, W.E.; Kalsbeek, H.C.; Wensing, T.; Elving, L.; Verbeek, B.; Wentink, G.H. BLAD: Bovine leucocyte adhesion deficiency. *Tijdschr. Diergeneeskd.* **1993**, *118*, 183–184. Available online: https://pubmed.ncbi.nlm.nih.gov/8096659/ (accessed on 19 January 2022).

60. Mirck, M.H.; Von Bannisseht-Wijsmuller, T.; Timmermans-Besselink, W.J.; Van Luijk, J.H.; Buntjer, J.B.; Lenstra, J.A. Optimization of the PCR test for the mutation causing bovine leukocyte adhesion deficiency. *Cell. Mol. Biol.* **1995**, *41*, 695–698. Available online: https://pubmed.ncbi.nlm.nih.gov/7580848/ (accessed on 19 January 2022). [PubMed]
61. Arrayet, J.L.; Oberbauer, A.M.; Famula, T.R.; Garnett, I.; Oltjen, J.W.; Imhoof, J.; Kehrli, M.E., Jr.; Graham, T.W. Growth of Holstein calves from birth to 90 days: The influence of dietary zinc and BLAD status. *J. Anim. Sci.* **2002**, *80*, 545–552. [CrossRef] [PubMed]
62. Lubieniecki, K.; Grzybowski, G. Diagnostyka molekularna wrodzonego niedoboru leukocytarnych czasteczek adhezyjnych [BLAD] u bydla. *Med. Weter.* **1997**, *53*, 214–217. Available online: http://agro.icm.edu.pl/agro/element/bwmeta1.element.agro-article-b700eed0-54f0-4e01-9630-f8a5be251758 (accessed on 19 January 2022).
63. Ruść, A.; Kamiński, S. Prevalence of complex vertebral malformation carriers among Polish Holstein-Friesian bulls. *J. Appl. Genet.* **2007**, *48*, 247–252. [CrossRef] [PubMed]
64. Ruść, A.; Kamiński, S. Detection of Brachyspina carriers within Polish Holstein-Friesian bulls. *Pol. J. Vet. Sci.* **2015**, *18*, 453–454. [CrossRef]
65. Marzanov, N.S.; Popov, A.N.; Zinovieva, N.A.; Polezhaeva, V.A.; Ignatiev, V.M.; Brem, G. Screening of the BLAD syndrome gene in Black-and-white root animals. *Vestnik Rossiyskoy Akademii Sel'skokhozyaystvennykh Nauk (Her. Russ. Acad. Sci.)* **1997**, *4*, 59–61. Available online: https://elibrary.ru/item.asp?id=21635211 (accessed on 19 January 2022).
66. Marzanov, N.S.; Eskin, G.V.; Turbina, I.S.; Devrishov, D.A.; Tokhov, M.K.; Marzanova, S.N. *Gene Diagnostics and Distribution of the Immunodeficiency Allele, or BLAD Syndrome, in Black-and-White Cattle*; Rosinformagrotech: Moscow, Russia, 2013. Available online: https://elibrary.ru/item.asp?id=21055161 (accessed on 19 January 2022).
67. Kalashnikova, L.A.; Dunin, I.M.; Glazko, V.I.; Glazko, G.V.; Ryzhova, N.V.; Golubina, E.P. *DNA Technologies in the Evaluation of Farm Animals*; All-Russian Research Institute of Animal Breeding: Lesnye Polyany, Moscow Oblast, Russia, 1999. Available online: https://www.elibrary.ru/item.asp?id=25478053 (accessed on 19 January 2022).
68. Turbina, I.S.; Fedorova, E.V.; Kertieva, N.M.; Yeskin, G.V.; Turbina, G.S.; Marzanov, N.S. Genealogy and some biological features in BLAD carrier and non-carrier bulls. In *Selektsiya, kormleniye, soderzhaniye sel'skokhozyaystvennykh zhivotnykh i tekhnologiya proizvodstva produktov zhivotnovodstva: Trudy VNIIPlem (Breeding, Feeding, Keeping Farm Animals and Technology for the Production of Livestock Products: Proceedings of VNIIPlem)*; All-Russian Research Institute of Animal Breeding: Lesnye Polyany, Moscow Oblast, Russia, 2004; pp. 3–7.
69. Kalashnikova, L.A.; Khabibrakhmanova, Y.A.; Prozherin, V.P.; Yaluga, V.L. Genotyping of brachyspina anomaly in Holsteinized cattle of the Kholmogory breed. In *Molecular Diagnostics 2021—Section 21: Technologies for the Detection of Infectious and Hereditary Animal Diseases, Proceedings of the 10th Anniversary International Scientific and Practical Conference, Moscow, Russia, 9–11 November 2021*; Center for Strategic Planning and Management of Biomedical Health Risks: Moscow, Russia, 2021; Volume 2, pp. 186–188.
70. Vătăşescu-Balcan, R.A.; Manea, M.A.; Georgescu, S.E.; Dinischiotu, A.; Tesio, C.D.; Costache, M. Evidence of single point mutation inducing BLAD disease in Romanian Holstein-derived cattle breed. *Biotechnol. Anim. Husb.* **2007**, *23*, 375–381. [CrossRef]
71. Grobet, L.; Charlier, C.; Hanset, R. Diagnosis of bovine leucocyte adhesion deficiency (BLAD) at the DNA level. *Ann. Med. Vet.* **1993**, *137*, 27–31. Available online: https://agris.fao.org/agris-search/search.do?recordID=BE9300687 (accessed on 19 January 2022).
72. Eggen, A.; Duchesne, A.; Laurent, P.; Grohs, C.; Denis, C.; Gautier, M.; Boichard, D.; Ducos, A. Controlling genetic disorders in the French dairy cattle population. In Proceedings of the 29th International Conference on Animal Genetics, Tokyo, Japan, 11–16 September 2004; International Society for Animal Genetics (ISAG), INT: Tokyo, Japan, 2004; p. 35. Available online: https://hal.inrae.fr/hal-02762175/file/ISAG_Proceedings_2004_1.pdf (accessed on 19 January 2022).
73. Valour, D.; Michot, P.; Eozenou, C.; Lefebvre, R.; Bonnet, A.; Capitan, A.; Uzbekova, S.; Sellem, E.; Ponsart, C.; Schibler, L. Dairy cattle reproduction is a tightly regulated genetic process: Highlights on genes, pathways, and biological processes. *Anim. Front.* **2015**, *5*, 32–41. [CrossRef]
74. Anon. Erbfehler—Was ist Blad? *Hann. Land. Forstwirtsch. Ztg.* **1991**, *11*, 91.
75. Kuczka, A.; Schwenger, B. BLAD im Griff [The right feeling for BLAD [bovine leukocyte adhesion deficiency]]. *Tierzüchter* **1993**, *45*, 32–35. Available online: https://agris.fao.org/agris-search/search.do?recordID=DE19930058362 (accessed on 19 January 2022).
76. Lienau, A.; Stöber, M.; Kehrli, M.E.; Tammen, I.; Schwenger, B.; Kuczka, A.; Pohlenz, J. Bovine Leukozyten-Adhäsions-Defizienz: Klinisches Bild und Differential-Diagnostik (Bovine leukocyte adhesion deficiency: Clinical picture and differential diagnosis). *Dtsch. Tierarztl. Wochenschr.* **1994**, *101*, 405–406. Available online: https://pubmed.ncbi.nlm.nih.gov/7851303/ (accessed on 19 January 2022).
77. Tammen, I. Weiterentwicklung des DNA-Tests auf BLAD (Bovine Leukozyten-Adhäsions-Defizienz) für den Einsatz in Rinderzucht und Klinischer Diagnostik. Ph.D. Thesis, Tierärztliche Hochschule Hannover, Hannover, Germany, 1994. Available online: https://agris.fao.org/agris-search/search.do?recordID=US201300291400 (accessed on 19 January 2022).
78. Duesmann, K.; Schmidt, W.; Görlach, A. Die PCR als Methode zur Gendiagnostik im Routinebetrieb einer Besamungsstation am Beispiel der Bovinen Leukozyten-Adhäsions-Defizienz (BLAD). *Reprod. Domest. Anim.* **1994**, *29*, 193.
79. Engelhardt, I. Inzucht, Bedeutende Ahnen und Wahrscheinlichkeit für BLAD-Merkmalsträger in der Deutschen Schwarzbuntzucht. Ph.D. Thesis, Tierärztliche Hochschule Hannover, Hannover, Germany, 1996. Available online: https://agris.fao.org/agris-search/search.do?recordID=US201300016803 (accessed on 19 January 2022).

80. Distl, O.T. The use of molecular genetics in eliminating of inherited anomalies in cattle. *Arch. Anim. Breed.* **2005**, *48*, 209–218. [CrossRef]
81. Buck, B.C.; Ulrich, R.; Wöhlke, A.; Kuiper, H.; Baumgaertner, W.; Distl, O. Missbildungen an der Wirbelsaule mit multiplen Organanomalien bei einem Kalb der Rasse Deutsche Holsteins (Vertebral and multiple organ malformations in a black and white German Holstein calf). *Berl. Munch. Tierarztl. Wochenschr.* **2010**, *123*, 251–255. [CrossRef] [PubMed]
82. Segelke, D.; Täubert, H.; Reinhardt, F.; Thaller, G. Considering genetic characteristics in German Holstein breeding programs. *J. Dairy Sci.* **2016**, *99*, 458–467. [CrossRef]
83. Glazko, V.I.; Lavrovsky, V.V.; Filenko, A.N.; Mariutsa, A.E. Intrabreed genetic differentiation and the presence of the BLAD mutation in Holstein . *Sel'skokhozyaistvennaya Biol. (Agr. Biol.)* **2000**, *4*, 45–47. Available online: https://scholar.google.com/scholar?cluster=6636562904833334721&hl=en&oi=scholarr (accessed on 19 January 2022).
84. Mariutsa, A.E. Population-Genetic Mechanisms of Adaptation and Spread of Semi-Lethal Recessive Mutations on the Example of BLAD in Cattle. Ph.D. Thesis, Institute of Agroecology and Environmental Management UAAS, Kyiv, Ukraine, 2005. Available online: http://www.disslib.org/populjatsyonno-henetycheskye-mekhanyzmy-adaptatsyy-y-rasprostranenye-poluletalnykh.html (accessed on 19 January 2022).
85. Birukova, O.D. The Population-Genetic Monitoring of Gene Pool Forming of the Ukrainian Black-and-White Dairy Breed. Ph.D. Thesis, Institute of Animal Breeding and Genetics UAAS, Chubinske, Kyiv Region, Ukraine, 2005. Available online: http://www.irbis-nbuv.gov.ua/cgi-bin/irbis_nbuv/cgiirbis_64.exe?C21COM=2&I21DBN=ARD&P21DBN=ARD&Z21ID=&Image_file_name=DOC/2005/05bodrmp.zip&IMAGE_FILE_DOWNLOAD=1 (accessed on 19 January 2022).
86. Meylan, M.; Abegg, R.; Sager, H.; Jungi, T.W.; Martig, J. Fallvorstellung: Bovine Leukozyten-Adhäsions-Defizienz (BLAD) in der Schweiz (Case presentation: Bovine leukocyte adhesion deficiency (BLAD) in Switzerland). *Schweiz. Arch. Tierheilkd.* **1997**, *139*, 277–281. Available online: https://pubmed.ncbi.nlm.nih.gov/9411734/ (accessed on 19 January 2022). [PubMed]
87. Citek, J.; Rehout, V.; Hajkova, J.; Pavkova, J. Monitoring of the genetic health of cattle in the Czech Republic. *Vet. Med.* **2006**, *51*, 333–339. [CrossRef]
88. Li, Y.; Han, G.; Zhang, S.; Li, N.; Sun, F. Detection of bovine leukocyte adhesion deficiency (BLAD) and haplotype analysis. *Acta Vet. Zootech. Sin.* **2008**, *39*, 1285–1288. Available online: http://www.xmsyxb.com/CN/Y2008/V39/I9/1285 (accessed on 19 January 2022).
89. Chu, Q.; Sun, D.; Yu, Y.; Zhang, Y.; Zhang, Y. Identification of complex vertebral malformation carriers in Chinese Holstein. *J. Vet. Diagn. Investig.* **2008**, *20*, 228–230. [CrossRef] [PubMed]
90. Zhang, Y.; Fan, X.; Sun, D.; Wang, Y.; Yu, Y.; Xie, Y.; Zhang, S.; Zhang, Y. A novel method for rapid and reliable detection of complex vertebral malformation and bovine leukocyte adhesion deficiency in Holstein cattle. *J. Anim. Sci. Biotechnol.* **2012**, *3*, 24. [CrossRef] [PubMed]
91. Fang, L.; Li, Y.; Zhang, Y.; Sun, D.; Liu, L.; Zhang, Y.; Zhang, S. Identification of brachyspina syndrome carriers in Chinese Holstein cattle. *J. Vet. Diagn. Investig.* **2013**, *25*, 508–510. [CrossRef] [PubMed]
92. Li, Y.; Zhai, L.; Fang, L.; Zhang, S.; Liu, L.; Zhu, Y.; Xue, J.; Xiaoqing, L.; Qiao, L.; Sun, D. A novel multiplex polymerase chain reaction method for the identification of brachyspina syndrome carriers in Chinese Holstein cattle. *J. Vet. Sci. Med. Diagn.* **2016**, *5*. [CrossRef]
93. Takahashi, K.; Miyagawa, K.; Abe, S.; Kurosawa, T.; Sonoda, M.; Nakade, T.; Nagahata, H.; Noda, H.; Chihaya, Y.; Isogai, E. Bovine granulocytopathy syndrome of Holstein-Friesian calves and heifers. *Nihon Juigaku Zasshi* **1987**, *49*, 733–736. [CrossRef] [PubMed]
94. Tajima, M.; Irie, M.; Kirisawa, R.; Hagiwara, K.; Kurosawa, T.; Takahashi, K. The detection of a mutation of CD18 gene in bovine leukocyte adhesion deficiency (BLAD). *J. Vet. Med. Sci.* **1993**, *55*, 145–146. [CrossRef] [PubMed]
95. Nagahata, H.; Miura, T.; Tagaki, K.; Ohtake, M.; Noda, H.; Yasuda, T.; Nioka, K. Prevalence and allele frequency estimation of bovine leukocyte adhesion deficiency (BLAD) in Holstein-Friesian cattle in Japan. *J. Vet. Med. Sci.* **1997**, *59*, 233–238. [CrossRef] [PubMed]
96. Nagahata, H. Bovine leukocyte adhesion deficiency (BLAD): A review. *J. Vet. Med. Sci.* **2004**, *66*, 1475–1482. [CrossRef]
97. Healy, P.J. Bovine leukocyte adhesion deficiency (BLAD)—Another genetic defect of Holstein/Friesians. *Aust. Vet. J.* **1992**, *69*, 190. [CrossRef]
98. Norouzy, A.; Nassiry, M.R.; Eftekhari Shahrody, F.; Javadmanesh, A.; Mohammad Abadi, M.R.; Sulimova, G.E. Identification of bovine leukocyte adhesion deficiency (BLAD) carriers in Holstein and Brown Swiss AI bulls in Iran. *Russ. J. Genet.* **2005**, *41*, 1409–1413. [CrossRef]
99. Rezaee, A.R.; Nassiry, M.R.; Valizadeh, R.; Tahmoorespour, M.; Javadmanesh, A.; Zarei, A.; Janati, H. Study of complex vertebral malformation disorder in Iranian Holstein bulls. *World J. Zool.* **2008**, *3*, 36–39. Available online: https://profdoc.um.ac.ir/paper-abstract-1014702.html (accessed on 19 January 2022).
100. Hemati, B.; Saberi, J.; Noshary, A.R. The study of complex vertebral malformation genetic defect in a population of Sistani cows. *J. Plant Anim. Environ. Sci.* **2014**, *4*, 485–487. Available online: https://www.cabdirect.org/cabdirect/abstract/20143288422 (accessed on 19 January 2022).
101. Öner, Y.; Keskin, A.; Elmaci, C. Identification of BLAD, DUMPS, citrullinaemia and factor XI deficiency in Holstein cattle in Turkey. *Asian J. Anim. Vet. Adv.* **2010**, *5*, 60–65. [CrossRef]

102. Patel, R.K.; Singh, K.M.; Soni, K.J.; Chauhan, J.B.; Sambasiva Rao, K.R. Low incidence of bovine leukocyte adhesion deficiency (BLAD) carriers in Indian cattle and buffalo breeds. *J. Appl. Genet.* **2007**, *48*, 153–155. [CrossRef]
103. Gholap, P.N.; Kale, D.S.; Sirothia, A.R. Genetic diseases in cattle: A review. *Res. J. Anim. Vet. Fish. Sci.* **2014**, *2*, 24–33. Available online: http://www.isca.in/AVFS/Archive/v2/i2/5.ISCA-RJAVFS-2014-005.php (accessed on 19 January 2022).
104. Nasreen, F.; Altaf Malik, N.; Naeem Riaz, M.; Anver Qureshi, J. Detection and screening of bovine leukocyte adhesion deficiency in Pakistan using molecular methods. *Hereditas* **2009**, *146*, 74–78. [CrossRef]
105. Stadler, P.; Van Amstel, S.R.; Van Rensburg, I.B.; Williams, M.C. Verdagte oorerflike granulositopatie in vier Holstein frieskalwers (Suspected inherited granulocytopathy in four Holstein Friesian calves). *J. S. Afr. Vet. Assoc.* **1993**, *64*, 172–177. Available online: https://pubmed.ncbi.nlm.nih.gov/8176699/ (accessed on 19 January 2022).
106. FAO. Gateway to Dairy Production and Products. Available online: https://www.fao.org/dairy-production-products/en/ (accessed on 19 January 2022).
107. Kosyachenko, N.M.; Abramova, M.V.; Ilyina, A.V.; Zyryanova, S.V.; Konovalov, A.V.; Kosourova, T.N. *Holstein Breed in the Creation of Improved Genotypes and Intrabreed Types of Cattle*; Kantsler: Yaroslavl, Russia, 2020; ISBN 978-5-907417-06-9. Available online: https://search.rsl.ru/ru/search#q=ISBN%20978-5-907417-06-9 (accessed on 19 January 2022).
108. Cole, J.B.; Null, D.J.; VanRaden, P.M. Phenotypic and genetic effects of recessive haplotypes on yield, longevity, and fertility. *J. Dairy Sci.* **2016**, *99*, 7274–7288. [CrossRef]
109. Marzanov, N.S.; Turbina, I.S.; Eskin, G.V.; Turbina, G.S.; Ignat'ev, V.M.; Popov, A.N.; Harlizius, B. Screening of the gene of the leukocytal adhesion deficiency in Black-and-white Holsteinized cattle. *Sel'skokhozyaistvennaya Biol. (Agr. Biol.)* **2003**, *38*, 23–30. Available online: https://www.elibrary.ru/item.asp?id=18100805 (accessed on 19 January 2022).
110. Grant, V. *The Evolutionary Process: A Critical Study of Evolutionary Theory*; Columbia University Press: New York, NY, USA, 1991; ISBN 0-231-07324-0. Available online: https://www.google.com/books/edition/The_Evolutionary_Process/qAt_QgAACAAJ?hl=en (accessed on 19 January 2022).
111. Kaminski, S.; Cieslinska, A.; Kostyr, E. Polymorphism of bovine beta-casein and its potential effect on human health. *J. Appl. Genet.* **2007**, *48*, 189–198. [CrossRef]
112. Woodford, K.B. *Devil in the Milk: Illness, Health and Politics of A1 and A2 Milk*; Chelsea Green Publishing: White River Junction, VT, USA, 2009; ISBN 978-1-60358-102-8. Available online: https://www.google.com/books/edition/_/QqD6btySu7AC?hl=en (accessed on 19 January 2022).
113. Marzanov, N.S.; Devrishov, D.A.; Abylkasymov, D.A.; Marzanova, S.N.; Konovalova, N.V.; Libet, I.S. Characteristics of Russian dairy cattle by the occurrence of genotypes and alleles in the beta casein locus. *Veterenariya Zootekhniya i Biotehnologiya (Vet. Med. Zootech. Biotechnol.)*, 2020; 1, 47–52. [CrossRef]
114. Marzanov, N.S.; Abylkasymov, D.A.; Devrishov, D.A.; Marzanova, S.N.; Libet, I.S. The characteristic of allelotype in cows of Black and multicolored breed of β- and k-casein locus and qualitative indicators of milk. In *Aktual'nyye Voprosy Molochnoy Promyshlennosti, Mezhotraslevyye Tekhnologii i Sistemy Upravleniya Kachestvom (Actual Issues of the Dairy Industry, Intersectoral Technologies and Quality Management Systems)*; All-Russian Dairy Research Institute: Moscow, Russia, 2020; pp. 368–376. [CrossRef]
115. Marzanova, S.N.; Alekseev, Y.I.; Konovalova, N.V.; Turbina, I.S.; Devrishov, D.A.; Sochivko, D.G.; Marzanov, N.S. Developing a method of diagnostics of complex vertebral malformation (CVM) by Real-Time PCR method in Black-and-white cattle. *Problemy Biologii Produktivnyh Životnyh (Probl. Product. Anim. Biol.)* **2011**, 79–82. Available online: https://elibrary.ru/item.asp?id=17644709 (accessed on 19 January 2022).
116. Marzanov, N.S.; Devrishov, D.A.; Marzanova, S.N.; Getokov, O.O.; Abylkasymov, D.A.; Libet, I.S. DNA diagnostics of populations of the Blackand-white breed by the beta-casein locus. *Veterenariya Zootekhniya i Biotehnologiya (Vet. Med. Zootech. Biotechnol.)* **2021**, *3*, 78–84. [CrossRef]
117. Wiggans, G.R.; Cole, J.B.; Hubbard, S.M.; Sonstegard, T.S. Genomic Selection in Dairy Cattle: The USDA Experience. *Annu. Rev. Anim. Biosci.* **2017**, *5*, 309–327. [CrossRef]
118. Matthews, D.; Kearney, J.F.; Cromie, A.R.; Hely, F.S.; Amer, P.R. Genetic benefits of genomic selection breeding programmes considering foreign sire contributions. *Genet. Sel. Evol.* **2019**, *51*, 40. [CrossRef]
119. Gutierrez-Reinoso, M.A.; Aponte, P.M.; Garcia-Herreros, M. Genomic Analysis, Progress and Future Perspectives in Dairy Cattle Selection: A Review. *Animals* **2021**, *11*, 599. [CrossRef]
120. Bouquet, A.; Juga, J. Integrating genomic selection into dairy cattle breeding programmes: A review. *Animal* **2013**, *7*, 705–713. [CrossRef]
121. Brito, L.F.; Bedere, N.; Douhard, F.; Oliveira, H.R.; Arnal, M.; Peñagaricano, F.; Schinckel, A.P.; Baes, C.F.; Miglior, F. Review: Genetic selection of high-yielding dairy cattle toward sustainable farming systems in a rapidly changing world. *Animal* **2021**, *15* (Suppl. S1), 100292. [CrossRef]
122. Smaragdov, M.G. Genomic selection of milk cattle. The practical application over five years. *Russ. J. Genet.* **2013**, *49*, 1089–1097. [CrossRef]
123. Plemyashov, K.V.; Smaragdov, M.G.; Romanov, M.N. Molecular Genetic Polymorphism in Animal Populations and Its Application in Intensive Breeding of Dairy Cattle—A Review. In *Materials of the 3rd International Scientific and Practical Conference on Molecular Genetic Technologies for Analysis of Gene Expression Related to Animal Productivity and Disease Resistance, Moscow, Russia, 30 September 2021*; Pozyabin, S.V., Kochish, I.I., Romanov, M.N., Eds.; Sel'skokhozyaistvennye tekhnologii: Moscow, Russia, 2021; pp. 368–378. Available online: https://elibrary.ru/item.asp?id=46668865 (accessed on 19 January 2022).

124. Plemyashov, K.V.; Smaragdov, M.G.; Romanov, M.N. Genomic Assessment of Breeding Bulls. In *Materials of the 3rd International Scientific and Practical Conference on Molecular Genetic Technologies for Analysis of Gene Expression Related to Animal Productivity and Disease Resistance, Moscow, Russia, 30 September 2021*; Pozyabin, S.V., Kochish, I.I., Romanov, M.N., Eds.; Sel'skokhozyaistvennye tekhnologii: Moscow, Russia, 2021; pp. 363–367. Available online: https://elibrary.ru/item.asp?id=46668864 (accessed on 19 January 2022).
125. Sharko, F.S.; Khatib, A.; Prokhortchouk, E.B. Genomic estimated breeding value of milk performance and fertility traits in the Russian Black-and-white cattle population. *Acta Nat.* **2022**, *14*, 109–122. [CrossRef]
126. Zinovieva, N.A.; Dotsev, A.V.; Sermyagin, A.A.; Deniskova, T.E.; Abdelmanova, A.S.; Kharzinova, V.R.; Sölkner, J.; Reyer, H.; Wimmers, K.; Brem, G. Selection signatures in two oldest Russian native cattle breeds revealed using high-density single nucleotide polymorphism analysis. *PLoS ONE* **2020**, *15*, e0242200. [CrossRef]
127. Krivonogova, A.S.; Bruter, A.V.; Makutina, V.A.; Okulova, Y.D.; Ilchuk, L.A.; Kubekina, M.V.; Khamatova, A.Y.; Egorova, T.V.; Mymrin, V.S.; Silaeva, Y.Y.; et al. AAV infection of bovine embryos: Novel, simple and effective tool for genome editing. *Theriogenology* **2022**, *193*, 77–86. [CrossRef]
128. Cziszter, L.-T.; Ilie, D.-E.; Neamt, R.-I.; Neciu, F.-C.; Saplacan, S.-I.; Gavojdian, D. Comparative study on production, reproduction and functional traits between Fleckvieh and Braunvieh cattle. *Asian-Australas. J. Anim. Sci.* **2017**, *30*, 666–671. [CrossRef]
129. Ernst, L.K.; Zhigachev, A.I. *Monitoring of Animal Genetic Diseases in the System of Large-Scale Breeding*; Russian Agricultural Academy Publishing House: Moscow, Russia, 2006; ISBN 5-85941-250-9. Available online: https://search.rsl.ru/ru/search#q=ISBN%205-85941-250-9 (accessed on 19 January 2022).
130. Marzanov, N.S.; Amerkhanov, H.; Yeskin, G.; Turbina, G.; Fedorova, E.; Turbina, I.; Samorukov, Y.; Kiyko, E.; Popov, N.; Shukurova, E.; et al. Genogeography of BLAD in populations of Black-and-white cattle in Russia and abroad. *Molochnoye i Myasnoye Skotovodstvo (Dairy Beef Cattle Breed.)* **2008**, *4*, 2–5. Available online: https://www.elibrary.ru/item.asp?id=11149171 (accessed on 19 January 2022).
131. Ussenbekov, E.S.; Terletskiy, V.P. On the absence of genetic defects BLAD, CVM, DUMPS and BC in stud bulls of the local Alatau cattle breed. *Veterinariya (Vet. Med.)* **2016**, *6*, 49–51. Available online: https://www.elibrary.ru/item.asp?id=26556053 (accessed on 19 January 2022).
132. Terletskiy, V.P.; Buralkhiyev, B.A.; Ussenbekov, Y.S.; Yelubayeva, M.; Tyshchenko, V.I.; Beyshova, I.S. Screening for mutations that determine the development of hereditary diseases in breeding cattle. *Aktual'nyye Voprosy Veterinarnoy Biologii (Actual Quest. Vet. Biol.)* **2016**, *3*, 3–6. Available online: https://www.elibrary.ru/item.asp?id=26644420 (accessed on 19 January 2022).
133. Marron, B.M.; Robinson, J.L.; Gentry, P.A.; Beever, J.E. Identification of a mutation associated with factor XI deficiency in Holstein cattle. *Anim. Genet.* **2004**, *35*, 454–456. [CrossRef]
134. Kuznetsov, V.M. Statistical analysis of pedigrees. *Zootekhniya (Zootechnics)*. 1998, 2, pp. 5–8. Available online: https://vm-kuznetsov.ru/files/book/Pap1998_2_pedig.pdf (accessed on 19 January 2022).
135. Zhigachev, A.I.; Suller, I.L. Hereditary anomalies and their control in cattle. *Praktik (Practitioner)*, 2002; 3–4, 46–53.
136. Zhigachev, A.I. Estimation of breeding male for invisible genetic defects. *Zootekhniya (Zootechnics)* **2001**, *2*, 10–12. Available online: https://www.elibrary.ru/item.asp?id=9124606 (accessed on 19 January 2022).

MDPI
St. Alban-Anlage 66
4052 Basel
Switzerland
Tel. +41 61 683 77 34
Fax +41 61 302 89 18
www.mdpi.com

*Agriculture* Editorial Office
E-mail: agriculture@mdpi.com
www.mdpi.com/journal/agriculture

www.ingramcontent.com/pod-product-compliance
Lightning Source LLC
Chambersburg PA
CBHW061626210926
51850CB00009B/2447
*9783036582467*